Advances in Construction ICT and e-Business

T0187837

This internationally conducted study of the latest construction industry practices addresses a broad range of Information and Communication Technology applications. Drawing on research conducted in the US and UK, this book presents the state of the art of various e-business processes, and examines BIM, virtual environments and mobile technologies.

Innovation is a theme that runs throughout this book, so in addition to the direct impact of these new technical achievements, it also considers the management styles that helped them to emerge. Examples from industry are illustrated with case studies and presented alongside research from some of the best known academics in this field.

This book is essential reading for all advanced students and researchers interested in how ICT is changing construction management and the construction industry.

Srinath Perera is Professor and Chair of Built Environment and Construction Management at the School of Computing, Engineering & Mathematics, Western Sydney University, Sydney, Australia.

Bingunath Ingirige is Professor of Project Management and Resilience at the Global Disaster Resilience Centre (GDRC), School of Art Design and Architecture, University of Huddersfield, West Yorkshire, UK.

Kirti Ruikar is a Senior Lecturer in Architectural Engineering at the School of Civil and Building Engineering, Loughborough University, UK.

Esther Obonyo is an Associate Professor of Engineering Design and Architectural Engineering at Penn State University, USA.

Advances in Construction ICT and e-Business

Edited by
Srinath Perera, Bingunath Ingirige,
Kirti Ruikar and Esther Obonyo

Routledge
Taylor & Francis Group

LONDON AND NEW YORK

First published 2017 by Routledge

2 Park Square, Milton Park, Abingdon, Oxon, OX14 4RN
605 Third Avenue, New York, NY 10017

Routledge is an imprint of the Taylor & Francis Group, an informa business

First issued in paperback 2020

British Library Cataloguing-in-Publication Data
A catalogue record for this book is available from the British Library

Library of Congress Cataloging in Publication Data
A catalog record for this book has been applied for

ISBN: 978-1-138-91458-2 (hbk)
ISBN: 978-0-367-73620-0 (pbk)

Typeset in Times New Roman
by codeMantra

Contents

Figures and tables

Figures

Tables

Contributors

Editors

Professor Srinath Perera PhD MSc IT BSc (Hons) QS MRICS AAIQS ICECA FAIB

Institution: Western Sydney University

Email: Srinath.perera@westernsydney.edu.au

Srinath Perera holds a personal chair in built environment and construction management at Western Sydney University, Sydney Australia, and leads the construction management research. He leads research in construction ICT and sustainability focusing on e-business, carbon estimating, construction management and the use of public private partnerships in construction. His expertise in dealing with multi disciplines have enabled him to be a lead partner in several EU-funded projects in disaster resilience, paving the way to apply construction economic techniques in building resilience. He is the coordinator of the International Council for Research and Innovation in Building and Construction (CIB) task group TG83: e-Business in Construction. He is a chartered surveyor and a member of the Royal Institution of Chartered Surveyors (RICS) and a Fellow of the Australian Institute of Building. He currently serves as an editorial board member of several international journals and is the chairman of the Editorial Panel's Procurement subcommittee of the Institution of Civil Engineers proceedings on Management Procurement and Law. He is also the co-author of the leading text book on construction management, *Cost Studies of Buildings, 6th Edition,* recently published by Routledge. He has over one hundred peer reviewed publications. He has supervised and examined several doctoral students worldwide.

Professor Bingunath Ingirige Bsc (Hons) MBA PhD MRICS FHEA

Institution: University of Huddersfield

Email: B.Ingirige@hud.ac.uk

Professor Bingunath Ingirige holds the chair in project management and resilience at the School of Art Design and Architecture, University of

Huddersfield, UK. He was previously a senior lecturer at the School of the Built Environment, University of Salford, UK. Prior to that, he worked at the Department of Building Economics of the University of Moratuwa, Sri Lanka as a Lecturer from 1994 to 2000. He has published and delivered keynotes in the area of web-enabled project management in many forums. He is also a member of the Royal Institution of Chartered Surveyors (RICS). Bingu is also interested in improving performance in the construction industry both from an angle of improving its overall resilience as well as improving the effectiveness and efficiency of project management in construction. Since 2011, he has held the position of the joint coordinator of the Conseil International du Bâtiment (CIB) Task Group TG83—eBusiness in construction. He teaches and leads the MSc in Advanced Project Management in Construction programme at the University of Huddersfield.

Dr Kirti Ruikar EngD MSc BArch FHEA

Institution: Loughborough University

Email: k.d.ruikar@lboro.ac.uk

Dr Kirti Ruikar is a senior lecturer in architectural engineering in the School of Civil and Building Engineering at Loughborough University. She is a fellow of the Higher Education Academy (FHEA). Her research interests are in strategic knowledge, information and technology management. Fields of active interest include e-business, e-readiness, BIM, fire-resilient building design, knowledge and information management, collaborative working environments and BPR. She has led several research projects in these areas and has supervised eight doctoral graduates to successful completion. Her research is widely published and she has over one hundred publications in these fields. She has co-edited a book on *e-Business in Construction* (Wiley-Blackwell 2008) and has co-authored a book on *Collaborative Design Management* (Routledge 2013). She has been a guest editor of the *Journal of IT in Construction*'s special issues on 'E-commerce in Construction' (2006); 'Technology Strategies for Collaborative Working' (2009); and 'Innovation in Construction e-Business' (2011). Dr Ruikar is also an associate editor of the internationally reputed *Journal of IT in Construction* (ITcon.org) and a joint coordinator of CIB Task Group, TG83 on 'e-Business in Construction'. She sits on the scientific committees of various international conferences and is a member of the editorial review boards of international journals.

Dr Esther Obonyo

Institution: The Pennsylvania State University

Email: eao4@psu.edu

Dr Esther Obonyo is associate professor of engineering design and architectural engineering at the Pennsylvania State University. She is also a

2015/2016 Jefferson Science Fellow. Between August 2004 and July 2015, she was a faculty member at the University of Florida's (UF) Rinker School of Construction Management, first serving as assistant professor until her promotion to the rank of associate professor in 2012. While at UF, she also served as a faculty entrepreneurship fellow at College of Business. In addition, for the past three years, she has been a visiting professor in TIP University Manila through the PhilDev IDEA Program (an initiative directed at enhancing the practice of technopreneurship in the Philippines). Beyond her academic credentials, Dr Obonyo has extensive industry experience, having worked as a construction engineer, project manager and innovations analyst in several engineering and construction companies in Kenya, the United Kingdom and the United States. Her research interest cuts across the following themes: sustainable structural materials, intelligent information and knowledge-based systems and entrepreneurship. Dr Obonyo has won several NSF awards. Her work has been disseminated through over a hundred journals, papers, conference proceedings and presentations.

Authors

Eric Adzroe

Institution: Evangelical Presbyterian University College, Ghana

Email: kofi.adzroe@epuc.edu.gh

Eric Kofi Adzroe is the director of physical development and estate management at the Evangelical Presbyterian University College Ho, Ghana. He holds a Ph.D. from the University of Salford. He is a chartered builder with many years of industrial experience in the Ghanaian construction industry. He had in the past several industrial engagements with professional bodies within the Ghanaian construction industry. He is actively involved in helping to bring e-business technology to micro enterprises SMEs within the Ghanaian construction industry. His area of expertise and interest is in e-business in construction and smart business.

Dr Aizul Harun BSc MSc PhD

Institution: Universiti Teknologi Malaysia

Email: aizulnahar.kl@utm.my

Dr Aizul is a senior lecturer at Universiti Teknologi Malaysia. During his career, he has worked in the areas of land administration and project development as well as in the management of the environments of the construction projects in the Malaysian private sector for nine years and now in the academic sector since 2009. His areas of research interest include the Internet of Things, environmental management system, building information modelling and sustainable development.

Professor Mustafa Alshawi BSc MSc PhD FCIOB

Institution: University of Salford

Email: M.A.Alshawi@salford.ac.uk

Professor Alshawi is the director of the BIM-Award–winning Unisearch Ltd (UK). He has over thirty years of academic and industrial experience in introducing new technologies to improve business performance of organizations. He was the associate dean at the University of Salford, and before that, he was the director of the Build and Human Environment Research Institute. He is also the editor-in-chief of the international journal *Construction Innovation* and the author of over a hundred and thirty publications in fields such as IT and process improvement, BIM and business process re-engineering.

Professor Chimay Anumba FREng DSc PhD Dr.h.c. CEng/PE FICE FIStructE FASCE

Institution: University of Florida

Email: anumba@engr.psu.edu

Chimay Anumba is professor and dean, College of Design, Construction and Planning at the University of Florida. He is a fellow of the Royal Academy of Engineering. He holds a Ph.D. in civil engineering from the University of Leeds, UK; a higher doctorate—doctor of science—from Loughborough University, UK; and an honorary doctorate from Delft University of Technology (Netherlands). His research interests include advanced engineering informatics, concurrent engineering, knowledge management, distributed collaboration systems, and intelligent systems. He has over four hundred and fifty scientific publications in these fields and more than $150 m in research support. He has also supervised forty-five doctoral graduates and mentored over twenty-three postdoctoral researchers.

Dr Zeeshan Aziz BEng MSc PhD FHEA ICIOB

Institution: University of Salford

Email: Z.Aziz@salford.ac.uk

Dr Aziz is a senior lecturer at University of Salford. He has led work on numerous research bids from funding agencies such as EU, British Council, HEA, Royal Academy of Engineering and Highways England. His areas of research interest include intelligent construction collaboration, building information modelling, building energy planning and modelling and mobile computing.

Dr Anas Bataw PhD

Company: KEF Holdings

Email: anas.bataw@kefholdings.com

Dr Anas Bataw is a certified BIM manager with proficient industrial and academic experience in building information modelling (BIM) and sustainable design in the UK, UAE and Malaysia. Anas has gained recognition as a unique BIM expert by providing high levels of BIM consultation to the public and private sectors towards establishing methodologies and best practices to support BIM implementation, adoption and management in the UK and UAE, while being actively involved with the BIM Academic Forum (BAF) contributing towards the adoption of BIM training and education at universities and training centres in the UK.

Liam Brady MRICS (lapse) MCIOB (lapse) Bsc (Hons) Building Surveying

Institution: Manchester City Council

Email: brady@manchester.gov.uk

Liam has worked at MCC for thirty-eight years, graduating from Salford University in 1997 BSc (hons) in building surveying. As client programme manager on the £100m Town Hall Complex completed in March 2014, he acted as the interface between the project team and Manchester's operational team, including working with Cabinet Office on Government Soft Landings and exploring how the project's digital technologies could add operational value and benefit post-construction.

Liam is currently leading a citywide energy-saving pilot project and is promoting the city to adopt soft landings and lessons learned on future large-scale MCC projects.

Samuel Brand BSc (QS) Hons

Institution: Northumbria University

Email: sam.brand@hotmail.co.uk

Sam Brand is a graduate quantity surveyor currently pursuing a professional cycling career with Team Novo Nordisk. Prior to this, he studied BSc quantity surveying at Northumbria University. He had a year of industry placement with Gardiner and Theobald LLP, subsequently working part time. He graduated with an honours degree in 2015. Before joining the university, he completed a HNC in construction and the built environment whilst working for a local quantity surveying firm. He is a keen sports enthusiast and a professional cyclist and has currently taken a career break in quantity surveying to pursue his sporting career.

Professor Patricia Carrillo CEng BSc MSc PhD FICE FCIOB

Institution: Loughborough University

Email: P.M.Carrillo@lboro.ac.uk

Professor Patricia Carrillo is the associate dean of teaching in the School of Civil and Building Engineering. Her research has focused on the exploitation of ICT on construction projects, knowledge management and lessons learned in UK construction organisations, PFI governance, and mergers and acquisitions in the construction sector. She was awarded the prestigious Royal Academy of Engineering Global Award to undertake research on knowledge management in Canada and the USA. To date, she has co-authored six books, including three on knowledge management, and has published over 170 journal papers, conference papers and reports.

George Charalambous MEng MPhil

Institution: CH2M

Email: giorgoscharalambous86@gmail.com

George Charalambous is assistant structural engineer at CH2M, Birmingham. Prior to that, he was graduate structural engineer at Amey, Birmingham. He obtained an MPhil from Loughborough University for which he conducted a three-year industry-based research project on BIM and collaboration tools, based at a renowned construction software-as-a-service provider. He holds a 1st Class MEng in civil engineering from the University of Bristol. His interests are BIM, digitally-mediated communication, social network analysis, semantic technology, knowledge management, systems thinking and biomimicry in engineering design. He is a member of the Institution of Civil Engineers.

Dr Yongjie Chen BSc MSc PhD PENG

Email: staryjchen@gmail.com

Institution: Independent/Freelance

Dr Yongjie Chen is currently an independent professional engineer in Kingston Ontario, Canada. Her background is a combination of civil/structural engineering and ICT with over ten years' working experience in the industry and eight years' research experience in the areas of electronic procurement, benchmarking and strategic implementation of electronic business in China, North America and the UK.

Joanna Chomeniuk MSc

Institution/organisation: North West Construction Hub

Email: j.chomeniuk@manchester.gov.uk

Joanna Chomeniuk is an organizational psychologist, urbanism enthusiast and framework lead for the North West Construction Hub (NWCH). Her areas of interest are focused on relationships between urban space and its users as well as on modern building technologies and the opportunities that they bring to renovations and revitalization projects. She is interested in social, economic and technological development of urban public space, heritage initiatives and new building technologies. She advises and works with clients who wish to procure building contractors to deliver much more than just a building and who are interested in BIM and achieving social value on construction projects across the North West. Joanna has a master's degree from University of Gdansk, Poland.

Professor Malcolm Cook

Institution: Loughborough University

Email: Malcolm.Cook@Lboro.ac.uk

Malcolm Cook is a professor of building performance analysis and associate dean of research for the School of Civil and Building Engineering in Loughborough University. His research interests include the use of computational fluid dynamics and dynamic thermal simulation programs for modelling natural ventilation in non-domestic buildings.

Dr Peter Demian BA/MA MEng MSc PhD MASCE MCIOB FHEA

Institution: Loughborough University

Email: P.Demian@lboro.ac.uk

Peter Demian is senior lecturer at Loughborough University. He teaches design and design management and conducts research on software for design and construction, particularly BIM. His postgraduate studies were at Stanford University and undergraduate studies at Cambridge University (all in civil engineering). He is a chartered construction manager, a member of the Chartered Institute of Building and the American Society of Civil Engineers, and a fellow of the Higher Education Academy.

Dr Robert Eadie BEng(Hons) MSc(DIS) PhD PGCertPD(Researchers) CEng FIEI MCIHT MAPM EURING FHEA

Institution: Ulster University

Email: r.eadie@ulster.ac.uk

Dr Robert Eadie is course director for the MSc Civil and Infrastructure Engineering course at Ulster University. His PhD was related to e-procurement in construction, and his research focuses on procurement and pedagogy. He spent twenty years in industry before moving into academia. A fellow of Engineers Ireland, he is currently northern region secretary and is on the Professional College of CIGNI. He is a chartered Northern Ireland committee member of CIHT.

Dr Gerald Feldman PhD MSc BSc PGCert LT FHEA

Institution: Birmingham City University

Email: gerald.feldman@bcu.ac.uk

Gerald Feldman is a postdoctoral researcher at the Centre for Enterprise Systems at Birmingham City University, UK. He received his PhD from Birmingham City University in 2015. His PhD research explored enterprise systems upgrade decision processes and drivers. Before joining academia, Gerald has worked in several industries, managing enterprise information systems for over seven years. His research and teaching interests focus on improving the integration of people, processes, and technology to support decision-making at all levels, mainly through the application of structured decision analysis, business process modelling and socio-technical systems theory.

Tristan Gerrish

Institution: BuroHappold Engineering and Loughborough University

Email: Tristan.Gerrish@BuroHappold.com

Tristan Gerrish is a research engineer with BuroHappold Engineering and Loughborough University, investigating how BIM can be used to help understand and improve building energy performance.

Professor David Greenwood MA MSc. FCIOB PhD

Institution: Northumbria University

Email: david.greenwood@northumbria.ac.uk

David Greenwood is professor of construction project management in the faculty of Engineering and Environment, Northumbria University, UK. He is director of BIM Academy (Enterprises) Ltd., an industry-academia joint venture spin-out company, a member of the Board of Construct-IT, and former chair of the Association of Researchers in Construction Management (ARCOM). He has published over one hundred academic journal and conference papers and has authored and co-authored several textbooks. He has over twenty years of experience in consulting, training and lecturing around the world for commercial and governmental organisations and is an active promoter of better practice in the construction industry.

Mark Johnson

Institution: BuroHappold Engineering (at time of writing—but now at Brentwood Design Partnership)

Email: Mark.Johnson@BuroHappold.com (see note above Mark.Johnson@ BrentwoodGroup.co.uk)

Mark Johnson is an associate director in BuroHappold Engineering, leading the Environments team in the north of England. With fifteen years' experience in the industry, he delivers client-focussed, high quality affordable engineering solutions across a wide range of projects.

Professor Patrizia Lombardi PhD MSc BA/MA

Institution: Politecnico di Torino, Interuniversity Department of Urban and Regional Studies and Planning, Viale Mattioli 39, 10125 Turin (IT)

Email: patrizia.lombardi@polito.it

Patrizia Lombardi (PhD, MSc, BA/MA) is full professor of urban planning evaluation and project appraisal of Politecnico di Torino and head of the Interuniversity Department of Regional and Urban Studies and Planning (DIST). She is an established figure in the field of evaluating smart and sustainable urban development for over twenty years, publishing widely in the subject area and coordinating, or serving as lead partner, in several pan-European projects related to smart cities, post carbon society and cultural heritage: BEQUEST, INTELCITY, INTELCITIES, ISAAC; SURPRISE; UNImetrics; MILESECURE-2050; POCACITO; DIMMER; KIC InnoEnergy/ EIT ICT Lab; EEB Cluster/MIUR, SHAPE-ENERGY.

Dr Eric Lou

Institution: University of Manchester

Email: eric.lou@manchester.ac.uk

Eric Lou spent a decade in construction, higher education, information technology and environmental industries in managerial positions; and completed over £60 million in building refurbishment and new build projects before joining academia. He is now the programme director for the MSc in project management (Professional Development Programme) for the Asia Pacific region. His research interest encompasses the trichotomy of people-process-technology in the areas of project management, sustainability, corporate responsibility, BIM and IT management. Dr Lou is also the principal investigator for the Newton-Ungku Omar Institutional Links fund with Malaysia researching into sustainability-led design through building information modeling (SuLeD-BIM).

Sara Moghadam MSc BSc

Institution: Politecnico di Torino, Interuniversity Department of Urban and Regional Studies and Planning, Viale Mattioli 39, 10125 Turin (IT)

Email: sara.torabi@polito.it

Sara Torabi Moghadam graduated *cum laude* from the Polytechnic university of Turin in 2014 with a BSc and MSc in sustainable architecture. She is currently a PhD student under the supervision of Professor Patrizia Lombardi at the Polytechnic University of Turin, where she works in the field of "zero energy buildings in smart urban districts" at the Interuniversity Department of Regional and Urban Studies and Planning. Her research focuses on spatial decision support systems for evaluating different energy retrofitting scenarios for built environment at the urban scale based on multiple criteria analysis, promoting effective and sustainable urban planning toward low carbon cities.

Dr Naif Alaboud BEng MSc PhD

Institution: Umm Al-Qura University

Email: nsaboud@uqu.edu.sa

Dr Naif is an assistant professor at Umm Al-Qura University. His areas of research interest include construction project management, building information modelling, ICT in construction, mobile computing.

Mark Phillip

Institution: BuroHappold Engineering

Email: Mark.Phillip@BuroHappold.com

Mark Phillip is a partner and regional director in BuroHappold Engineering. With over twenty years' experience as a consulting engineer, Mark has delivered many high profile and award winning projects, engaging with clients to deliver technical solutions with value and elegance.

Dr Pathmeswaran Raju PhD MSc BCompSc PGCert FHEA

Institution: Birmingham City University

Email: path.raju@bcu.ac.uk

Dr Pathmeswaran Raju is a reader in knowledge engineering in the Knowledge-Based Engineering Lab at the Birmingham City University (BCU). His research interests centre around investigating the model-based engineering approach for solving complex problems in the areas of knowledge-based engineering and system engineering. He led the development of knowledge models for Rolls-Royce and EU Clean Sky-funded Platform Independent Knowledge Model project and also knowledge-based decision support tools for the Energetic Algae project. Dr Raju joined BCU in September 2011 from the University of Salford, where he worked for six years in an EPSRC-funded Innovative Manufacturing Research Centre.

Dr Alan Redmond PhD MRICS MCSCE PGCE

Institution: UCIrvine Division of Continuing Education

Email: redmonda@uci.edu

Dr Alan Martin Redmond received his PhD in late 2013 from the School of Real Estate & Construction Economics and Management at Dublin Institute of Technology. He is currently completing a professional credit course at University of California, Irvine studying system engineering; a combination of product development and service delivery fundamentals including program and project management techniques and keen business skills. His professional certifications feature both finance and engineering principles, such as a chartered member of the Royal Institution of Chartered Surveyors (MRICS) and member of The Canadian Society of Civil Engineers (MCSCE) and CFIRE (Council Member of Finance, Investment, and Real Estate—National Institute of Building Science, USA) and member of Association Française d'Ingénierie System (AFIS—Recherches et Innovations en IS, France).

Dr Anushi Rodrigo PhD BSc (QS)

Institution: Faithful and Gould, USA

Email: Anushi.Rodrigo@fgould.com

Anushi Rodrigo recently joined Faithful and Gould, USA and currently working with the Koch Pipeline Company (Flint Hills Resources) team, providing large capital cost management and project control services. She recently completed her PhD at Northumbria University, UK majoring in construction e-business. Her PhD focused on e-business capability and maturity in construction processes. She graduated as a quantity surveyor with a first-class degree from University of Moratuwa, Sri Lanka in 2009 and moved to UK in 2010 to complete her postgraduate studies. She currently lives in Minnesota, USA and work and research in the area of construction e-business.

Dr Emine Thompson BA MA MSc PhD FHEA AoU

Institution: Northumbria University

Email: emine.thompson@northumbria.ac.uk

Emine Mine Thompson is a senior lecturer in the Department of Architecture and Built Environment, Faculty of Engineering and Environment, Northumbria University, and the programme leader for MSc Future Cities. She has substantive expertise in areas related to digital place-making activities, in particular to smart and future cities, virtual city modelling and city information modelling, with a focus on smart/future cities with a public engagement and involvement perspective. Emine is also a manager of the Virtual Reality and Visualisation group that runs the Virtual Newcastle-Gateshead project at Northumbria University, and has skills, knowledge and numerous publications relating to BIM, VR, AR and architectural and urban visualisation.

Professor Tony Thorpe CEng CITP BSc MSc PhD FICE FBCS FCIOB FRICS MIMgt

Institution: Loughborough University

Email: A.Thorpe@lboro.ac.uk

Tony Thorpe is dean of the School of Civil and Building Engineering at Loughborough University, and professor of construction information technology. He graduated in civil engineering at Nottingham University, followed by a masters and doctorate at Loughborough. He was seconded for periods to both industry and the Building Research Establishment developing advanced IT systems for construction organisations. His current research interests are in information and communications modelling and systems, innovative construction technologies, and site-based computing. He is a chartered engineer and a fellow of the Institution of Civil Engineers, the British Computer Society, the Chartered Institute of Building and the Royal Institution of Chartered Surveyors.

Dr Jacopo Toniolo PhD MSc BSc

Institution: Politecnico di Torino, Interuniversity Department of Urban and Regional Studies and Planning, Viale Mattioli 39, 10125 Turin (IT)

Email: jacopo.toniolo@polito.it

Jacopo Toniolo, born in 1981, research fellow at Politecnico di Torino since 2008, PhD in energy engineering, is an expert on the energy consumption of building. He has designed more than fifty heating ventilation and air conditioning systems, and he worked in specific research project on district heating and HVAC systems. He developed a specific knowledge in biomass heating systems, building management systems and district energy systems with practical design and European-funded projects (milesecure2050.eu, dimmer.eu, iservcmb.info).

Professor Jason Underwood BEng MSc PhD MCInstCES MBPsS FHEA

Institution: University of Salford

Email: j.underwood@salford.ac.uk

Professor in the digital built environment and programme director of the MSc in building information modelling and integrated design within the School of the Built Environment at the University of Salford. Director of construct IT for business industry-led non-profit–making collaborative membership-based network. Background in civil/structural engineering and construction ICT. Over twenty years' research experience in the area of concurrent engineering, integrated and collaborative computing in construction, product and building information modelling, and organisational e-readiness towards delivering strategic value from ICT investment through both UK- and EU-funded research on which he has published extensively. Editor-in-chief of the *Journal of 3D Information Modelling* (IJ3DIM), specifically focused on BIM along with 3D GIS and their integration.

Michele Victoria BSc (Hons) QS

Institution: Northumbria University

E-mail: michele.f.victoria@northumbria.ac.uk

Michele is currently undertaking a PhD in embodied carbon estimating in buildings at Northumbria University. She also worked as a demonstrator on a part-time basis and involved in other research activities. Before she joined Northumbria University, she worked as a lecturer in the Department of Building Economics, University of Moratuwa, Sri Lanka. She graduated from the same department with a first-class honours degree in quantity surveying. She has published several book chapters, conference papers and research reports in the areas of sustainable development, cost management, social media in construction and organisational studies.

Dr Paul Wilkinson BA PhD DipPR FCIPR

Institutions: University of Westminster; pwcom.co.uk Ltd

Email: paul.wilkinson@pwcom.co.uk

Paul Wilkinson has been working in the UK construction industry since 1987 and researches and writes about construction collaboration technology platforms, SaaS and related developments in fields such as BIM, mobile technologies and social media. He is deputy chair of the information systems panel at the Institution of Civil Engineers, and on the management team at Construction Opportunities for Mobile IT. A fellow of the CIPR, he chairs its policy and campaigns committee and its construction and property special interest group (CAPSIG), and was a member of its social media panel. He also co-founded built environment social media group Be2camp.

David Woodcock

Organisation: Manchester City Council (MCC)

Email: d.woodcock@manchester.gov.uk

David started out working as a project and technical support officer in 1999. After years of sitting behind a computer monitor, he took a more physical role in facilities management with MCC. This helped to shape his knowledge on building issues and the need for accurate building information and data. In 2011, he took a new position within Capital Programmes to investigate how digital technologies used during construction could bring added value to the day-to-day building operations. David now works in Corporate Property to enhance their property systems. He now sits behind two monitors instead of one!

Dr Steven Yeomans BSc(Hons) MSc(Dist) EngD ICIOB AFHEA MInstLM

Institution: Loughborough University

Email: S.G.Yeomans@lboro.ac.uk

Steven Yeomans is the centre research manager at the Centre for Innovative and Collaborative Construction Engineering at Loughborough University, where he completed his own doctorate of engineering in ICT-enabled collaborative methodologies. His subsequent research and teaching focuses on collaborative building information modelling underpinned by cloud based digital technologies. Steven was formerly head of collaboration for an international engineering consultancy, and a consulting board member of the Avanti programme. He has been an advocate of digital collaboration in construction for more than fifteen years.

1 Introduction

Srinath Perera, Kirti Ruikar,
Bingunath Ingirige, and Esther Obonyo

1.1 Background

The construction industry of most countries is considered the growth engine of the economy often acting as the regulator of economic activity. This makes it important to continually innovate its practice and processes. One of the primary ways of infusing innovation in to the construction industry is through the adoption of Information Communication Technologies (ICT) in to its endemic processes. Many previous reports that analysed the level of adoption of ICT by sectors (e-Business W@tch, 2007, 2008, 2010) and its performance (Kang *et al.*, 2013) reported that the construction industry as one of the sectors that is most IT backward in terms of ICT adoption and use.

These issues of low ICT adoption in the construction industry is further exacerbated due to the fact that it is heavily fragmented in its operations in terms of design and construction, through the involvement of different distinct professions (architecture, civil engineering, structural engineering, mechanical engineering, quantity surveying, construction management, planning, among others) multi-layered supply chains and other silos of separation. This provides a greater role for ICT in the construction industry to act as a catalyst for integration and collaboration creating a platform for innovative applications and process re-engineering. ICT offers a plethora of enabling technologies that create the platform for these developments to take hold. This book aims to explore the use of numerous e-business enabling technologies within the construction industry.

This chapter first defines the term construction e-business and introduces the different types of enabling technologies that are explored in the various chapters of this book. Finally, it provides a guide to the structure and the layout of the book.

1.2 Defining construction e-business

This section is an attempt in defining construction e-business by first analysing the definition of e-business and projecting it to application in the construction sector. The term e-business first originated from the work of

IBM (Gerstner, 2002; Chaffey, 2011) and introduced the term e-business as a way of transforming key business processes by using Internet technologies (IBM, 1997). Subsequently they revised the definition giving it a much broader outlook (IBM, 2001: p. 5) as;

> the process of using web technology to help businesses streamline processes, improve productivity, and increase efficiency. It is about using the internet infrastructure and related technologies to enable business anywhere and anytime.

However, in the UK, the Department of Trade and Industry (DTI) developed a much broader view of e-business in their international benchmarking study analysing the adoption of e-business by Small and Medium Enterprises (SMEs). They termed e-business as

> the integration of Information and Communication Technologies (ICTs) for business processes (DTI, 2000).

Here, ICT is defined as any technology used to support information gathering processing, distribution and use which includes hardware, software and network systems (Baynon-Davies, 2013). This broader definition of e-business as adoption of ICT to business processes indicates that it has transformative potential to reinvent business processes through integration and redefining of business processes (DTI, 2004).

Li (2007) takes a similar approach and defines e-business as

> e-business is about developing new ways of working by innovatively exploiting the new capabilities of Information and Communication Technologies in general and the Internet and related technologies in particular.
>
> (Li, 2007: p. 2)

More recently, Chaffey (2011) used a more information centric approach to define e-business activities as

> All electronically mediated information exchanges, both within an organisation and with external stakeholders supporting the range of business processes.
>
> (Chaffey, 2011: p. 12)

These definitions identify several central characteristics in defining e-business. Information and its process of communication, the technologies used are primary constructs of the definition. These processes are considered within the context of business applications. This broader approach to e-business definition is supported by many authors (Laudon and Laudon, 2002;

Aranda-Mena and Stewart, 2005; Anumba and Ruikar, 2008; Xu and Quaddus, 2010; Goncalves *et al.*, 2010).

Therefore, the application of e-business with this broader definition in mind, to the processes in the construction industry is termed as construction e-business. In this book the term construction e-business if defined as 'the application of information communication technologies to construction business processes'. These include business processes utilised by construction companies (be it construction contractors, sub-contractors, suppliers, or consultants such as designers, engineers, managers or surveyors). The process of adoption of ICT to construction processes would naturally lead to creating new ways of conducting business activities which enable transformation of business processes with added value. This is considered as innovation in construction business processes, thus ICT led innovation.

1.3 e-Business–enabling technologies

e-Business–enabling technologies play a vital role in organisations when e-business activities blend with their day to day business practices. The technologies that enable organisations to adopt ICT and reinvent business processes are considered as e-business enabling technologies. Organisations should carefully consider the selections and choices of hardware, software, human resource and IT services as they persuade functional, financial and technical requirements of an organisation (Perera and Karunasena, 2008). The IT solutions generated through the use of e-business enablers are the driving forces behind the e-business adoption that would pronounce potential competitive advantages for adopting businesses. They have changed not only the way in which businesses communicate and interact, but also the way in which information is stored, exchanged and viewed (Anumba and Ruikar, 2008). The construction e-business enabling technologies discussed in this chapter are briefly explored below.

1.3.1 Internet technologies

Internet refers to the global information system that is logically linked together by a globally unique address space based on the Internet Protocol (IP) or its subsequent extensions (Leiner *et al.*, 2009). It is able to support communications using the Transmission Control Protocol/Internet Protocol (TCP/IP) suite or its subsequent extensions, and other compatible protocols and provides, users or makes accessible, either publicly or privately, high level services layered on the communications and related infrastructure. The Internet is implemented in three primary methods (Ashworth and Perera, 2015: chap. 23). Firstly, as the World Wide Web where information is provided to the masses through the use of Internet technologies. Secondly, Intranet is its implementation as a private network used exclusively within a company or organisation. It uses the Internet technologies, but does not necessarily function through the Internet (Vlosky *et al.*, 2000). These are private computing networks, internal to an organisation, allowing access only to authorised users. The third implementation is the Extranets. It is a network that links business partners to one another over the Internet (Volsky *et al.*, 2000) and this linkage is usually occurring by companies allowing their partners to access certain areas of their intranet (Greengard, 1997). It is private to a group of users defined by membership of the group. Chapter 8 explores the use of Extranets for project collaboration. Although there are no other dedicated chapters on the Internet and its variations most enabling technologies discussed are often implemented either fully or partially using Internet technologies.

1.3.2 e-Commerce and its infrastructure

e-Commerce involves use of electronic transactions in commercial process-ing related activities of businesses. There are a number of definitions that explains e-commerce in many ways. This book considers e-commerce as a subset of broadly defined e-business. Chapter 2 of this book explores the e-commerce infrastructure, its classifications, technologies and drivers and barriers.

1.3.3 Cloud computing

Cloud computing (CC) can be defined as 'a model for enabling convenient, on demand network access to a shared pool of configurable computing resources that can be rapidly provisioned and released with minimal man-agement effort or service provider interaction' (Mell and Grance, 1999). Organisations can use CC as a service to obtain software, platform, infra-structure or data storage (Sultan, 2010). CC makes e-business concept more accessible by providing a foundation and cost effective infrastructure for e-business activities. e-Business services can be borrowed through public clouds, private clouds, community clouds or hybrid clouds. There are many advantages of using CC as an enabling technology for e-business within an organisation as on-demand self-service, broad network access, resource pooling, rapid elasticity and measured service (Dillon *et al.*, 2010). In con-struction context, CC can be utilised to develop cost effective collaborative and data sharing solutions. Chapters 9, 10 and 11 explore the use of BIM within a cloud-based environment for collaborative design.

1.3.4 e-Procurement

e-Procurement originated with the use of various digital media such as CDs for media common formats such as EDI for data exchange, but found its nat-ural residence in Internet technologies. Chaffey (2009) defines e-procurement as 'electronic integration and management of all procurement activities including purchase request, authorisation, ordering, delivery and payment between purchaser and supplier'. e-Procurement includes sourcing, tendering, invoicing, auctions and transactions completed utilising electronic means (Ashworth and Perera, 2015). In the construction industry e-procurement includes on the one hand sourcing of labour, material and plant by contractors, and on the other hand the e-tendering processes followed to procure buildings and other structures as well as for materials procurement. Chapter 3 explores e-procurement in details followed by Chapter 4 dealing with e-procurement within a BIM environment.

1.3.5 BIM

Building information modelling (or management) is one of the fastest growing enabling technologies that have high relevance and applicability to the construction industry. In the UK and in many other parts of the world there is significant government led promotion of use of BIM in the construction industry. BIM represents a building or structure as a hierarchical computer based object model often using object oriented modelling techniques. It enables to define buildings and its constituent elements and components in a hierarchical object structure that represent the characteristics of each object in terms of attribute–value pairs. This information rich platform has great potential for integrating the fragmented construction industry provided it is used in the right way as a collaborative tool. These aspects of BIM are explored in Chapters 5, 6 and 7 with industry case studies explaining successful applications of the technology. Chapters 9, 10 and 11 further explore use of BIM as collaboration tools both at project level and district level modelling.

1.3.6 Social media

Social media in one technology that has surpassed the scope of its own original purpose of social interactions to become one of the greatest applications of Internet technologies for business purposes. The advent of Web 2.0 technology has propelled social media beyond mere social interactions to marketing, recruitment, sourcing to many other business applications (Ashworth and Perera, 2015). Chapter 15 provides the background knowledge related to social media in construction with Chapter 16 providing a case study analysis followed by Chapter 17 exploring mobile computing.

1.3.7 AI-agents

Artificial Intelligence grew from its origins of Turing experiments to the popular research paradigm in the 1980s and 1990s to becoming a major part in modern hardware to software applications. The potential for application of AI is limitless. Some construction industry specific applications are based on the paradigm of Multi-Agent Systems (Vermeulen and Pyka, 2015).

Chapter 19 reviews the use of multi agent systems in construction e-business operations taking APRON (Obonyo *et al.*, 2005) a prototype for the specification and procurement of construction products as one such example.

1.4 Structure of the book

This book is an attempt to provide a coherent monograph of the state of construction e-business and its advancements following from Anumba and Ruikar (2008) fulfilling a knowledge gap in construction ICT literature. The book consists of 20 chapters structured in to seven sections:

1   Introductions: Chapter 1 defining the scope of the subject and the text.
2   Procurement: chapters dealing with ICT applications enhancing construction procurement (Chapters 2, 3 and 4).
3   Building information modelling (BIM): reviewing the state of application of BIM (Chapters 5, 6 and 7).
4   Cloud and Collaboration Technologies: reviewing the use of many collaboration technologies such as extranets, could computing and BIM (Chapters 8, 9, 10 and 11).
5   Process Issues: analysing the methodologies and status of ICT adoption related process issues and their management (Chapters 12, 13 and 14).
6   Social and Media Technologies: reviewing the state of Internet based technologies such as social media (Chapters 15 and 16), communication technologies such as mobile computing (Chapter 17) and visualisation technologies such as virtual reality (Chapter 18).
7   Conclusions: providing trends and development for the future with a review of agent technology (Chapter 19) and with conclusions and future trends (Chapter 20).

The following paragraphs provide a quick overview of the individual chapters in the book.

Chapter 1 provides an introduction to the book providing the background context within which this book has been conceived. It defines construction e-business and provides an overview of the e-business enabling technologies that have been reviewed within this book.

Chapter 2 discusses the use of e-commerce with the construction industry exploring e-commerce classifications and technologies used based on features of e-commerce. The chapter provides a detailed account of drivers and barriers for e-commerce in construction.

Chapter 3 analyses the advances in e-procurement in the construction industry taking a step by step detailed look at the various stages of the procurement process. It reviews the European e-advertising requirements for construction contracts. The electronic awarding and e-auctions are analysed in detail explaining the stages involved in each of these e-procurement methods. It also tracks major government backed initiatives that promote e-procurement.

Chapter 4 reviews the drivers for e-procurement and BIM in the construction industry. It compares and contrasts the drivers and barriers to BIM with those of e-procurement. It also looks at the interoperability between the two systems.

Chapter 5 sets the background for building information modelling/ management (BIM) within the context of UK government and other countries taking active steps towards implementation and adoption of BIM within the construction industry. BIM has been reviewed as a radical, disruptive and fast-moving phenomenon. It provides a state of the art account with a full coverage of BIM from its origins to current level of development and beyond, to its future potential.

Chapter 6 is an attempt to showcase the implementation of BIM in the construction industry taking three case studies of BIM applications. Each case study describes the project and its uses of BIM in place of conventional design and construction processes, and then looks at the challenges and solutions developed as part of the BIM adoption throughout these projects.

Chapter 7 provides a detailed single case study of a public sector organisation implementing BIM as a test case. It provides an account of the BIM journey for implementing BIM for the Manchester Central Library and Town Hall Extension Project. The influences of key decisions during this journey and on behaviours that enhance collaboration and cooperation in understanding of BIM requirements are discussed in the chapter.

Chapter 8 provides the latest highlights on extranets for web-enabled project management. Extranet software use in the construction industry enables its participants to communicate, exchange information, data storage and collaborate across a standard platform. The chapter provides details on how extranets have changed everyday business operations and the challenges in the future if such technologies are to be implemented within day to day practice in the construction industry.

Chapter 9 describes the use of business rules associated with semantic knowledge in order to identify appropriate environmental and health policies on the Web. The authors discuss techniques for semantically enhancing policy documents within an open BIM model by exchanging information via BIM XML and representational state transfer (REST) 'systems-of-systems' they adopted to realise their objective of creating a 3D virtual representation model connected to policy documents.

Chapter 10 discusses the development of a context-specific conceptual-model ontology, which can support the discourse of requirements engineering while also providing a robust and universally applicable framework for evaluating the communication capabilities of BIM. The authors outline their use of this approach to develop a model for waste in BIM process interactions in their 'WIMBIM' project.

Chapter 11 discusses the district information modelling concept (DIM) and presents it as a feasible approach in efforts directed at scaling up the impact of energy efficiency models from the individual building to neighbourhood level. DIM is presented in this chapter as a new concept based on integration building information modelling (BIM) and geospatial information systems (GIS) with real-time data. The chapter includes a description of an open platform being developed as part of ongoing research for real-time data processing and visualization at the district level based on information about buildings, the energy distribution grid and user behaviour.

Chapter 12 presents capability maturity modelling (CMM) of construction e-business processes as a ICT process-management methodology that enables organisations to seek improvements to their e-business processes. It reviews CMM concepts and different approaches used in various sectors and evaluates their applicability in the construction industry. It also provides a detailed account of a construction e-business capability model that has been developed for use within the construction industry.

Chapter 13 discusses the development of a novel strategic e-business framework that highlights the aspects to be considered internally (within an organisation) and externally (within projects) to enhance collaboration and derive business benefits from the implementation. It focuses on the development and implementation of the framework. The chapter starts from the discussion of the crucial needs of industry organisations when formulating their e-business strategies, and the he introduction of the adopted methodology for developing the framework. This is followed by a review of the different relevant approaches for strategy formulation and framework development. The main body of the chapter presents a detailed description of the framework and its evaluation. Future implementation consideration is also discussed in the concluding section of the chapter.

Chapter 14 presents SMEs in construction as a very good source for e-business, thereby demonstrating that e-business is not just a tool for larger organisations but also a good tool that will be well received if positioned well among smaller businesses. This chapter reinforces the position of e-business use not only by examining e-business use that are appropriate for SME and micro organisations in construction but it also reports from a developing country case of the Ghananian construction industry to demonstrate the importance of e-business within their industry.

Chapter 15 provides a brief introduction to and history of the emergence of social media, highlighting the evolution from the first-generation 'Web 1.0' technologies of the early 1990s to more user-friendly and interactive 'Web 2.0' Internet platforms. It also discusses the use of social media in the construction industry and uses examples and case studies for added context.

Chapter 16 defines social media, classifies the various social media platforms, discusses the emergence of the social media platforms against a timeline and discusses the benefits of, drivers of and barriers to social media usage across various industries. It then specifically discusses the application of social media in the construction industry using two case studies. Conclusions and discussions highlight the most popular social media platforms among construction organisations, including the drivers and barriers of social media implementation in construction organisations.

Chapter 17 introduces mobile computing in construction as a technology, which is widely being publicised as a means to cope with the problems associated with the hazardous nature of the construction industry. Innovation is generated in this area to link various facets of construction with the power of emerging mobile computing technologies. The chapter provides a very good literature review and a synthesis and also discusses a case of application of mobile services for on-site environmental surveillance. The discussion here will be beneficial for many upcoming practitioners in the construction industry.

Chapter 18 provides a state-of-the-art review on visual communication tools based on an analysis of published literature and selected case studies. The chapter discuss trends that are continue to enhance the interactive and customization features within visualization tools such as 3D modelling, animation/walkthrough, virtual reality, augmented reality, building information modelling, and others.

Chapter 19 discusses the benefits of using agent-based systems in information and knowledge management for construction. In described first generation agent-based models that were generally speaking, agent-centred multi-agent systems' (ACMAS). The authors discuss the potential for scaling up their impact through the design and use of organization-centred multi-agent systems' (OCMAS). They also examine feasibility of adapting solutions-driven agent models to enhance data-driven decision support tools.

Chapter 20 provides a conclusion to the book. It first provides a brief summary of the key issues covered in each chapter before following on to advancements and enablers in ICT in construction. It then provides an account of key barriers to e-Business in construction and explores the benefits of adoption of ICT in construction processes and practices. Finally it provides an overview of future directions in e-Business in Construction.

References

Anumba, C. J. and Ruikar, K. (2008). *e-Business in Construction*. Oxford, UK: Wiley-Blackwell.

Aranda-Mena, G. and Stewart, P. (2005). 'Barriers to e-business adoption in construction: international literature review', QUT Research Week 2005 Conference. Brisbane 4–8 July 2005.

Ashworth, A. and Perera, S. (2015). *Cost Studies of Building*. Abingdon and New York: Routledge.

Beynon-Davies, P. (2013). *eBusiness*. 2nd ed. New York: Palgrave Macmillan.

Chaffey, D. (2011). *e-Business and e-commerce management: strategy, implementation and practice*. 5th ed. Essex: Pearson Education Limited.

————. (2009) *e-Business and e-commerce management.* 4th ed. Essex: Pearson Education Limited.

Dillon, T., Wu, C. and Chang, E. (2010). Cloud computing: issues and challenges. 24th IEEE International Conference on Advanced Information and Applications, pp. 27–33.

DTI (2000). Business in the Information Age—International Benchmarking Study 2000. UK Department of Trade and Industry.

———— (2004). Business in the Information Age—International Benchmarking Study 2004, UK Department of Trade and Industry. https://www.google.co.uk/url?sa=t&rct=j&q=&esrc=s&source=web&cd=4&ved=0ahUKEwiChvrX04bRAhWDVZQKHd8_BE0QFggpMAM&url=http%3A%2F%2Fwww.knowledgebusiness.com%2Fknowledgebusiness%2FTemplates%2FViewAttachment.aspx%3FhyperLinkId%3D2254&usg=AFQjCNGiE2-vw4M1xL1OdnG-ZJSN_dFpAg&sig2=Vyn5ptmPV4F5fgNQrSMk_A&cad=rja. Accessed on 22 December 2016.

e-Business W@tch (2006–7). The European e-business report, A portrait of e-business in 10 sectors of the EU economy, 5th Synthesis Report of the e-Business W@tch, Retrieved January 28, 2012, from http://bookshop.europa.eu/en/the-european-e-business-report-pbNBAU06001/

———— (2008). The European e-Business Report 2008 The impact of ICT and e-business on firms, sectors and the economy 6th Synthesis Report of the Sectoral e-Business Watch, Office for Official Publications of the European Communities, 2008 ISBN 978-92-79-09355-5, Retrieved March 21, 2016, from http://aei.pitt.edu/54205/1/2008.pdf

e-Business W@tch (2010). ICT and e-Business for an Innovative and Sustainable Economy, 7th Synthesis Report of the Sectoral e-Business Watch (2010). Retrieved March 21, 2016, from http://www.aimme.es/archivosbd/observatorio_oportunidades/ICT_and_e-busuness_for_an.pdf

Gerstner, L. V. (2002). Who says elephants can't dance. New York: Harper Collins Publishers Inc.

Goncalves, R. M., Santos, S. S. and Morais, E. P. (2010). E-business maturity and information technology in Portuguese SMEs, IBIMA Publishing [Online]. [Viewed April 2015]. Available from: http://www.ibimapublishing.com /journals/CIBIMA/2010/303855/303855.pdf.

Greengard, S. (1997). Extranets – Linking Employees With Your Vendors. Workforce. Nov. 1997 28–33.

IBM (1997). IBM100 – e-business. [Online]. IBM's 100 Icons of Progress. [Viewed April 2015]. Available from: http://www-03.ibm.com/ibm/history/ibm100/us/en/icons/ebusiness.

———— (2001). IBM @server iSeries e-business handbook: a V5R1 technology and product reference. IBM Corporation, International Technical Support Organization.

Kang, Y., O'Brien, W. J. and Mulva, S. P. (2013). Value of IT: indirect impact of TI on construction project performance via best practices. *Automation in Construction*, 35, pp. 383–396.

Laudon, K. C. and Laudon, J. P. (2002). Management Information systems: managing the digital firm. 7th ed. USA: Prentice-Hall Inc.

Leiner, B. M., Cerf, V. G., Clark, D. D., Kahn, R. E., Kleinrock, L., Lynch, D. C., Postel, J., Roberts, L. G. and Wolff, S. (2009). A brief history of the Internet. *ACM SIGCOMM Computer Communication Review*, 39 (5), pp. 22–31.

Mell, P. and Grance, T. (2010). The NIST Definition of Cloud Computing, NIST, viewed 21 June 2011, http://www.nist.gov/itl/cloud/upload/cloud-def-v15.pdf

Obonyo, E. A., Anumba, C. J. and Thorpe, A. (2005). Specification and procurement of construction products using agents. In Anumba, C. J., Ugwu, O. O and Ren. Z., eds, *Agents and Multi-Agent Systems in Construction*. London and New York: Taylor & Francis Group.

Perera, S. and Karunasena, G. (2008). A decision support model for the selection of best value information technology procurement method. *Journal of Information Technology in Construction*, 13, pp. 224–243.

Sultan, N. (2010). Cloud computing for education: A new dawn?. *International Journal of Information Management*, 30, pp. 109–116.

Vermeulen, B. and Pyka, A. (2015). Agent-based Modeling for Decision Making in Economics under Uncertainty. Economics Discussion Papers, No. 2015–45, Kiel Institute for the World Economy. http://www.economics-ejournal.org/economics/discussionpapers/2015-45.

Vlosky, R. P., Fontenot, R., and Blalock, L. (2000). Extranets: Impacts on Business Practices and Relationships. *Journal of Business and Industrial Marketing*, 15(6).

Xu, J. and Quaddus, M. (2010). E-business in the 21st century: realities, challenges and outlook. Singapore: World Scientific Publishing Co. Pte. Ltd., Intelligent Information Systems – Vol. 2.

2 Exploiting e-commerce in construction

Pathmeswaran Raju and Gerald Feldman

2.1 Introduction

The advancement in technology has provided organisations with a platform that offers an alternative manner to deliver their services; as such, many organisations now rely on the World Wide Web (WWW) and the Internet as a medium to offer their services and conduct business. e-Commerce plays an important role in construction organisations' day-to-day business practices. It is considered that e-commerce solutions would have a greater impact on the fragmented construction industry by unifying the different stakeholders, including clients, regulatory authorities, consultants, contractors and the supply chain. The use of e-commerce in the construction industry is steadily growing when compared to other industrial sectors, and, in some cases, it is becoming a mainstream method of conducting transactions such as procurement and payment of commodities between companies. This chapter provides an overview of e-commerce and e-commerce technologies, presents extant e-commerce literature in construction and explores the drivers and barriers to the adoption of e-commerce within the construction industry.

2.2 e-Commerce

2.2.1 History of e-commerce and definitions

The use of e-commerce in the construction industry is steadily growing when compared to other industrial sectors, however, in some cases it is becoming a mainstream method of transactions such as procurement and payment of commodities between companies. The emergence of e-commerce can be associated with electronic data interchange (EDI), which was used mainly for data transfer between organisations (Bhutto *et al.*, 2005a). The development of EDI led the beginning of e-commerce revolution in 1960s with the digital transfer of documents from one computer to another, reducing the need for sending traditional mails and faxes. Businesses including construction and building related companies transferred the order forms, invoices, and delivery information electronically from their computers to other company's computers. According to Bhutto *et al.* (2005b) EDI created the foundation for e-commerce, yet EDI cannot be considered as an e-commerce solution because of its configuration. The development of electronic funds transfer (EFT) also enabled businesses to do the electronic transactions over the computer networks. In 1979, the American National Standards Institute (ANSI) introduced a data exchange standard, ASC X12 for sharing business documents over the computer networks. The development of communications networks such as Advanced Research Projects Agency Computer Network (ARPANet) and the terminal interface processor (TIP) helped to revolutionise the network communications in late 1970s.

With the advent of Transmission Control Protocol (TCP) and Internet Protocol (IP) and the resulting ARPAnet's switch over to TCP/IP in 1982 enabled all computers to transmit information equally using an approved standard. This created the next phase in computer networking and e-commerce. In 1990, Tim Berners Lee proposed the hypertext project to build World Wide Web using an interface called 'browser'. He also developed the hypertext mark-up language (HTML), with specifications for uniform resource locators (URLs). In 1991, National Science Foundation lifted the ban on commercial businesses operating over the Internet, paving the way for e-commerce revolution. With the arrival of graphical user-friendly browsers such as Mosaic and the HTML webpages, the web-based e-commerce shifted to its next phase of development.

Websites such as BidCom and Cephren attempted to provide web portals for design and construction and were effective for typical project management tasks (Johnson *et al.*, 2002). Johnson and Xia (2000) discussed a number of early e-commerce case studies in construction. One of the large architecture/engineering firms, 3D/International attempted to improve its competitive position by providing customised e-commerce tools to clients (Johnson and Xia, 2000). Conoco, one of the global energy companies introduced standardised use of the off-the-shelf software and e-commerce concepts to enhance productivity, allowing the organisation to easily connect with all employees and reducing the cost of both implementation and training (Johnson and Xia, 2000). Web portals such as HomeWire that provided services and tools for homeowners, builders, and contractors, and FreeMarkets that conducted online auctions for industrial parts, raw materials, and commodities allowed construction industry to embrace the web-based e-commerce technology.

Bhutto *et al.* (2005b) suggest that there are many variations to the definition for e-commerce in the literature and it is based on the perspective it is applied. For example Kalakota and Whinston (1997) categorise the definition of e-commerce into three perspectives: firstly as means for delivering information, buying and selling products and (or) services. Secondly, as an approach to simplify workflows and automate processes. Thirdly, as a tool to manage and reduce operational costs and improve service delivery along with customer satisfaction. Garrett and Skevington (1999) cited in (Bhutto *et al.*, 2005b) explain e-commerce as 'trading by means of new communication technology (everything beyond voice telephony, fax and telex)'. It includes all aspects of trading, including commercial market making, ordering, supply chain management, and the transfer of money'. Anumba and Ruikar (2002) suggest that e-commerce is 'doing business by electronic means, typically over the Internet'. This was reiterated by Kamaruzaman *et al.* (2010) suggesting that e-commerce is the act of selling or purchasing services or products over the Internet.

On the other hand, Baladhandayutham and Venkatesh (2010) claim that e-commerce is part of a broader term 'e-business', which includes buying and selling online along with inventory tracking, managing production, supply chain management, customer support services, and collaborative engineering. Stewart (2001) supports this explanation of e-business; however, he suggests that it is necessary to review the processes, cultivate new skill sets, and building new relationships with customers. While Ruikar and Anumba (2009) explanation of 'e-business' bears similarity to the broader definition for e-commerce offered by Kalakota and Whinston (1997), which can be summarised as the application of technology for automation of business transactions, the delivery of information, products/services and other online services over the Internet. Drawing from these various definitions, it is apparent that e-commerce is not only about conducting business via the Internet, but also about online communication, collaboration, and commercial transactions between multiple organisations, individuals, and government agencies.

2.2.2 *Classification of e-commerce*

e-Commerce can be classified into the following categories based on the parties involved in a transaction (Bhasker, 2013). Laudon and Traver (2014) included peer-to-peer (P2P) and mobile commerce (m-commerce) as additional classifications of e-commerce.

- Business-to-business (B2B)
- Business-to-consumer (B2C) or consumer-to-business (C2B)
- Consumer-to-consumer (C2C)
- Business-to-government (B2G) or government-to business (G2B)
- Business-to-employee (B2E)
- Peer-to-peer (P2P) e-commerce
- Mobile commerce (m-commerce)

2.2.2.1 *Business-to-business (B2B) e-commerce*

Business-to-business (B2B) e-commerce is about facilitating business transactions electronically between two or more businesses, directly or via an intermediary. B2B e-commerce includes all types of interactions among businesses. This includes order management, invoice and payments, and inventory management. Bhasker (2013) defines three types of B2B e-commerce models such as buyer-centric, supplier-centric, and intermediary-centric. In a supplier-centric model, which is usually the business model for dominant suppliers, supplier creates an electronic market place for other businesses with customised pricing and solutions. In a buyer-centric model, major businesses with high volume purchasing power set-up market places for purchases of goods and services. In an intermediary-centric model, a third party creates a market place for buyers and sellers to carry out e-commerce.

2.2.2.2 *Business-to-consumer (B2C) or consumer-to-business (C2B) e-commerce*

Business-to-consumer (B2C) e-commerce facilitates business transactions between businesses and consumers. In B2C, businesses sell goods and services directly to consumers without any need for intermediaries and the consumers have the capabilities to choose the businesses based on the quality and pricing. Businesses benefit from direct marketing and more profit, while consumers benefit from lower prices and timely delivery. On the other hand, consumer-to-business (C2B) e-commerce offers the opportunity for consumers to specify the requirements and the budget for particular goods or services and then businesses match the requirements with best offers.

2.2.2.3 *Consumer-to-consumer (C2C) e-commerce*

Consumer-to-consumer (C2C) e-commerce facilitates business transactions among consumers. It provides market places for consumers to trade products and services to other consumers through Internet. In this model, both buyers and sellers perform transactions over the third party trading platforms such as auction sites, and have the ability to negotiate the prices and communicate directly with other consumer. The buyers often benefit from lower prices, while the sellers have the responsibility for packaging and shipping. In some cases such as blogs and reviews that target other consumers, no business transaction is carried out between consumers; however, the publishers may benefit from online advertisements and web traffic.

2.2.2.4 *Business-to-government (B2G) or government-to business (G2B) e-commerce*

Business-to-government (B2G) and government-to business (G2B) e-commerce refers to transactions that are performed between businesses and governments. B2G e-commerce covers services provided to government departments by businesses in order to provide public access to government services such as tax affairs, vehicle and transport services, and local taxes. G2B e-commerce covers the business activities that are aimed at businesses to fulfil government service requirements such as e-tendering. B2G and G2B e-commerce activities increased in the last decade due to governments looking to reduce the cost and increase the use of Internet to provide public services.

2.2.2.5 Business-to-employee (B2E) e-commerce

Business-to-employee (B2E) e-commerce covers intra-organisational business activities between the employees and organisation. This is mainly carried out through the company intranet and could include human resource functions such as sharing and updating personal information, financial transactions such as expenses claims and personal development activities such as booking training events. B2E e-commerce also facilitates the communication between the management and the employees and helps to improve the organisation-employee relationships.

2.2.2.6 Peer-to-peer (P2P) e-commerce

In peer-to-peer (P2P) e-commerce, the data and information are shared by users directly with other users without any intermediary or central web server. Users who are both suppliers and consumers form the nodes of the peer-to-peer network, share their computer resources such as storage space and Internet bandwidth to access, and share the data. The P2P networks have been used to share music files, videos and software, and introduced new model of user interaction, however due to copyright issues of sharing files, some of the popular P2P networks have been shut down. Since then, hybrid models have been introduced to include both peer-to-peer and client-server models.

2.2.2.7 Mobile commerce (m-commerce)

Mobile commerce (m-commerce) covers the use of mobile devices to carry out business transactions over the wireless Internet. Businesses and consumers use mobile devices such as smart phones and tablet computers and connect to use other businesses and consumers through Wi-Fi hotspots and 3G/4G mobile Internet to conduct business operations such as money transfer, stock trading, and buying and selling product and services. With the arrival of smart phones and the mobile apps, the use of m-commerce has increased significantly in recent years.

2.3 e-Commerce in construction

2.3.1 Related studies

With the success of e-commerce businesses such as Amazon and eBay, many businesses including construction organisations began to embrace the potentials of e-commerce in their business operations. The construction industry is regarded as an essential contributor to the country's economy development (Bhutto *et al.*, 2005b), according to Rhodes (2015) 'in 2014 the construction industry in the UK contributed £103 billion in economic output, 6.5% of the total'. While the use of e-commerce is also contributing towards the economy for example from 2006 up to 2010, the revenue is US$10 trillion, with United States and Europe contributing 79% to this global revenue (Kamaruzaman *et al.*, 2010). While in the UK, according to Office for National Statistics (2015) e-commerce sales were around £1.5 billion, which has shown 24% growth rate between 2008 and 2014. Hence, it can be argued that e-commerce has transformed the world of business and introduced new business models. Although, the construction industry has often been slow to adopt new technologies, there are indication that organisations tend to adopt a small subset of the current e-commerce technologies, such as e-procurement, e-tendering, e-collaboration, e-auctions, and e-payment (Stewart, 2001; Anumba and Ruikar, 2009). Various research studies have been undertaken to exploit the potentials of e-commerce in construction, as shown in Table 2.1.

Due to the surging popularly of building information modelling (BIM) in industry and academia, Ren *et al.* (2012) attempted to combine e-commerce and BIM approaches and proposed a framework to integrate BIM and e-commerce in a material procurement process. BIM integrates key building data in a 3D model that can be used in collaborative environment to manage product information throughout a project lifecycle. It has been considered as a game-changing technological development for the construction industry. However, the benefit of integrating e-commerce with BIM is not yet fully realised and therefore it is unlikely to have any real impact on the e-commence development in the near future.

On the other hand, the retail industry is fast adopting the Internet of Things (IoT) technology, which is a network of connected objects that can communicate with each other wirelessly. The IoT has been predicted as a technology to push e-commerce to the next level and therefore could prove to be a significant technology for e-commerce development in construction industry. Manyika *et al.* (2015) predict that B2B applications of IoT have greater economic potential than consumer applications and estimate that the economic impact of IoT applications could be from $3.9 trillion to $11.1 trillion per year in 2025 (Manyika *et al.*, 2015). However, in order to realise the full potential of e-commerce and any emerging technologies, construction sector has to innovate itself by improving its processes, up-skilling its workforce, investing in research and development, taking calculated risks and embracing new technologies.

2.3.2 *Technologies for e-commerce in construction*

The e-commerce has challenged the traditional business models and introduced new ways of conducting businesses. The popularly of e-commence models and their successes are made possible by its unique features as discussed by Laudon and Traver (2014). These eight dimensions of e-commerce technologies are listed in Table 2.2.

Table 2.1 Some of the e-commerce research studies in construction

Construction activity	e-Commerce applications in construction	Studies
Procurement	Construction material procurement	Kong and Li (2001)
	Construction material procurement using Internet-based agent system	Hadikusumo et al. (2005)
	e-Procurement in construction Sector SMEs	Vitkauskaite and Gatautis (2008)
	Information sharing between e-commerce systems for construction material procurement	Kong et al. (2004)
	Intelligent agents for construction procurement negotiation	Dzeng and Lin (2004)
	Geographical Information Systems System for e-commerce application in construction material procurement	Li et al. (2003)
Construction alternatives	On-line decision support system for construction alternatives	Kaklauskas et al. (2007)
Project Management	Web-enabled Project Management	Alshawi and Ingirige (2003)
Supply chain	Construction supply chain coordination	Xue et al. (2007)
	Supply chain optimization in B2B construction marketplaces	Castro-Lacouture et al. (2007)
Bidding	e-Bidding proposal preparation system	Arslan et al. (2006)
Process improvement	Construction business process improvement through e-commerce	Ruikar et al. (2003)
Collaboration	Project-based web platform for collaboration in social e-business model	Costa and Tavares (2012)
	e-Commerce application system for collaboration and information sharing	Wang et al. (2007)
	Intranet and extranet to generate a co-operative virtual workplace and business environment	Wong (2007)
Waste management	e-Commerce based waste exchange platform for demolition waste exchange	Chen et al. (2006)
Information management	A web-based system for facilitating construction information management and communication	Chassiakos and Sakellaropoulos (2008)

Table 2.2 e-Commerce technologies for construction

Unique features of e-commerce	e-Commerce technologies
Ubiquity	• Internet and Web technology such as TCP/IP and HTML • Devices such as personal computers, laptops, smart phones and tablet computers • Broadband, 3G/4G mobile technologies
Global Reach	• Internet and web technology such as HTTP, SSL, web browsers • Web servers and clients • Internet telephony—Voice Over Internet Protocol (VOIP)
Universal Standards	• Internet and web standards and languages such as HTML, XML, CSS, DOM, ECMAScript, URLs and SSL • Web accessibility standards
Richness	• Video, audio and text via Internet and Web • Streaming media delivering audio/video • Databases and database servers
Interactivity	• Web and interaction technologies such as 3D and animation • Video conferencing and teleconferencing • Tele-immersion such as virtual and augmented reality
Information density	• Communication and collaboration technologies such as e-mails and web chats • Intelligent databases and data mining techniques • Search engines and intelligent agents • Intranets (located within a single organisation) and Extranets (can be accessed by outsiders)
Personalisation/customisation	• Communication technology, standards and tools such as personalised emails, instant messaging and chats • Customisable user interfaces with touch-screen capabilities
Social technology	• Communication and mobile technologies • Big Data and analytical technologies • Cloud computing and storage technologies

Source: Adopted from Laudon and Traver (2014).

The construction organisations can benefit from e-commerce technologies in various ways. The benefits are discussed according to the identified e-commerce features.

- **Ubiquity:** Construction organisations are able to access the Internet and the World Wide Web from anywhere, anytime, through personal computers and mobile devices. This enables them to conduct commerce, communicate, and collaborate with other business, consumers and governments around the clock from any part of the world.
- **Global reach:** Construction organisations can communicate, collaborate, conduct business operations and carry out commercial transactions with other businesses, consumers and governments across the world without any boundaries. This provides them with global market for their products and services and offers access to global supply chain of manufacturers and suppliers.
- **Universal standards:** Construction organisations can conduct business transactions and communicate across the globe seamlessly and securely without any modification. This reduces the cost and time of business operations and improves productivity and efficiency. Businesses can also develop web-based business applications that can be used by anyone and are built with forward and backward compatibilities in evolving web standards.
- **Richness:** Rich media including video and audio can be produced by the construction businesses and streamed to partners and clients for purposes such as marketing and communication. This increases the consumer experience and can reach out to wider audiences. Rich media can also be used to provide training to employees through videos and facilitate face-to-face meetings with video conferencing.
- **Interactivity:** Construction businesses can benefit from rich and personalised interaction with other businesses and consumers through online mediums. Organisations are able to provide face-to-face experience to their partners and clients on a global scale through 3D and animated graphics. Businesses can also deliver visualisation and interaction capabilities such as 3D building models about the products and services through virtual and augmented reality techniques.
- **Information density:** Construction organisations benefit from reduced cost and time to process and store complex business and financial data by using intelligent databases and data mining techniques. Businesses can also improve the quality and accuracy of information and provide up-to-date information to employees, partners and clients. Using the internal networked information such as Intranets, organisations are able to supply data and information to the right people at the right time using information retrieval capabilities.
- **Personalisation/customisation:** Construction businesses can benefit from personalised products and services delivered to consumers and partners.

This will provide them with effective communication and collaboration mechanisms with other businesses and consumers. Manufacturers and builders are able to supply customised products, services, and orders based on individual specifications according to clients' budgets.

- **Social technology:** Construction organisations can utilise the social content created by consumers to support the business planning and business intelligence. Social technology can be used to understand consumer demand, preferences, and market trends. Peer reviews and ratings that are provided by users can be used to improve the business processes and promote the products and services. Business can also benefit from the access to social networks for potential consumers and businesses and for e-marketing purposes.

2.4 Drivers and barriers for e-commerce adoption in construction

2.4.1 Drivers for e-commerce adoption

Bhutto *et al.* (2005b) argue that the construction industry cannot afford to lag behind in the adoption of e-commerce technologies, which can be utilised to address competitive market conditions, constant price fluctuation and the need to provide sustainable and innovative project delivery strategies along with procurement practices. There are several initiatives to promote the adoption of technology solutions within the construction industry (Smyth, 2010), as a means to improve performance and make the industry more competitive. Some of these initiatives within the UK construction industry date back to 1998, for example the Egan *et al.* (1998) report, which proposed several measures to 'advance the knowledge and practice of construction best practice', developing trust and respect with the construction supply chain, measuring performance, etc. Despite this earlier initiative, there have been little progress, as reported in the HM Government (2013) construction 2025 report, which sets a vision that 'Construction in 2025 is no longer characterised, as it once was, by late delivery, cost overruns, commercial friction, late payment, accidents, unfavourable workplaces, a workforce unrepresentative of society or as an industry slow to embrace change'. This implies there is still a lag in the construction industry when it comes to the adoption of technology that can improve the construction industry's efficiency and effectiveness.

In context of this chapter, drivers are regarded as the factors that encourage the adoption of e-commerce technologies within the construction industry. There are several studies (Love *et al.*, 2001; Stewart, 2001; Ruikar, 2004; Eadie *et al.*, 2007; Eadie *et al.*, 2010; Lou and Goulding, 2010; Isikdag *et al.*, 2011; Laryea and Ibem, 2014a) that have explored the drivers and barriers for e-commerce technologies adoption in the construction industry. These drivers, as in Table 2.3, have been identified from the above-published work and are used to provide context to the discussion of promoting e-commerce adoption.

Table 2.3 Drivers of e-commerce

Drivers of e-commerce	
Service/process trends	Streamlined supply chain Integrated service
Organisational	Increased process transparency and visibility Continuous innovation
Technology advancements	Enterprise applications: connect the organization Infrastructure convergence: merge voice, data and video Information access and connectivity

It is understood that the construction industry is more project-driven and have various stakeholders working towards a single project (Goulding and Lou, 2013), with the possibility spanning multiple clients, suppliers, contractors etc. Optimising the business processes and providing a flexible fulfilment and delivery, which could result in integrated services that would potentially offer consistent and reliable information and effective communication channels. In addition, process re-engineering can lead to transparency and visibility across all business sector, which could lead to more co-operation and collaboration, along with establishing business relationships. According to Lou and Alshawi (2009), having the ability to collaborate and share information with stakeholders is a sought after trait in any industry. This reflects on the organisation's stability, which in return could lead to the organisation gaining a competitive edge.

With the advancements and technology infrastructure, it is regarded as a significant enabler of e-commerce and an indicator for growth. Taking advantage of the technological trends could lead to high market diffusion, which could open-up multiple channels for conducting business, this could be through provide end-to-end connectivity with all stakeholders, which could improve company's image to external environment (community and corporate). Thus, it is postulated that the adoption of e-commerce would result in cost reduction, performance improvement, competitive edge and efficient procurement process, improve project planning, supply chain management, logistics and document management (Anumba and Ruikar, 2009; Aboelmaged, 2010; Baladhandayutham and Venkatesh, 2012). These factors could be utilised as the baseline for assessing the need for change, especially when considering adopting e-commerce technologies to support the construction industry.

2.4.2 Barriers for e-commerce adoption

Barriers refer to the commonly found obstructions to e-commerce within the construction industry. The hesitation to adopt the e-commerce solution can be associated with the stakeholders' mind-set (Lindsley and Stephenson,

2008). This has been a main concern that has been highlighted in previous studies, for example (Ruikar, 2004) indicated the need for construction industry to embrace change and move away from its traditional working practices. In fact, Ruikar (2006) suggest that when it comes to stakeholders' opinion about adoption of e-commerce in constructions industry, there are two school of thought: those who are in support and those who are against e-commerce technologies. Those in support of e-commerce face stiff competition to provide detailed reasoning to support the change from traditional methods that most people are used to. Possibly this could be associated with lack of understanding of the benefits that can be brought about by embracing new approaches or maybe fear of being replaced with technological solutions (Love *et al.*, 2001).

Although Baladhandayutham and Venkatesh (2012) suggest e-commerce solutions could aid in potential performance improvement, Lou and Alshawi (2009) list several factors that can hinder the adoption of e-commerce. These are: the industry's structure; a fragmented supply chain; and limited support from few 'champions'. Another study conducted by Ruikar (2004) establishes the barriers for the adoption of e-commerce technologies within the UK construction sector, these are: technology capabilities, telecommunication networks, trust and reliability and regulatory issues. Love *et al.* (2001) explored barriers for e-commerce adoption in Australian construction industries. They established several barriers and categorised these drivers into four main categories: organisational (such as lack of employee knowledge), financial (such as the cost of system requirements and maintenance), technical (such as risks associated with security and authentication) and behaviour (such as fear of losing one's job). Several other studies (Eadie *et al.*, 2007, 2010; Isikdag *et al.*, 2011; Laryea and Ibem, 2014a) have examined the barriers of e-commerce adoption, extending the literature by providing alternative barriers to the adoption of e-commerce.

Most of these studies categorised these drivers into either external environment, internal environment, organisation or technology to provide a more holistic perspective to the effect of these drivers. Although there are numerous studies and variation of the different barriers classification, these barriers are interrelated and are interdependent to each other. For example, technology capabilities could be influenced by the external environment such as suppliers and customers. Al-Somali *et al.* (2015) utilises T-O-E framework to explore e-commerce adoption and suggest the framework would assist in explaining the inter-relationship between the different barriers. Additionally Rowe *et al.* (2012) suggests that the T-O-E framework an established framework to study adoption of technology within organisation. In the Technology-Environment-Organisation (T-O-E) framework, Tornatzky and Fleischer (1990) suggest that the adoption of technology is influenced by external, internal and technology factors. The T-O-E framework offers three contexts: technology, organisation, and environment. The technology context represents existing and new technologies relevant to the organisation, including their benefits, compatibility, and complexity. Organisational context describes the internal measures such as scope, size, managerial support, and availability of resources. Environmental context refers to the field in which the organisation operates; this includes elements such as government legislation. Although in this section the T-O-E constructs are not adopted, the categorises are utilised as a means to offer a more generic categorisation of the barriers identified from the different studies mentioned above, in order to establish commonality between the barriers and their interactions. Table 2.4 summarises the different barriers identified from previous studies into technology, organisation, and environment context.

2.4.2.1 Technology barriers

This categorisation refers to the role of technology in the adoption of e-commerce. While the advancement of technology is always considered to be an advantage to the any industry, Flanagan and Marsh (2000) suggest the important element is how organisations leverage these technologies. The advancement of technology introduces its own challenges as such as integration of the different systems such as e-procurement, e-tendering, e-collaboration, e-auctions, e-payment etc. As there is no holistic e-commerce solution, integration becomes an on-going challenge that organisations need to consider and support with relevant knowledge of the

Table 2.4 Summary of e-commerce adoption barriers

Studies	Categories	Barriers
Love *et al.* (2001); Ruikar (2004); Rankin *et al.* (2006); Eadie *et al.* (2007); Isikdag *et al.* (2011); Oyediran and Akintola (2011); Laryea and Ibem (2014a)	Technology	Lack of widely accepted e-commerce solutions
		Compatibility issues
		Integration of e-commerce with other systems
		Interoperability of e-commerce systems
		Lack of Flexibility
		Security issues
		Infrastructure
	Organisation	Management support
		Cultural Issues
		Lack of technical expertise
		Low or lack of awareness
		Operational cost Issues
	Environment	Loss of business relationship with costumers
		Legislative/government support
		Complicated procedures and extended relationships
		Ownership of information

technologies. The challenges of integration may also result in interoperability issues when different application are utilised. However, Eadie *et al.* (2007) highlight the availability of approve packages for construction industry trading electronically (CITE) and some supporting open source software, as means to address this barrier. However, they still argue that interoperability between applications to be a concern for e-commerce adoption. This also may result in aligning the processes with the technology offerings; in fact, according to Love *et al.* (2001), the identification of which technology to be used to match the organisation's requirements is one of the major barriers. This was reiterated in Lou and Alshawi (2009), where it was suggested that the barriers to e-commerce are process alignment with technology functionality. For example, Love *et al.* (2001) cite one of the respondents suggesting that when technology adopted are not aligned to the requirements it may lead to added costs and losses to the business.

The frequent technology advancement may also lead to systems being incompatible with each other for example, if one of the stakeholders within the supply chain uses the latest technology, and the other stakeholders use outdated systems, this may result in compatibility issues, which could lead to information failing to be transferred appropriately. Additionally, there is growing concern for security and confidentiality of the information during transmission (Eadie *et al.*, 2007); as with any Internet system, the data is transferred to multiple network providers. Thus, there is an extra effort required to ensure no tempering of information is done and that the e-commerce infrastructure is secured and not easily hacked. Therefore, it is essential for organisations to evaluate the benefits of technology before adopting it; however, each organisation perceives benefits differently, and what one organisation considers valuable may be different to another organisation. Therefore, most organisations gauge the benefits of e-commerce based on its usefulness and effectiveness to streamline the processes, improve performance, and reduce operational costs.

2.4.2.2 *Organisation barriers*

The organisation barriers refer to internal factors that hinder the adoption of e-commerce; these include management support, cultural issues, lack of expertise and operational costs. Normally, the adoption of e-commerce would result in processes change (Ruikar, 2004), that aims to align the technology to the organisation requirements. This ensures the organisation adapts to the new medium of delivering their services and market conditions. In order to support these transformations, there is a need to eliminate redundant processes and re-engineering some of the existing processes or formulate new processes. These results in organisation requiring invest in experts or re-training their staff in order to ensure they can take full advantage of the new technology (Love *et al.*, 2001; Eadie *et al.*, 2007). Additionally, when contemplating adoption of any technology, it is important to gain top management involvement and support as it plays a significant role in the decision. For example, when the top management are in support, these projects will be given priority and assigned realistic timelines and resources. The cases where management support is not attained, there is very little possibility for such projects to be given any considerations. Therefore management buy-in is very critical, as it shows the level of encouragement to use new technology to support the organisations operations (Al-Somali *et al.*, 2015). However, in order to gain this buy-in a clear justification which outlines the benefits and return on investment needs to be provided in order to persuade the top management (Stewart, 2001). According to Ruikar *et al.* (2006) and Chen *et al.* (2013) people play an important role in supporting change, as they could be resist technology changes in fear of their jobs or simply because they have no expertise to use the technology. In their study Kheng and Al-Hawamdeh (2002) report that 60% of their respondent preferred other solutions when compared to e-procurement. According to Eadie *et al.* (2007) the resistance to change can be addressed providing effective leadership and creating an organisational culture that is in favour of change. Additionally, to promote change, there is a need to understand the need for change, the training requirements, along with communicating these to all levels of employees and stakeholder, in order to create inclusiveness and identifying change champions within the organisation.

2.4.2.3 Environmental barriers

According to Chaffey *et al.* (2009) environment barriers include consideration of both macro-environment and micro-environment that hinder the adoption of e-commerce. These normally include the various stakeholders, such as suppliers, consumers, partners, and legal entities. The interaction and collaboration of these stakeholders are vital for the success of these projects that these stakeholders work in cohesion (Goulding and Lou, 2013). Thus, it can be argued that the stakeholder's can hinder the adoption of e-commerce, especially if some of them are still operating using traditional methods. For example, if an organisation implements an e-commerce solution there is a possibility to hinder collaboration with others, as the information would still be required to be submitted manually to those using traditional methods. Thus, for an effective e-commerce adoption, all the partners need to be convinced to adopt the technologies. This could explain why organisations avoid changing from the traditional ways of operating or why they avoid just utilising some aspect of e-commerce technologies that assist with the daily routine, such as e-tendering or e-procurement (Stewart, 2001; Anumba and Ruikar, 2009). In addition, e-commerce technologies rely on the Internet as a medium to conduct business; hence, the variety of national approaches and strategies, or the lack thereof (Carayannis and Popescu, 2005), could hinder the adoption within the global collaboration environment. Therefore, in order to be effective and truly borderless, there is a need for a global standardisation of the strategies, unique standards to be followed globally in order to ease the adoption of e-commerce solutions. However, the manner in which construction industries operate is very complex and is governed and informed by regulatory bodies and institutional policies to become competitive. This implies that there is a need for evaluating the legal risks, according to Ismail and Kamat (2006), these legal risks include contract formation, attribution, privacy, conflict of laws, authentication etc., prior to adopting any technology in order to comply with the regulations and standards. For example, Eadie *et al.* (2007) report in their finding that 'only 26% of respondents agreed that ICT was acceptable as admissible written proof during construction'. This indicates that the level of awareness of the advantages offered by e-commerce and technology as a whole is still limited, specifically when it comes to regulatory and legislative issues, which may result in some stakeholders being reluctant to adopt e-commerce solutions.

2.4.3 *Exploiting e-commerce in construction*

The use of e-commerce in the construction industry could lead the industry in the right direction of fulfilling the construction 2025 vision and be in line with the recommendations of the Egan report. As the adoption and use of e-commerce technologies could result in performance improvement, integrated supply chain and lean construction (Laryea and Ibem, 2014b). However, the adoption of e-commerce is a complex undertaking, and the decision is dependent on balancing the interaction of numerous technological, organisational and environmental barriers with an understanding of the benefits. One key observation from previous studies (for example Bhutto *et al.*, 2005b; Ruikar, 2006; Laryea and Ibem, 2014a), is that the construction industry is still lagging behind in the adoption of e-commerce technologies. Probably creating an awareness of the e-commerce benefits and its application could be vital in promoting its adoption. Some of the barriers can be overcome by readily available solutions. For example, the issue of security can be addressed with the implementation of firewalls, and other security measures to reduce the risks of security breaches. Hence, it can be argued that more attention should be given to identify strategies to enable and promote the realisation and leveraging of the benefits of e-commerce.

To date, there are several studies proposing models and frameworks, such as Ruikar *et al.* (2006) proposed VERDICT, Chen *et al.* (2013) proposed a strategic e-business framework and Al-Somali *et al.* (2015) advocated a stage-oriented model (SOM) for e-commerce adoption; all these studies aimed to support organisations to evaluate their e-readiness. According to Lou (2010) the evaluation of e-readiness is a global agenda; however, it is still not fully explored. Ruikar *et al.* (2006) and Chen *et al.* (2013) focused their efforts on the construction industry, and it is argued that the use of these e-readiness tools would provide an in-depth understanding on the need to change and identify the skills and knowledge gaps, along with exploring the external environment factors that could hinder the adoption of e-commerce solution. Thus, such an approach would allow the organisation to demonstrate the level of its readiness to adopt and establish the opportunities and benefits for adopting e-commerce. This could be a step forward in promoting e-commerce adoption. However, it should be noted that the adoption of e-commerce varies depending on the company size and type of service offered. Therefore, any model for adoption should take into consideration the business needs and the organisation ability of leveraging the potential of e-commerce technologies.

The IoT applications could potentially revolutionise the construction sector, from applications in construction sites for equipment maintenance and health and safety of workers, to energy management in buildings, to human productivity applications using the augmented-reality devices (Manyika *et al.*, 2015). Bitcoin, which is a digital currency and a peer-to-peer payment system that allow payments to be sent directly from one party to another without going through a central authority could revolutionise IoT based e-commerce in construction (Noyen *et al.*, 2014). By combining e-commerce with IoT, the management and ordering of construction materials and supplies could be automated, the building energy usage could be reduced with smart thermostats and appliances, and the status of construction equipment can be monitored, maintenance and repair services can be arranged and the invoices can be settled automatically. Turber *et al.* (2014) and Porter and Heppelmann (2014) proposed new business models based on IoT and construction organisations have to embrace the possibilities offered by the IoT in their current and future business models. However extant literature in construction discipline have not yet provided effective methods for such considerations or implementations.

2.5 Summary and conclusions

It is evident from the success of e-commerce in retail and other sectors that e-commerce technology has the potential to provide several benefits to construction industry. However, the adoption and implementation of e-commerce technologies in construction industry are limited compared to other sectors. This is due to several challenges and issues as discussed in terms of technology, organisation and environment factors. This chapter discussed e-commerce and the related benefits to construction businesses. Through e-commerce, construction organisations can communicate, collaborate, conduct business and carry out commercial transactions with other businesses, consumers, and governments across the world. Businesses are able to provide face-to-face experience to their partners and clients on a global scale through visualisation and interaction capabilities. However, the decision to adopt e-commerce is dependent on balancing the interaction of numerous technological, organisational and environmental barriers. This chapter argued that more attention should be given on identifying strategies to promote the realisation and leveraging of the benefits of e-commerce in construction sector. It is also argued that the use of e-readiness tools would provide necessary understanding on the need to change and explore the external factors that could hinder the adoption of e-commerce solution.

References

Aboelmaged, M.G. (2010). Predicting e-procurement adoption in a developing country: an empirical integration of technology acceptance model and theory of planned behaviour. *Industrial Management and Data Systems*, 110(3), 392–414.

Alshawi, M. and Ingirige, B. (2003). Web-enabled project management: an emerging paradigm in construction. *Automation in Construction*, 12(4), 349–364.

Al-Somali, S.A., Gholami, R. and Clegg, B. (2015). A stage-oriented model (SOM) for e-commerce adoption: a study of Saudi Arabian organisations. *Journal of Manufacturing Technology Management*, 26(1), 2–35.

Alshawi, M. and Ingirige, B. (2003). Web-enabled project management: an emerging paradigm in construction. Automation in Construction, 12(4), pp. 349–364.

Anumba, C.J. and Ruikar, K. (2002). Electronic commerce in construction—trends and prospects. *Automation in Construction*, 11(3), 265–275.

Anumba, C.J. and Ruikar, K. (2009). *e-Business in construction*. Wiley-Blackwell, Oxford, UK.

Arslan, G., Tuncan, M., Birgonul, M.T. and Dikmen, I. (2006). E-bidding proposal preparation system for construction projects. *Building and Environment*, 41(10), 1406–1413.

Baladhandayutham, T. and Venkatesh, S. (2010). B2B e-Commerce: an integrated approach for construction industry. In *2010 International Conference on Financial Theory and Engineering (ICFTE)*, Proceedings, IEEE, Computer Society, Dubai, United Arab Emirates, 18–20 June, pp. 127–131.

Baladhandayutham, T. and Venkatesh, S. (2012). Construction industry in Kuwait—an analysis on e-procurement adoption with respect to supplier's perspective. *International Journal of Management Research and Development*, 2(1), 1–17.

Bhasker, B. (2013). *Electronic commerce: framework, technologies and applications.* Tata McGraw-Hill Education, New Delhi.

Bhutto, K., Thorpe, A. and Stephenson, P. (2005a). E-commerce in UK construction. In *Proceedings of the 3rd International Conference on Innovation in Architecture, Engineering and Construction*, 2, pp. 989–998, Rotterdam, the Netherlands.

Bhutto, K., Thorpe, T. and Stephenson, P. (2005b). E-commerce and the construction industry. In *Proceedings of the 21st Annual Association of Researchers in Construction Management* (Khosrowshahi, F., ed.), pp. 1345–1353, SOAS, University of London, London.

Carayannis, E.G. and Popescu, D. (2005). Profiling a methodology for economic growth and convergence: learning from the EU e-procurement experience for central and eastern European countries. *Technovation*, 25(1), 1–14.

Castro-Lacouture, D., Medaglia, A.L. and Skibniewski, M. (2007). Supply chain optimization tool for purchasing decisions in B2B construction marketplaces. *Automation in Construction*, 16(5), 569–575.

Chaffey, D., Ellis-Chadwic, F., Mayer, R. and Johnston, K. (2009). *E-business and e-commerce management*, 4th ed., Prentice Hall, London.

Chassiakos, A. and Sakellaropoulos, S. (2008). A web-based system for managing construction information. *Advances in Engineering Software*, 39(11), 865–876.

Chen, Y., Ruikar, K. and Carrillo, P.M. (2013). Strategic e-business framework: a holistic approach for organisations in the construction industry. *Journal of Information Technology in Construction*, 18, 306–320.

Chen, Z., Li, H., Kong, S.C., Hong, J. and Xu, Q. (2006). E-commerce system simulation for construction and demolition waste exchange. *Automation in Construction*, 15(6), 706–718.

Costa, A.A. and Tavares, L.V., 2012. Social e-business and the Satellite Network model: Innovative concepts to improve collaboration in construction. Automation in Construction, 22, pp. 387–397.

Dzeng, R.-J. and Lin, Y.-C. (2004). Intelligent agents for supporting construction procurement negotiation. *Expert Systems with Applications*, 27(1), 107–119.

Eadie, R., Perera, S. and Heaney, G. (2010). Identification of e-procurement drivers and barriers for UK construction organizations and ranking of these from the perspective of quantity surveyors. *Journal of Information Technology in Construction*, 15, 23–43.

Eadie, R., Perera, S., Heaney, G. and Carlisle, J. (2007). Drivers and barriers to public sector e-procurement within Northern Ireland's construction industry. *Journal of Information Technology in Construction*, 12, 103–120.

Egan, J., Raycraft, M., Gibson, I., Moffatt, B., Parker, A., Mayer, A., Mobbs, N., Jones, D., Gye, D. and Warburton, D. (1998). Rethinking Construction. Available at: http://constructingexcellence.org.uk/wp-content/uploads/2014/10/rethinking_construction_report.pdf [Accessed 23rd January 2016].

Flanagan, R. and Marsh, L. (2000). Measuring the costs and benefits of information technology in construction. *Engineering, Construction and Architectural Management*, 7(4), 423–435.

Goulding, J.S. and Lou, E.C.W. (2013). E-readiness in construction: an incongruous paradigm of variables. *Architectural Engineering and Design Management*, 9(4), 265–280.

Hadikusumo, B., Petchpong, S. and Charoenngam, C. (2005). Construction material procurement using Internet-based agent system. *Automation in Construction*, 14(6), 736–749.

HM Government. (2013). Construction 2025, Industrial Strategy: Government and Industry in Partnership. (URN BIS/13/955). Available at: https://www.gov.uk/government/uploads/system/uploads/attachment_data/file/210099/bis-13-955-construction-2025-industrial-strategy.pdf [Accessed 23rd January 2016].

Isikdag, U., Underwood, J., Ezcan, V. and Arslan, S. (2011). Barriers to e-procurement in Turkish AEC industry. In Proceedings of the CIB W78-W102 2011: International Conference, Sophia Antipolis, France.

Ismail, I.A. and Kamat, V.R. (2006). Evaluation of legal risks for e-commerce in construction. *Journal of Professional Issues in Engineering Education and Practice*, 132(4), 355–360.

Johnson, R., Clayton, M., Xia, G., Woo, J.H. and Song, Y. (2002). The strategic implications of e-commerce for the design and construction industry. *Engineering Construction and Architectural Management*, 9(3), 241–248.

Johnson, R.E. and Xia, G. (2000). The impact of e-commerce on the design and construction industry. In Proceedings of the 22nd Annual Conference of the Association for Computer-Aided Design in Architecture, Washington, D.C.

Kaklauskas, A., Zavadskas, E.K. and Trinkunas, V. (2007). A multiple criteria decision support on-line system for construction. *Engineering Applications of Artificial Intelligence*, 20(2), 163–175.

Kalakota, R. and Whinston, A.B. (1997). *Electronic commerce: a manager's guide*. Addison-Wesley, Reading, MA.

Kamaruzaman, K.N., Handrich, Y.M. and Sullivan, F. (2010). e-Commerce adoption in Malaysia: trends, issues and opportunities. In Ramasamy, R. and Ng, S. (eds.) *ICT Strategic Review 2010/11 E-commerce for Global Reach PIKOM*. The National ICT Association of Malaysia, Putrajaya, Malaysia.

Kheng, C.B. and Al-Hawamdeh, S. (2002). The adoption of electronic procurement in Singapore. *Electronic Commerce Research*, 2(1), 61–73.

Kong, C. and Li, H. (2001). E-commerce application for construction material procurement. *International Journal of Construction Management*, 1(1), 11–20.

Kong, S.C., Li, H., Hung, T.P., Shi, J.W., Castro-Lacouture, D. and Skibniewski, M. (2004). Enabling information sharing between e-commerce systems for construction material procurement. *Automation in Construction*, 13(2), 261–276.

Laryea, S. and Ibem, E.O. (2014a). Barriers and prospects of e-procurement in the South African construction industry. In 7th Annual Quantity Surveying Research Conference, Pretoria, South Africa.

Laryea, S. and Ibem, E.O. (2014b). Patterns of technological innovation in the use of e-procurement in construction. *Journal of Information Technology in Construction*, 19, pp. 104–125.

Laudon, K.C. and Traver, C.G. (2014). *E-commerce: business, technology, society*, 10th ed., Pearson, Boston, MA.

Li, H., Kong, C., Pang, Y., Shi, W. and Yu, L. (2003). Internet-based geographical information systems system for e-commerce application in construction material procurement. *Journal of Construction Engineering and Management*, 129(6), 689–697.

Lindsley, G. and Stephenson, P. (2008). E-tendering process within construction: a UK perspective. *Tsinghua Science and Technology*, 13(S1), 273–278.

Lou, E.C.W. (2010). E-readiness: how ready are UK construction organizations to adopt IT. In *Proceedings of the 26th Annual ARCOM Conference*, pp. 6–8, Leeds, UK.

Lou, E.C.W. and Alshawi, M. (2009). Critical success factors for e-tendering implementation in construction collaborative environments: people and process issues. *Journal of Information Technology in Construction,* 14, 98–109.

Lou, E.C.W. and Goulding, J.S. (2010). The pervasiveness of e-readiness in the global built environment arena. *Journal of Systems and Information Technology*, 12(3), 180–195.

Love, P.E.D., Irani, Z., Li, H., Cheng, E.W.L. and Tse, R.Y.C. (2001). An empirical analysis of the barriers to implementing e-commerce in small-medium sized construction contractors in the state of Victoria, Australia. *Construction Innovation: Information, Process, Management*, 1(1), 31–41.

Manyika, J., Chui, M., Bisson, P., Woetzel, J., Dobbs, R., Bughin, J. and Aharon, D. (2015). The Internet of Things: Mapping the Value Beyond the Hype. [Accessed 23rd January 2016]. Available at: http://www.mckinsey.com/insights/business_technology/the_internet_of_things_the_value_of_digitizing_the_physical_world.

Noyen, K., Volland, D., Wörner, D. and Fleisch, E. (2014). When Money Learns to Fly: Towards Sensing as a Service Applications Using Bitcoin. CoRR abs/1409.5841. [Accessed 15 March 2016]. Available at: http://arxiv.org/pdf/1409.5841v1.pdf.

Office for National Statistics. (2015). E-commerce and ICT Activity, 2014. [Accessed 15 January 2016]. Available at: http://www.ons.gov.uk/ons/dcp171778_425690.pdf.

Oyediran, O.S. and Akintola, A.A. (2011). A survey of the state of the art of e-tendering in Nigeria. *Journal of Information Technology in Construction*, 16, 557–576.

Porter, M.E. and Heppelmann, J.E. (2014). How smart, connected products are transforming competition. *Harvard Business Review*, 92(11), 64–88.

Rankin, J.H., Chen, Y. and Christian, A.J. (2006). E-procurement in the Atlantic Canadian AE industry. *Journal of Information Technology in Construction*, 11, 75–87.

Ren, Y., Skibniewski, M.J. and Jiang, S., 2012. Building information modeling integrated with electronic commerce material procurement and supplier performance management system. Journal of Civil Engineering and Management, 18(5), pp. 642–654.

Rhodes, C. (2015). Construction Industry: Statistics and Policy. [Accessed 6 October 2015]. Available at: http://www.parliament.uk/commons-library, House of Commons Library.

Rowe, F., Truex, D. and Huynh, M.Q. (2012). An empirical study of determinants of e-commerce adoption in SMEs in Vietnam: an economy in transition. *Journal of Global Information Management*, 20(3), 23–54.

Ruikar, K. (2004). Business process implications of e-commerce in construction organisations. Doctor of Engineering (EngD) Thesis, Loughborough University, Loughborough, UK.

Ruikar, K. (2006). E-commerce in construction. *Journal of Information Technology in Construction*, 11(Special Issue), 73–74.

Ruikar, K., Anumba, C. and Carrillo, P. (2003). Reengineering construction business processes through electronic commerce. *The TQM magazine*, 15(3), 197–212.

Ruikar, K. and Anumba, C.J. (2009). e-Business: the construction context. In Anumba, C.J. and Ruikar, K. (eds.) *e-Business in construction*. Wiley-Blackwell, Oxford, UK.

Ruikar, K., Anumba, C.J. and Carrillo, P.M. (2006). VERDICT—an e-readiness assessment application for construction companies. *Automation in Construction*, 15(1), 98–110.

Smyth, H. (2010). Construction industry performance improvement programmes: the UK case of demonstration projects in the 'Continuous Improvement' programme. *Construction Management and Economics*, 28(3), 255–270.

Stewart, P. (2001). The role of e-commerce systems for the construction industry. *The Australian Journal of Construction Economics and Building*, 1(2), 24–36.

Tornatzky, L.G. and Fleischer, M. (1990). *The processes of technological innovation*. Lexington Books, Lexington, MA.

Turber, S., Vom Brocke, J., Gassmann, O. and Fleisch, E. (2014). Designing business models in the era of Internet of things. In Hutchison, D., Kanade, T. and Kittler, J. *et al.* (eds.) *Advancing the impact of design science: moving from theory to practice*. Springer International Publishing, Cham, Switzerland, pp. 17–31.

Vitkauskaite, E. and Gatautis, R. (2008). e-Procurement perspectives in construction sector SMEs. *Journal of Civil Engineering and Management*, 14(4), 287–294.

Wang, Y., Yang, J. and Shen, Q. (2007). The application of electronic commerce and information integration in the construction industry. *International Journal of project management*, 25(2), 158–163.

Wong, C.H. (2007). ICT implementation and evolution: case studies of intranets and extranets in UK construction enterprises. *Construction Innovation*, 7(3), 254–273.

Xue, X., Wang, Y., Shen, Q. and Yu, X. (2007). Coordination mechanisms for construction supply chain management in the Internet environment. *International Journal of project management*, 25(2), 150–157.

3 Advances in electronic procurement for the construction industry

Robert Eadie and Srinath Perera

3.1 Introduction

This chapter traces e-procurement from its inception to its current status. It provides the background to its development, definition and description as to what it is and why it is needed. It focuses on e-procurement in the construction industry and indicates all of the stages it impacts on during the procurement process. It shows how to complete electronic advertisements for European Contracts and traces the main tendering process electronically and indicates how electronic awarding can take place. It also reviews the use of electronic auctions in construction before indicating the impact that e-procurement has had on the timelines used in the tender process. It concludes that at all stages electronic procurement has had a positive impact on the construction procurement process.

3.2 Background

The advent of computers in the construction industry brought great advantages in efficiency and effectiveness. This has now increased through interface management using building information modelling (BIM) at all stages of the construction process (Lim, 2015). Processes originally carried out on paper can now be carried out electronically. The letter 'e' started to be put in front of these processes to denote that they were now electronic. e-Business and e-procurement started to be recognised. e-Business can therefore be defined as the application of information communication technologies (ICT) to business activities and processes (Eom *et al.*, 2015). e-Business is a very wide term encompassing numerous activities and processes, one of which is procurement. In the construction sector, the procurement function is defined as the activities undertaken by a client or employer who is seeking to bring about the construction or refurbishment of a building (JCT, 2015). In looking at the associated cost of construction procurement activities it accounts for around 8%–10% of the gross domestic product (GDP) (Miller *et al.*, 2015; Lopes, 2012). The purchasing of construction activities therefore has a huge impact on the economies of countries. It is imperative therefore to get the procurement function correct.

3.2.1 Fundamentals of e-procurement

An e-procurement system may include an e-tendering system with web services that replaces the complete manual paper-based system (Hung *et al.*, 2013). The use of the Internet for collaboration and document management purposes (Kähkönen and Rannisto, 2015) also allows greater opportunity for the exchange of electronic contract documents (e-procurement) through the initial phase to contract administration. e-Procurement therefore provides a viable electronic alternative to the more traditional paper-based processes, such as tendering, associated with the procurement process. Electronic procurement relates to the complete buying and purchasing process and in construction relates mainly to the purchase of materials and equipment. The different elements of e-procurement are e-tendering, e-sourcing, e-invoicing, e-auctions, and e-transactions.

Minahan and Degan (2001) categorised these different elements of procurement into three categories: indirect procurement, direct procurement and sourcing. In the construction industry, indirect procurement can be purchasing day to day essentials such as office supplies, raw materials and sundries. This contains the elements e-transactions, e-invoicing and at times e-sourcing and e-auctions. e-Transactions relates to 'transaction-based' elements of purchasing, and electronically enables through live databases links to CD web-based interfaces, that allow purchasing. It incorporates the various methods indicated in Figure 3.1.

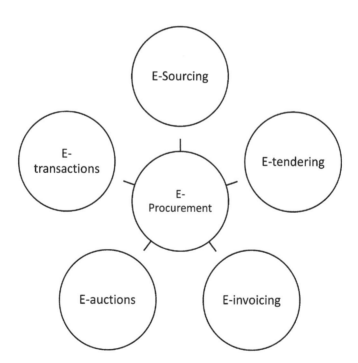

Figure 3.1 Methods of electronic transactions.

e-Invoicing can be defined as payment of invoices electronically. e-Sourcing relates to the provision of product information, negotiation on its price, purchasing and where necessary maintenance. Hung *et al.* (2013) states that e-procurement can make all these elements more efficient and effective. Lu *et al.* (2014) shows that there are two different types of procurement outsourcing: direct procurement outsourcing and indirect procurement outsourcing. Lu *et al.* (2014) state indirect procurement outsourcing is increasing as many companies have to procure services ranging from office services to IT services sourcing and results in increased efficiency in procurement. Direct procurement, sometimes called supply chain management, involves purchasing goods and organising activities to manufacture a completed product or products. In the construction industry, the purchase of materials, plant and labour services are examples of areas where direct procurement is used. Sourcing is defined according to Kim and Shunk (2003) as pertaining equally to indirect and direct procurement and is in the form of a four phase models (information, negotiation, settlement, and after-sales). It can be made electronic through to e-sourcing. The main element of construction procurement is the procurement of projects, carried out through the tender process. Making it electronic is known as e-tendering.

3.2.2 e-Tendering in construction

The tender process in construction involves all elements of the Kin and Shunk (2003) four phase model: information – advertising, negotiation – the tender process, settlement – contract award and after-sales – contract management. European Dynamics (2014b) in Figure 3.2 provides a schematic of these various stages. It can be seen that they call their e-advertising stage, e-notification, followed by e-tendering. Their e-awarding stage includes e-auctions. In this chapter, the initial stage described in the 'Stages of e-procurement section' equates to the European Dynamics (2014b) e-notification stage and is known as e-advertising because it contains the notifications at each stage of the process.

e-Tendering relates mainly to the purchase of construction projects, services or maintenance contracts. Brook (2008) states the construction industry initially used the 'Traditional Tendering' process to achieve fixed price lump-sum contracts. However, this strategy changed in 2001, when the UK government decided to move to multi-stage, quality price tendering for all public procurement (CPDNI, 2012a). This is a two stage process the initial supply of the documentation and then the assessment and award process of the contract. In this chapter, both of these stages are described in the 'Process/stages of e-procurement' section.

Traditional tendering outlined by Brook (2008) involves a single stage process, where tender documentation is released for economic operators to price according to Griffith, Knight and King (2003). As part of this process, the interested parties submit a 'tender' or price, via a bill of quantities or priced activity schedule, with lump sum prices for each of the activities, and the lowest of the prices (or bids) would usually be successful (Brook, 2008).

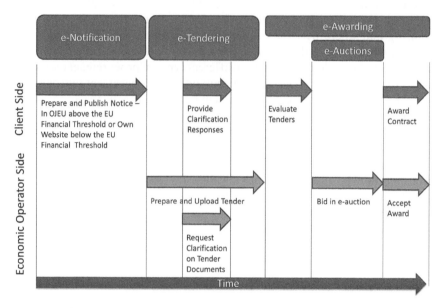

Figure 3.2 Stages of e-tendering.
Source: After European Dynamics, 2015b.

After determining the lowest bid documents are exchanged and a contract is entered into (Laryea and Hughes, 2009). This exchange of documents allows the contract to be evidenced in writing (Murdoch and Hughes, 2008, p. 117). Loosemore and Richard (2015) indicate that lowest price awards remain the dominant selection criterion in tenders.

However, Loosemore and Richard (2015) prove the bidder who submits the lowest price may not produce the same quality of construction due to lacking the experience or expertise. The quality assessment for selection takes into account other factors such as company employee experience, health and safety record, and performance on previous projects (CPDNI, 2012a). The UK government decided to assess quality at both 'prequalification' or selection stage and at a Quality/Price second or award stage by adopting the 'Restricted Procedure' in the Public Contracts Regulations (The Stationery Office Limited (TSO), Draft 2015, 2006, Amended 2009, Amended 2011). The 2015 Public Contracts Regulations came into force on 26 February 2015 (Cabinet Office, 2015). However, many longer term contracts were awarded under the previous regulations and the timescales for these have been included for completeness in Figure 3.3.

3.2.3 Process/stages of e-procurement

As described in the previous section the process of e-procurement for the tender process is split into four stages: e-advertising, e-document management, e-awarding and e-auctions. This process and how it has been adapted for e-tendering is shown in Figure 3.3. This indicates the stages that e-procurement takes in the restricted e-tendering procedure. The first stage relates to electronic advertising. This is necessary before the tender process begins for the initial advertisement notifying that the project is available to tender for and also after it concludes due to the need to advertise the award of the project for contracts over the European Threshold. The second stage relates to provision and submission of tender documentation. This stage is necessary for both the supply of the first stage prequalification documentation and also the invitation to tender and tender documentation. Once completed the tenders need to be submitted electronically. The third stage is analysis of the submitted documentation and correct selection and award. The fourth and final section relates to electronic auctions and is not mentioned in Figure 3.3 but is allowed for in the EU directives. Each one of these will be dealt with in depth in the sections following.

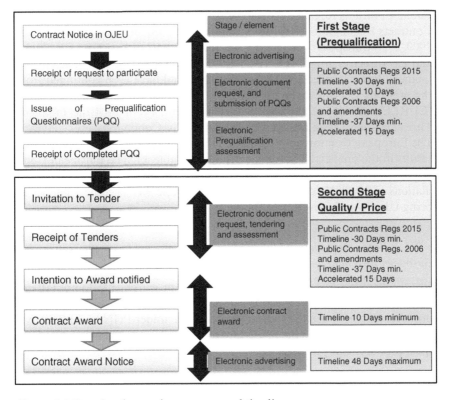

Figure 3.3 Restricted procedure process and timelines.

Table 3.1 2016 European financial thresholds above which the EU regulations apply

Public contracts regulations financial thresholds—from 1 January 2016

Financial threshold for:	Supply, services and design contracts	Works contracts	Social and other specific services new in 2016
Central Government	£106,047	£4,104,394	£589,148
	€135,000	€5,225,000	€750,000
Other contracting authorities	£164,176	£4,104,394	£589,148
	€209,000	€5,225,000	€750,000
Small lots	£62,842	£785,530	n/a
	€84,000	€1,000,000	

3.2.3.1 European procedures required for the e-tendering process

The Restricted Procedure is mainly used for contracts above the European Threshold. However, it can be used and is being used for many contracts under the European Threshold as stipulated in PGN 05-12 (CPDNI, 2012a). The Public Contracts Regulations 2015 (TSO, 2015) which superseded the Public Contracts Regulations 2006 (TSO, 2006) and subsequent amendments (TSO, 2009, 2011) transpose the current EU directive 2014/24/EU (EURLEX, 2014a) and EU directive 2014/25/EU (EURLEX, 2014b) into UK Law. The current financial limits applied from 1 January 2014 and have been continued from the previous EU directives 2004/17/EC (EURLEX, 2004a), and 2004/18/EC (EURLEX, 2004b) of the European Parliament. From 2004 the Financial Thresholds have been changed by the European Community every 2 years. The European Commission Regulation (EU) No 1336/2013 (EURLEX, 2013) results in the financial Thresholds defined in Table 3.1. The European Commission provides Sterling equivalents (Table 3.1). The Public Contracts Regulations (TSO, 2015) defines those to whom Table 3.1 Schedule 1 applies as being UK Central Government Bodies.

3.2.3.2 Procedures required for the e-tendering process below the European threshold

The thresholds mean that smaller government procurements do not fall under the scope of this European legislation. However, many of these smaller procurements still use the two stage procedure under PGN 05-12 (CPDNI, 2012b). Therefore this is the procedure that this chapter will use to illustrate the e-procurement process.

3.3 Stages of e-procurement

3.3.1 Stage 1 electronic advertising

Figures 3.2 and 3.3 indicate where electronic advertising fits into the standard overall e-procurement sequence. Advertising of government contracts is required in many locations worldwide. In the United States advertising of government contracts Clause 6101 b(1) states 'Unless otherwise provided in the appropriation concerned or other law, purchases and contracts for supplies or services for the federal government may be made or entered into only after advertising for proposals for a sufficient time' (Justia US Law, 2012). In Canada, the MERX system allows electronic advertisement to a group of prequalified suppliers and contractors or an unrestricted broad audience (MERX, 2015). AusTender (2015) allows tendering and then takes it to the next stage with delivery of documentation. Jetro in Japan (Jetro, 2015) provides a similar system.

3.3.1.1 Electronic advertisements UK background

The UK Government brought in legislation in 1988 named the Local Government Act for England and Wales (HMSO, 1988) which was the first to relate to advertising Government Projects. This related to Local Government. In this document, Clause 4(4) stated advertisements for public sector work had to be inserted into at least one newspaper. Furthermore, advertisement of Local Government contracts were required through Clause 7 of this Act (HMSO, 1988) and this was the first time advertisements for public sector Construction Works tenders had to be published by law. This was a paper based system and Clause 4(4) required the placing of advertisements in at least one newspaper. On 30 March 1999, a much wider remit on providing all public services electronically followed in the publication of the 'Modernising Government' white paper (Stationary Office, 1999). When published the document suggested the Internet was just an additional medium that could be used to further information provision with procurement advertisements under indicator BV157.

3.3.1.2 Electronic advertisements European background

Within the European Union, contracts above the European Threshold need to be advertised in the Official Journal of the European Union (OJEU). The OJEU has three sections L, C and S. The L Series contains EU legislation including directives and regulations. The C Series contains EU Case Law decisions and calls for expression of interest for EU programmes. The final section, called the supplementary S series, contains all the advertisements relating to contracts above the EU threshold. In the UK listings of advertisements of tender opportunities can be found at Tenders Direct (Tenders Direct, 2015). Advertisements consist of a number of parts from the very initial stages through to contract award, namely

- Prior information notice
- Contract notice
- Contract award notice
- Periodic indicative notice – utilities
- Contract notice – utilities
- Contract award notice – utilities
- Qualification system – utilities

- Notice on a buyer profile
- Simplified contract notice on a dynamic purchasing system
- Public works concession
- Contract notice – concession
- Design contest notice
- Results of design contest
- Notice for additional information, information on incomplete procedure or corrigendum
- Voluntary ex ante transparency notice
- Prior information notice for contracts in the field of defence and security
- Contract notice for contracts in the field of defence and security
- Contract award notice for contracts in the field of defence and security
- Subcontract notice

Templates for these are found on the SIMAP (2015) website. The two most important of these are the contract notice and the award notice, which are required by European and UK law (TSO, 2015). For this reason, the contract notice will be covered in detail in the following paragraphs.

Contract notices published in the supplementary S series must follow a strict format set out in and information requested by Annex II of Regulation (EC) 1564/2005. Failure to provide this information can lead to a financial penalty of up to 25% of the contract value as the advertisement is deemed to have broken EU law (DCAL, 2014). As part of the e-procurement process the OJEU notice is completed, submitted and advertised electronically. A standard format for OJEU adverts exists which must be followed. A step by step guide for completing an OJEU advertisement is provided in Annex 1.

Once completed the advertisement is then submitted electronically and displayed on the website. It can be seen from Figure 3.3 that contract award notices, part of the electronic advertising process also need to be submitted and displayed on the Tenders Direct website (Tenders Direct, 2015) within 48 days of contract award. Information from the contract notice should be replicated in the Award notice with the additional information requested in the template. The additional information relates to the winning bidder's details and the final value of the contract.

3.3.1.3 *Below threshold advertisements*

Procurement below the European Thresholds still uses advertisements in newspapers, among other means. DCAL (2014) suggests the following methods of advertising:

- Advertisement on the contracting authority's own website
- Advertisement on a portal website specifically created for contract advertisements

- Publication in national newspapers, specialist journals or national journals
- Publication of a voluntary OJEU notice in the official journal
- Local advertising (suitable for very small contracts)

As a pilot study for the rest of the UK, Northern Ireland was one of the first areas to examine tender advertising for construction contracts. As a result of this examination the Office of the First Minister and Deputy First Minister (2005) published a review of government advertising in Northern Ireland. Within this document, Clause 3.1.6. identified the Internet as an appropriate substitute to classified advertising for procurement. However, this did not provide a format for advertisements.

An ad-hoc approach followed as a result with each government department designing its own means of advertising projects. The result was that by 2008 over 20 systems in Northern Ireland were available from a variety of councils and central government departments, producing a fragmented and difficult to navigate system. The remainder of the UK had identified this issue and started producing tendering portals. These provide a single location allowing interested parties to access adverts in an easier way.

3.3.1.4 Benefits of an electronic advertisement system

One of the first publications in the area detailed research carried out by Eadie *et al.* (2007). This suggested a format for a web-based advertising website based on satisfaction ratings of a pilot scheme implemented by Roads Service, Northern Ireland. This showed that 82% of respondents preferred the web-based advertisement system to a paper based one. What came across strongly was that the Economic Operators favoured a single reference point where they could go to access all tenders. Classification of tenders was deemed a positive development leading to changes to the Roads Service website. This single source could produce time savings for companies trying to locate and keep informed of government contracts. Support was found from economic operators who considered that the system would decrease the possibility of advertisements being missed. There was 80% support for a system that divided the advertisement websites into future, current, closed and awarded tenders. These subdivisions were adopted by the Roads Service system and are still in place today even though the system now feeds directly into a central portal. While not required by law, Roads Service Northern Ireland were one of the first to publish the names of the winning tenderers on their website for under threshold tenders. Eadie *et al.* (2007) had shown 84% support for this. Further findings showed that savings up to £28,500 per year could be realised as the result of moving to the Internet. As well as cost savings, further benefits of e-advertising detailed by Eadie *et al.* (2007) were

- Time savings
- Central place for contract advertisements
- Not limited to when an external source – newspaper – can publish
- Improved visibility through universal twenty-four-hour access
- Improved communication
- Links to Freedom of Information (FOI) site
- Moderated by a senior officer
- Guaranteed to go live when required

Eadie *et al.* (2012) carried out further work into below threshold advertisements. This found that five years after the introduction of such systems that 101 economic operators considered 32% of contracts above £20,000 and below the European Threshold were still not being advertised electronically. This percentage increased to 44% in relation to award notices. This shows that significant financial saving could still be accrued from adopting an electronic system all of the time. This section shows the importance of electronic advertising and its advantages. However, it also shows that work needs to be completed on the central portal to improve user-friendliness. This would allow adoption for all contract sizes producing extra savings in these times of austerity.

3.3.2 Stage 2 electronic document provision and submission

3.3.2.1 Electronic procurement document provision and submission worldwide

Figure 3.3 indicates that after the advertisement that under the restricted procedure that the prequalification questionnaires need to be disseminated. In addition to its use across Europe Prequalification questionnaires are used worldwide. Liao (2002) indicates that the Taiwanese 'Electronic Tender Obtaining and Submitting System' streamlines the previous paper based administration procedures. This results in savings of US$40 million each year. A further example of automated downloading and uploading tender documentation was introduced by the Hong Kong government. This system named the 'Electronic Tendering System (ETS)' allows downloading and submission via the Internet (Chu *et al.*, 2004). Similar systems for the dissemination of documents exist worldwide – in Canada, the MERX system (MERX, 2015), in Australia, the AusTender system (AusTender, 2015) and Jetro in Japan (Jetro, 2015).

3.3.2.2   Electronic procurement document provision and submission in the UK

In the UK procurement is one of the issues for the devolved administrations. Therefore England, Scotland, Wales and Northern Ireland all have their own systems for delivering procurement documentation. In the UK in order to solve the issue of fragmentation Contracts Finder was launched to bring all Government Contracts for the whole of the UK together on 26 February 2015 (Gov.UK, 2015). Scotland has its own e-tendering portal (Scottish Government, 2015). This has an additional iPhone app to allow browsing and downloading of documents. Wales in Sell2Wales (2015) and Northern Ireland in eSourcingNI (2015) also have portal systems that deliver and accept submissions of contract documentation. For contracts outside central government, England has a number of e-procurement systems providing documentation fragmenting the landscape. In local government and the health service there are a number of smaller procurements relating to construction. They cover areas such as Health, for example the National Health Service (NHS, 2015), Food and Rural Affairs, for example the Department of the Environment Food and Rural Affairs (DEFRA, 2015), County Councils, for example Staffordshire County Council (2015) among others. However, these different bodies have adopted a number of systems badged for the various organisations. There are four main systems used for local government: Bravosolutions, European Dynamics, PROContract (PC)/PROACTIS (PA) and In-Tend.

3.3.2.3   Government systems for provision and submission

The initial framework agreement awarded in the UK, for an e-procurement solution to provide and receive tender documentation was awarded by the trading division of the Office of Government Commerce (OGC) known as OGCbuyingsolutions. The award of this contract for e-procurement services in 2003 dealt with provision of e-auctions (Bravosolution, 2013). A follow on framework agreement was notified as being awarded for e-sourcing solutions in the Official Journal of the European Union on 14 April 2009 (BravoSolution, 2013) to BravoSolution. BravoSolution was called off the framework to provide 'A framework utilising web-hosted technology to support the procurement process throughout the procurement life cycle, providing electronic tendering, evaluation, contract management and auctions'. This was deemed to be a 'pan government' initiative to try to ensure adoption of the same electronic tendering process across UK Central government departments. Many of the local government procurement departments also adopted the BravoSolution system. The framework for Northern Ireland needed renewed and was awarded on 25 February 2014 to European Dynamics (European Dynamics, 2014a). Northern Ireland's Central

Government site will soon move across. Scotland has also awarded its renewal to European Dynamics (European Dynamics, 2014c).These solutions have been awarded for the next five years. Private sector systems also exist allowing procurement to take place electronically. These systems work in much the same way.

3.3.2.4 *Process of document provision and submission*

Figure 3.4 provides a schematic indicating in a simplified manner the way in which such a system works.

3.3.3 *Stage 3 electronic awarding*

3.3.3.1 *Prequalification awarding*

In the restricted procedure two sections need to be analysed: The prequalification questionnaire (PQQ) for selection and the second stage for award. Eadie (2014) investigated the structure and content of PQQ's above the European threshold. Each PQQ contained the following sections- General Information, Economic and Financial Standing and compliance with EU/UK Procurement, Health and Safety and Minimum Standards

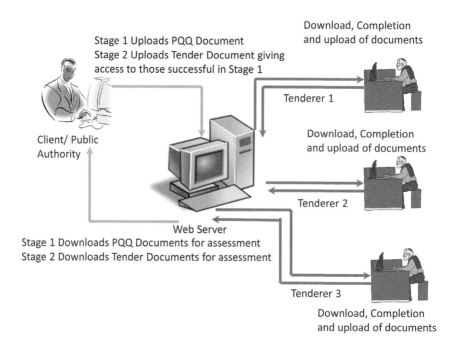

Figure 3.4 Simplistic e-tendering document supply and submission.

for Professional/Technical Ability. These sections are still mirrored in the CCS (2016) document setting out the format for central Government pre-qualification questionnaires. This follows the layout in PAS91:2013 (BSI, 2013). Two types of question are asked. These were Pass/Fail and Qualitative questions. Most e-procurement systems analyse the Pass/Fail questions Yes/No questions automatically. However, there are some PQQ submissions in Construction where the Pass/Fail requirements are evaluated against a Red, Amber and Green (RAG) status such as the Highways Agency PQQ (HA, 2016). These questions and the qualitative questions require input from the assessors. Different processes are carried out by personnel from different government departments. However, in essence the electronic system for this section of the selection process allows marks to be input by the assessors and a consensus to be arrived at. The complete process is detailed in Figure 3.5. In the restricted procedure 5–7 organisations go forward to the second stage.

3.3.3.2 Second stage awarding

A similar process looking forward is used to assess the second stage quality for award purposes. At the second stage after the quality is marked, the prices will be allowed to be viewed on the electronic system. The system analyses the prices. These are then combined with the quality marks already calculated and the preferred bidder identified. This preferred bidder is then placed on the e-advertising system to progress through the different stages of the Monetary Standstill period if above the European Threshold before the contract is awarded.

3.3.3.3 Awarding contracts under the European threshold

For contracts under the European Threshold Northern Ireland has introduced a further stage to cut down on the cost of assessment. This was due to the selection process using the prequalification questionnaires being

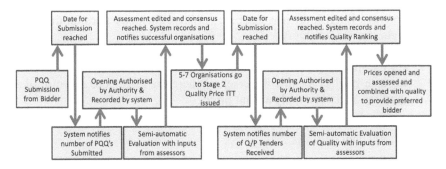

Figure 3.5 E-awarding electronic system.

expensive in terms of staff time. Northern Ireland Government set up a task group in 2008 to reduce costs and increase the speed of award (Northern Ireland Assembly, 2008). This resulted in PGN 05-12 being published by CPDNI (2012b) for a simplified approach to procurements over £30,000 and below the EU threshold. For the first time a 'Random Selection' (RS) process was brought into Construction Contracts. Following submission of documentation, names of six contractors going to compete in the second stage of the tender process were selected using a computer algorithm for three bands: (1) Construction Works over £30,000 to £500,000, (2) Construction Works over £500,000 to £2,000,000 and (3) Construction Services £30,000 to £111,676 (CPDNI, 2012b). These economic operators, having already passed the basic exclusion criteria and quality requirements, are then sent an invitation to tender. This system reduces the amount of assessment required saving time and money. Further examination of the system can be found in Eadie and McCavigan (2015).

3.3.4 Stage 4 electronic auctions

Auctions have been used in history since the start of the current era. Historically one of the first recorded auctions was in AD 193. when, after killing the emperor the Praetorian Guard auctioned the entire Roman Empire to the highest bidder (Klemperer and Temin, 2001). This would be an example of an English Auction where the bids are raised as the auction proceeds. A further auction type is the reverse auction with bids going down rather than up. This re-emerged as a procedure to reduce costs in the mid-1990s according to Kumar and Chang (2007). Arrowsmith (2001) defines these as *'a tendering procedure in which tenderers are provided with information on the other tenders, and are permitted to amend their own tenders on an on-going basis to beat those other tenders'*. This is sometimes known as a Dutch auction.

Arrowsmith (2001) and Teich *et al.* (2004) show two further types of electronic variations of auctions, bringing the total to four:

- Forward Dutch (or just Dutch) – Where the price goes down until the lowest price is reached
- Forward English (or just English) – Where the price goes up until the highest price is achieved
- Reverse Dutch – Where the price goes up until a supplier offers to supply at that price
- Reverse English (or just Reverse) – Where the price goes down until a supplier offers to supply at that price

3.3.4.1 Auctions and cost

Auctions drive down costs and have been promoted as providing best value for money (Soudry, 2004). Raghavan and Prabhu (2004) state that during reverse auctions, bidding carries on until the pre-established bidding period ends or no bidder is willing to reduce their bid further, whichever comes first. Auctions have reduced costs, in some cases by millions of dollars through the online reverse auction format but some are now suggesting that it is not a panacea and auctions have been discontinued after a few years (Gupta *et al.*, 2012). The difficulty is that the quality options are not taken into consideration going against John Ruskin's philosophy on procurement that has permeated through European Procurement Philosophy. The philosophy of assessing the quality of the product in addition to the price was first linked to John Ruskin (1819–1900) in the Washington Post in 1913. He is attributed with the following quotation '*It is unwise to pay too much, but it's worse to pay too little. When you pay too little, you sometimes lose everything because the thing you bought was incapable of doing the thing you bought it to do. The common law of business balance prohibits paying a little and getting a lot – it can't be done*'. (Bergeron and Lacinski, 2000, p. 66).

This has led to a rethinking of the initial thinking regarding the three most common types of auction used. Teich *et al.* (2004, p. 2) states '*The roots of electronic auction and negotiation mechanisms are in the auction and negotiation theory. Economists have investigated isolated, single good auctions. The ascending price English auction, the descending price Dutch auction and the Vickrey second-price sealed bid auction have been the most commonly studied auction mechanisms*'. Consideration needed to be taken into consideration as, on the surface, this goes against the ethos of the Latham (1994) and Egan (1998) reports into the UK Construction Industry. In order to be successful construction has used the 'iron triangle'; Cost, Time and Scope/Quality as a measure of project success (Atkinson, 1999). Pham *et al.* (2015) present an iteration of the Dutch auction, called two stage 'A+B' auctions. These take place with bids on cost (A) and time to deliver (B). However, on their own they do not address the quality issue. So the issue in construction is how to get the quality aspect into the equation.

3.3.4.2 *Auctions and quality*

Teich *et al.* (2004) suggests that the quality aspect has led to the development of a further type of Auction known as the 'Two Stage' auction. Quality was addressed through a sealed bid system and other issues followed leading to the name 'multiple-issue reverse auctions'. While the advent of the 'Two Stage' auction addressed the quality issue other issues are still outstanding. Smeltzer *et al.* (2003) see the auction as a short term fix resulting in a lack of loyalty as a buyer as the buyer is really only interested in a low price. Furthermore, Arminas (2002) supports this assertion in stating that as no long term relationship is built up leading ultimately to adversarial relationships developing. Ray *et al.* (2013) suggest that for this reason incentives should be used to keep bidders interested in participation in future procurements while maintaining a healthy level of competition. Even with this,

Tassabehji *et al.* (2006) state that the relationship aspect is the main reason for reverse auctions damaging supplier-buyer relationships and as a result being 'antithetical to what is currently regarded as good supply chain management'. The European Construction Institute magazine ECI news (2003, p. 5) further supports this stating 'Reverse on-line auctions of subcontracts may drive down prices in the short-term, but they will do nothing to improve long-term industry performance and they are totally alien to an open book approach to alliancing and supply chain integration'. Therefore, the short term relationship achieved through the auction, results in the ethos of continuous improvement, investment in quality of work, the training and investment of staff, and value management being often absent.

3.3.4.3 Auctions and legislation

However, since the Public Contract Regulations (TSO, 2006) onwards, dynamic purchasing systems and auctions have been authorised as part of European Union procurement. OGC (2008) guidance states that when stated in the Official Journal of the European Union (OJEU) Contract Notice, an eAuction may be used in conjunction with the open or restricted procedures (Figure 3.3). The auction comes between the receipt of tenders and intention to award. The initial price received through the tenders. This becomes the starting bid for the auction. Therefore the auction forces the price element down. Therefore, electronic auctions are best used to procure standardised products and services where price alone is the governing criteria. However, OGC (2008) do allow for an element of quality stating 'In the case of the most economically advantageous offer, any 'quality' features of the bid (e.g. terms of delivery or warranty) carried forward to the e-auction stage must be capable of being expressed as a value (figure or percentage) which can be incorporated within the formula which will be used to rank bids'.

In construction, price alone is normally not used as the sole award criterion due to the difference between it and the manufacturing industries. The different approach required between the two industries has been recorded in the findings of several court cases. One such court case in construction determined this difference as 'I think the most important background fact which I should keep in mind is that building construction is not like the manufacture of goods in a factory. The size of the project, site conditions, the use of many materials and the employment of various kinds of operatives make it virtually impossible to achieve the same degree of perfection that a manufacturer can'. (Eastern v. EME Developments 1991 55 BLR 114, 1991, p. 125). In procurement these issues are also present as Eadie *et al.* (2010b) shows that the drivers and barriers for e-procurement in construction act differently than those in the goods and services industries. These will be further examined later in the chapter.

3.3.4.4 Process and problems with auctions

The process for an electronic auction is indicated in Figure 3.6.

Figure 3.6 shows the process of the auction. The prices can be forced down by the auction, however, the quality element and the percentage costs for changes or dayworks in construction is vital to the success of a project. According to Sambhara *et al.* (2011), the low prices in the auction event are often '*accompanied by large switching costs which nullify any expected advantage*'. Therefore extras and dayworks need to be built into the quality assessment in construction. This is a common argument for restricting the use of auctions in construction to materials only where the specification has few unknowns or possible variations. In support of this, Katok and Roth (2004) state that electronic auctions are not suitable for every eventuality, but items with an exact specification such as plastic resin and personal computers can be procured successfully. Therefore, it can be seen that auctions for construction products should only be used when there is little by way of innovation or design required, the long term relationship with the supplier is not important and the items are clearly specified. However, Tassabehji *et al.* (2006) state that as operations are different with an e-auction that suppliers '*dislike the reverse e-auction format, which they perceive to lack openness and transparency*'. Reasons cited for this relate to knowledge of commercially sensitive information such as bid values being visible to all competing bidders. This means it is technically possible to work out the quality marks from watching the bids.

Furthermore, Smeltzer *et al.* (2003) conclude in an auction there is a possibility of a 'race' among those bidders in the 'challenge'. Instead of commercial interests being top of the agenda, there is the potential for bidders to get emotionally involved and continue bidding the process down in price

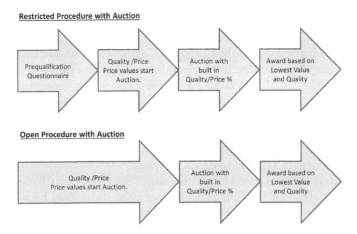

Figure 3.6 European procedures incorporating auctions.

to values that are not financially viable. Bidders have submitted bids below cost price in order to win and then failed to deliver. Sambhara *et al.* (2011) rank the following risks for reverse auctions, most important first: Inadequately specified requirements, Singular focus on price does not factor in total cost of ownership, Lack of competition in the auction, Inadequate supplier qualification, Resistance by internal clients within buying organization to reverse auction procedures and outcomes, Lack of top management support, Award terms not clearly communicated prior to auctions, Reluctance of suppliers to participate, Improper lot structuring, Buyer not faithful to the auction process, Omission of non-price criteria limits buyer's understanding of suppliers' full capabilities, Market conditions not conducive for reverse auctions, Quality of service and support could be reduced by suppliers to achieve offered price, Failure to honour award terms deters future supplier participation, Inclusion of suppliers who will not be awarded the business and Suppliers lack adequate knowledge of reverse auction process. It can be seen from this list that the major element in e-auction failure is inadequately specified requirements. Therefore it is vital in all construction electronic auctions to ensure that the requirements are specified correctly.

3.4 Government initiatives promoting e-procurement

One of the first initiatives relating to electronic procurement was 'Best Value' published in the 'Modernising Government' White Paper (Cabinet Office, 1999) on 30 March 1999. This specifically mentioned that procurement was one of the activities to be made electronic. The report measured progress through indicators; among these was Best Value Indicator 157 (BVI157) which set a target for public sector e-procurement to be complete by 2008 (Dacorum Borough Council, 2006). This kick-started the move to e-procurement and as a result most central and local government departments moved to electronic means of procuring services and works.

Work was carried out by the European Commission (2007) to identify the major driving forces in the implementation of e-procurement in construction which concluded that large European construction enterprises would adopt it and force it down through the supply chain. E-Business Watch (European Commission, 2007) reported the second lowest level of adoption of e-business was in construction in a comparison study of 10 major industries across Europe. The European Union had already incentivised the procurement procedures in relation to e-procurement.

The initial Public Contract regulations, TSO (2006) in Clauses 18, 19, 20 and 21 states that the four procurement procedures may be reduced in timescale if e-procurement is implemented. These are summarised in Table 3.2. The incentive was that reductions in time between 23% and 16% are achievable through using electronic documentation.

Table 3.2 Timings for e-procurement

Open	Restricted	Negotiated	Competitive dialogue
52 days or 36 days with Prior Information Notice (PIN)	First Stage 37 days Second Stage 40 days or 36 days with PIN	First Stage 37 days Second Stage time not specified	First Stage 37 days Second Stage time not specified
52 days reduced to 45 days for electronic contract notice or 29 days with PIN	First Stage reduced to 30 days for electronic contract notice	First Stage reduced to 30 days for electronic contract notice	First Stage reduced to 30 days for electronic contract notice
e-tendering means further reduction to 40 days	e-tendering means further reduction to 35 days	Negotiation time not specified	Time not specified
23% Overall time reduction—52 days to 40 days	16% Overall time reduction—37 + 40 = 77 days to 30 + 35 = 65 days	19% Overall time reduction- 1st stage—37 days to 30 days	19% Overall time reduction- 1st stage—37 days to 30 days

Table 3.3 Time reductions due to e-procurement

Open	Restricted	Other procedures
35 days or 15 days with Prior Information Notice (PIN)	First Stage 30 days Second Stage 30 days or 10 days with PIN	First Stage 30 days Second Stage time not specified

The European Commission has further enforced e-procurement use by making it mandatory across the European Union for Public Sector Contracts by June 2016 (Eurlex, 2012). According to the following section analysing the UK construction industries use of e-procurement further implementation needs to be completed to meet this target. However, this follows EU targets set in 2005 that failed to meet a deadline of 2010. It has however, allowed a reduction in the times for the various procedures to be reduced in the Public Contracts Regulations 2015 (TSO, 2015). These are shown in Table 3.3. These are provided in Clauses 28–32.

3.5 Conclusions

This chapter sets out the background, status and method for e-procurement in construction. It indicates how e-procurement provides a viable electronic alternative to the more traditional paper-based processes, such as tendering, associated with the procurement process. It uses the Minahan and Degan (2001) categorisation of the different elements of procurement: indirect procurement, direct procurement and sourcing. It shows that Hung *et al.* (2013) state that e-procurement makes all these elements more efficient and effective. It traces its evolution through from the Modernising Government White Paper in 1999 to the target for all public sector procurement across the European Union to be electronic by June 2016. It shows how the advent of e-procurement has allowed the time taken for tendering to be greatly reduced in the Public Contracts Regulation 2015 (TSO, 2015). Methods and stages of e-procurement are discussed through the different stages for construction operations: electronic advertising, the electronic tendering process, electronic assessment, through electronic auctions to electronic contract award. Each of these elements are discussed in detail indicating the improvements that e-procurement has delivered. Important steps and format for each of these steps for European and UK contracts are provided. The following chapter will look at the drivers and barriers to e-procurement and BIM adoption.

References

Arminas, D. (2002). Are relationships going, going, gone? Supply Management. March 14th 2002. Retrieved March, 2015, from https://www.cips.org/supply-management/news/2002/march/are-relationships-going-going-gone/.

Arrowsmith, S. (2001). E-commerce policy and the EC procurement rules: the chasm between rhetoric and reality, *Kluwer Law International*, 38, 1447–1477.

Atkinson, R. (1999). Project management: cost, time and quality, two best guesses and a phenomenon, it's time to accept other success criteria, *International Journal of Project Management*, 17(6), 337–342.

AusTender. (2015). AusTender Homepage. Retrieved March 2015 from https://www.tenders.gov.au/.

Bergeron, M. and Lacinski, P. (2000). *Serious Straw Bale: A Home Construction Guide for All Climates*. White River Junction, VT: Chelsea Green Publishing.

Bravosolution. (2013). BravoSolution Service Definition for SaaS Supply Management Services. Retrieved March 2015 from http://assets-production.govstore.service.gov.uk/Giii%20Attachments/BRAVOSOLUTION%20UK%20LTD/Bids/BravoSolution%20Service%20Definition%20for%20SaaS%20Supply%20Management%20Services%20G-Cloud%20III.pdf.

Brook, M. (2008). *Estimating and Tendering for Construction Work* (4th ed.). Burlington, MA: Butterworth-Heinemann.

BSI. (2013). PAS91:2013 Construction Prequalification Questionnaires. Retrieved March 2015 from http://shop.bsigroup.com/upload/PASs/PAS91-2013.pdf.

Cabinet Office. (1999). Modernising Government. White Paper. HMSO. Retrieved March 2015 from http://webarchive.nationalarchives.gov.uk/20140131031506/http://www.archive.official-documents.co.uk/document/cm43/4310/4310.htm.

Cabinet Office. (2015). Procurement Guidance Note PGN02/15. Retrieved March 2015 from https://www.gov.uk/government/publications/procurement-policy-note-0215-public-contracts-regulations-2015.

CCS. (2016). Standard Selection Questionnaire. Retrieved December 2016, from https://www.gov.uk/government/uploads/system/uploads/attachment_data/file/564780/Standard_Selection_Questionnaire_v3_Nov_16.docx.

Chu, P., Hsiao, N., Lee, F. and Chen, C. (2004). Exploring success factors for Taiwan's government electronic tendering system: behavioural perspectives from end users, *Government Information Quarterly*, 21, 219–234.

CourtsNI. (2008). McLaughlin and Harvey Limited -v- Department of Finance and Personnel [No. 2], Neutral Citation no.: [2008] NIQB 91, available at http://www.bailii.org/nie/cases/NIHC/QB/2008/91.html.

CPDNI. (2012a). Public Procurement: A Guide for Small and Medium Sized Enterprises PGN 02/12. Retrieved March 2015 from https://www.finance-ni.gov.uk/publications/procurement-guidance-note-0212-guide-public-procurement-small-and-medium-sized.

CPDNI. (2012b). PGN 05-12 Simplified Approach to Procurements Over £30,000 and Below EU Thresholds. Retrieved March 2015 from https://www.finance-ni.gov.uk/publications/procurement-guidance-note-0512-simplified-approach-procurement-over-%C2%A330000-and-under.

Craven, R. and Olson-Welsh, J. (2008). Northern Ireland Court Brings Down Framework Agreement, 33–35. Retrieved March 2015 from http://www.mayerbrown.com/files/Publication/5d183b5c-5ab8-4474-9bd1-5f27a26b1bf5/Presentation/PublicationAttachment/b6293233-cd25-4480-b055-c9e6b5a17501/0169con_Northern_Ireland_court_Article.pdf.

Dacorum Borough Council. (2006). IEG5 Report. Retrieved March 2015 from www.bipsolutions.com/docstore/doc/12757.doc.

DCAL. (2014). ERDF National Procurement Guidance ERDF-GN-1-004 Version 3, 3 July 2014. Retrieved March 2015 from https://www.google.co.uk/url?sa=t&rct=j&q=&esrc=s&source=web&cd=1&cad=rja&uact=8&ved=0CCIQFjAA&url=https%3A%2F%2Fwww.gov.uk%2Fgovernment%2Fuploads%2Fsystem%2Fuploads%2Fattachment_data%2Ffile%2F326610%2FERDF-GN-1-004_ERDF_National_Procurement_Requirements_V3.doc&ei=f5AKVe-VJsmwUZnTgaAI&usg=AFQjCNEHu5RpjmduIwaSAqchvCfTLZyZqg&bvm=bv.88528373,d.d24.

DEFRA. (2015). Procurement at Defra. Retrieved December, 2016, from https://www.gov.uk/government/organisations/department-for-environment-food-rural-affairs/about/procurement.2

Eadie, R. (2014). Does Europe need a specific prequalification system for highway projects? In: Proceedings of the 9th International Conference on Civil Engineering Design and Construction (Science and Practice), Varna, Bulgaria. Научно-технически съюз по строителство в България (*Scientific-Technical Union of Construction in Bulgaria*).

Eadie, R. and McCavigan, M. (2015). Random Selection: Winning the lottery in construction contracts, *International Journal of Procurement Management*, Vol. 9, No. 2, pp. 185–204.

Eadie, R., Perera, S., Heaney, G. and Carlisle, J. (2007). Construction contractor's perspective on web-based advertising of public sector construction contracts. In Boyd, David (ed.) *Proceedings of ARCOM*, September 2007, Belfast, Northern Ireland, 767–776.

Eadie, R., Perera, S. and Heaney, G. (2010). A cross discipline comparison of rankings for e-procurement drivers and barriers within UK construction organisations, *Journal of Information Technology in Construction*, 15, 23–43, http://www.itcon.org/2010/17.

Eadie, R., Millar, P. and McMorran, G. (2012) Public and private sector views of electronic government advertising in light of the Glover report in Northern Ireland. In: 7th International Conference on Innovation in Architecture, Engineering and Construction (AEC), Sao Paulo, Brazil. Escola Politecnica, University of Sao Paulo Brazil.

Eastern v. EME Developments (1991) Emson Eastern Ltd (in receivership) v EME Developments Ltd QUEEN'S BENCH DIVISION (OFFICIAL REFEREES' BUSINESS) His Honour Judge John Newey QC 55 BLR 114, 1991, p. 125 Retrieved December, 2016 From Lexus Library https://www.lexisnexis.com/uk/legal/search/enhRunRemoteLink.do?A=0.7840076570694092&service=citation&langcountry=GB&backKey=20_T25266237069&linkInfo=F%23GB%23CONLR%23vol%2526%25page%2557%25sel2%2526%25&ersKey=23_T25266237059.

ECI News. (2003). ECI Newsletter—Achieving Continuous Improvement and Value. Autumn 2003. Retrieved March 2015 from http://www.researchandmarkets.com/reports/1530394/long_term_partnering_achieving_continuous.

Egan, J. (1998) Rethinking Construction. Retrieved December 2016, from http://constructingexcellence.org.uk/wp-content/uploads/2014/10/rethinking_construction_report.pdf.

Eom, S., Kim, S. and Jang, W. (2015). Paradigm shift in main contractor-subcontractor partnerships with an e-procurement framework, *KSCE Journal of Civil Engineering*, DOI 10.1007/s12205-015-0179-5, 1–11.

E-SourcingNI. (2015). E-sourcing NI. Retrieved March 2015 from https://e-sourcingni.bravosolution.co.uk/web/login.shtml.

Eurlex. (2004a). European Union Directive 2004/17 Coordinating the Procurement Procedures of Entities Operating in the Water, Energy, Transport and Postal Services Sectors. Retrieved March 2015 from http://eur-lex.europa.eu/LexUriServ/LexUriServ.do?uri=CELEX:32004L0017:en:HTML.

Eurlex. (2004b). European Union Directive 2004/18 on the Coordination of Procedures for the Award of Public Works Contracts, Public Supply Contracts and Public Service Contracts. Retrieved March 2015 from http://eur-lex.europa.eu/LexUriServ/LexUriServ.do?uri=CELEX:32004L0018:En:HTML.

Eurlex. (2012). A Strategy for e-Procurement. Retrieved March 2015 from http://eur-lex.europa.eu/legal-content/EN/TXT/?uri=CELEX:52012DC0179.

Eurlex. (2013). Commission Regulation (EU) No 1336/2013 of 13 December 2013 Amending Directives 2004/17/EC, 2004/18/EC and 2009/81/EC of the European Parliament and of the Council in Respect of the Application Thresholds for the Procedures for the Awards of Contract (Text with EEA Relevance). Retrieved March 2015 from http://eur-lex.europa.eu/legal-content/EN/TXT/?qid=1398238103896&uri=CELEX:32013R1336.

Eurlex. (2014a). Directive 2014/24/EU of the European Parliament and of the Council of 26 February 2014 on Public Procurement and Repealing Directive 2004/18/EC (Text with EEA Relevance). Retrieved March 2015 from http://eur-lex.europa.eu/legal-content/EN/TXT/?uri=uriserv:OJ.L_.2014.094.01.0065.01.ENG.

Eurlex. (2014b). Directive 2014/25/EU of the European Parliament and of the Council of 26 February 2014 on Procurement by Entities Operating in the Water, Energy, Transport and Postal Services Sectors and Repealing Directive 2004/17/EC (Text with EEA Relevance). Retrieved March 2015 from http://eur-lex.europa.eu/legal-content/EN/TXT/?qid=1426711583200&uri=CELEX:32014L0025.

European Commission. (2007). The European e-business report, a portrait of e-business in 10 sectors of the EU economy, 5th Synthesis Report of the e-Business W@tch. Retrieved March 2015 from http://aei.pitt.edu/54204/.

European Dynamics (2014a). Newsroom. Retrieved March 2015 from http://www.e-pps.eu/default/page-view_category/catid-89/id-23.html.

European Dynamics (2014b). Pioneering the e-procurement era, Retrieved December, 2016, from http://www.eurodyn.com/ePPS_broch_portrait_02_high.pdf.

European Dynamics (2014c). Press Releases. Retrieved December 2016, from http://www.eurodyn.com/default/page-view_category/catid-16/id-408/type-press.html.

Gov.UK. (2015). Contracts Finder. Retrieved March 2015 from https://www.gov.uk/contracts-finder.

Griffith, A., Knight, A. and King, A. (2003). *Best Practice Tendering for Design and Build Projects.* London: Thomas Telford, Ltd.

Gupta, A., Parente, S. and Sanyal, P. (2012). Competitive bidding for health insurance contracts: Lessons from the online HMO auctions, International Journal of Health Care Finance and Economics, Vol. 12, Iss 4, 303–22.

HA (2016). Highways Agency PQQ Assessment Questionnaire, Retrieved December, 2016, from http://assets.highways.gov.uk/about-us/procurement-pre-qualification-pack/Pre%20Qual%20Annex%20C%20Min%20H%26S%20Requirement%20.doc.

HMSO. (1988). Local Government Act 1988. London: Her Majesty's Stationary Office Limited.

Hung, P., Dickson K., and Chiu, L. (2013). 'e-Tendering with Web Services: A Case Study on the Tendering Process of Building Construction', In 2013 IEEE International Conference on Services Computing, pp. 582–588.

Hung, W., Lin, C., Tai, Y. Ho, C. and Jou, J. (2014). Exploring the impact of web-based e-procurement on performance: organisational, interorganisational and systems perspectives, International Journal of Logistics Research and Applications, 17(3), 200–215.

JCT. (2015). Procurement. Retrieved March 2015 from http://corporate.jctltd.co.uk/products/procurement/.

Jetro. (2015). Japanese Government Procurement. Retrieved March 2015 from http://www.jetro.go.jp/en/database/procurement/.

Justia US Law. (2012). Title 41—Public Contracts. Retrieved March 2015 from http://statecodesfiles.justia.com/us/2012/title-41/subtitle-ii/chapter-61/section-6101/section-6101.pdf.

Kähkönen, K. and Rannisto, J. (2015). Understanding fundamental and practical ingredients of construction project data management, *Construction Innovation*, 15(1), 7–23.

Katok, E. and Roth, A. (2004). Adoption of electronic commerce tools in business procurement: enhanced buying center structure and processes, Management Science, 50(8), 1044–1063.

Kim J. and Shunk D. (2003). Matching indirect procurement process with different B2B e-procurement systems, Computers in Industry Vol. 53. 153–164.

Klemperer, P. and Temin, P. (2001). An early example of the 'Winner's Curse' in an auction, *Journal of Political Economy*, 109(6). Retrieved March 2015 from http://www.nuff.ox.ac.uk/users/klemperer/earlyexample.pdf.

Kumar, S. and Chang, C. (2007). Reverse auctions: how much total supply chain cost savings are there? A conceptual overview, *Journal of Revenue and Pricing Management*, 6(2), 77–85.

Laryea, S. and Hughes, W. (2009). Commercial reviews in the tender process of contractors, *Engineering, Construction and Architectural Management*, 16(6), 558–572.

Latham, Sir M. (1994). Constructing the Team, *Joint Government of Industry Review of Procurement and Contractual Arrangements in the U.K. Construction Industry*, HMSO, London.

Liao, T., Wang, M. and Tserng, H. (2002). A framework of electronic tendering for government procurement: a lesson learned in Taiwan, *Automation in Construction*, 11(6), 731–742.

Lim, Y. (2015). Use of BIM approach to enhance construction interface management: a case study, *Journal of Civil Engineering and Management*, 21(2), 201–217.

Loosemore, M. and Richard, J. (2015). Valuing innovation in construction and infrastructure: Getting clients past a lowest price mentality, Engineering, Construction and Architectural Management, Vol. 22 Iss: 1, 38–53.

Lopes, J. (2012). Construction in the economy and its role in socio-economic development. In Ofori, G. (ed.) *New Perspectives on Construction in Developing Countries*. Abingdon: Spon, 40–71.

Lu, Q., Meng, F. and Goh, M. (2014). Choice of supply chain governance: Self-managing or outsourcing?, International Journal of Production Economics, Vol. 154, August 2014, 32–38.

MERX. (2015). Construction Procurement. Retrieved March 2015 from http://www. merx.com/English/NonMember.asp?WCE=Show&TAB=3&PORTAL=MERX& State=42&TEMPLATE_NAME=PrivateBuyerConstructionProcurement&hcode= rRUg6dgcJeejzMQ4uH7uBw%3d%3d.

Miller, D., Doh, J. and Mulvey, M. (2015). Concrete slab comparison and embodied energy optimisation for alternate design and construction techniques, *Construction and Building Materials*, 80, 329–338.

Minahan, T. and Degan, G. (2001). *Best Practices in e-Procurement*. The Abridged Report, Boston: Aberdeen Group. Retrieved March 2015 from http://www. inkoopportal.com/inkoopportal/download/common/e-procurement_1_.pdf.

Murdoch, J. and Hughes, W. (2008). *Construction Contracts* (4th ed.). London: Taylor & Francis.

NHS. (2015). Welcome to the NHS Commercial Solutions Tendering Site. Retrieved March 2015 from https://commercialsolutions.bravosolution.co.uk/web/login.shtml.

Northern Ireland Assembly. (2008). Appendix 4: Memoranda and Papers from Department of Finance and Personnel. Retrieved March 2015 from http:// archive.niassembly.gov.uk/finance/2007mandate/reports/Report_19_08_09R_ memoranda.htm.

Office of the First Minister and Deputy First Minister. (2005). *A Review of Government Advertising*. Belfast, Northern Ireland: Office of the First Minister and Deputy First Minister.

OGC. (2008). eAuctions. Retrieved March 2015 from http://webarchive.nationalarchives.gov.uk/20110601212617/http:/ogc.gov.uk/documents/OGC_Guidance_on_ eAuctions.pdf.

Pham, L., Teich, J., Wallenius, H. and Wallenius, J. (2015). Multi-attribute online reverse auctions: recent research trends, *European Journal of Operational Research*, 242(1), 1–9.

Raghavan, N. and Prabhu, M. (2004). Object-oriented design of a distributed agent-based framework for e-procurement, *Production Planning & Control*, 15(7), 731–741.

Ray, A., Jenamani, M. and Mohapatra, P. (2013). Relationship preserving multi-attribute reverse auction: a web-based experimental analysis, *Computers & Industrial Engineering*, 66(2), 418–430.

Sambhara, C., Keil, M., Rai, A. and Kasi, V. (2011). Buyers' perceptions of the risks of Internet enabled reverse auctions. In Baumann, C., Blasum H., Bormer, T. and Tverdyshev, S. (eds.) *Proceedings of Architectures and Applications for*

Mixed-Criticality Systems (AMCIS) 2011, Newport Beach, March 2011, Paper 352. Retrieved March 2015 from http://aisel.aisnet.org/amcis2011_submissions/352/.

Scottish Government. (2015). Public Contracts Scotland. Retrieved March 2015 from http://www.publiccontractsscotland.gov.uk/search/search_mainpage.aspx.

Sell2Wales. (2015). Welcome to Sell2Wales. Retrieved March 2015 from http://www.sell2wales.gov.uk/.

SIMAP. (2015). NUTS. Retrieved March 2015 from http://simap.ted.europa.eu/web/simap/nuts.

Smeltzer, L. and Carr, A. (2003). Electronic reverse auctions promises, risks and conditions for success, *Industrial Marketing Management*, 32(6), 481–488.

Soudry, O. (2004). Promoting economy: electronic reverse auctions under the EC directives on public procurement, *Journal of Public Procurement*, 4(3), 340–374.

Staffordshire County Council. (2015). Welcome. Retrieved March 2015 from https://www.staffordshire.gov.uk/Homepage.aspx.

Stationary Office. (1999). Modernising Government White Paper. Retrieved March 2015 from http://webarchive.nationalarchives.gov.uk/20140131031506/http://www.archive.official-documents.co.uk/document/cm43/4310/4310.htm.

Tassabehji, R., Taylor, W., Beach, R. and Wood, A. (2006). Reverse e-auctions and supplier–buyer relationships: an exploratory study, *International Journal of Operations and Production Management*, 26(2), 166–184.

Teich, J., Wallenius, H., Wallenius, J. and Koppius, O. (2004). Emerging multiple issue e-auctions, *European Journal of Operational Research*, 159(1), 1–16.

Tenders Direct. (2015). Tender Alert Service for the Public Sector. Retrieved March 2015 from http://www.tendersdirect.co.uk/?source=OJEC&lgid=OJEC.

TSO. (2006). Public Contracts Regulations 2006. HMSO, London. Retrieved March 2015 from http://www.opsi.gov.uk/si/si2006/uksi_20060005_en.pdf.

TSO. (2009). Public Contracts (Amendment) Regulations 2009. Retrieved March 2015 from http://www.legislation.gov.uk/uksi/2009/2992/contents/made.

TSO. (2011). The Public Procurement (Miscellaneous Amendments) Regulations 2011. Retrieved March 2015 from http://www.legislation.gov.uk/uksi/2011/2053/pdfs/uksi_20112053_en.pdf.

TSO. (2015). The Public Contracts Regulations 2015. Retrieved March 2015 from http://www.legislation.gov.uk/uksi/2015/102/pdfs/uksi_20150102_en.pdf.

Annex 1 Completing an OJEU Notice

A step by step guide is provided below for completing an OJEU advert.

Section I: Contracting entity

I.1) Name, addresses and contact point(s) –
This should contain the name and address of the contracting authority and the contact person's name, address (if different), telephone number, fax and e-mail address.

Four further sub-sections complete this element of the OJEU notice –

1 Internet address(es): This entry is normally the website of the contracting entity;
2 Further information can be obtained from: This is normally completed by putting a dot in the button stating the contact(s) named above;

3 Specifications and additional documents (including documents for a dynamic purchasing system) can be obtained from: This can be completed using the same information as the first section above and;

4 Tenders or requests to participate must be sent to: This is to allow the naming of a different person than the person in the initial part if required. Normally the person in the original part is reentered.

I.2) Type of the contracting authority –
Choice can be made between

1 Ministry or any other national or federal authority, including their regional or local sub-divisions,
2 National or federal agency/office,
3 Regional or local authority,
4 Regional or local agency/office,
5 Body governed by public law,
6 European institution/agency or international organisation
7 Or the catchall option named 'Other' where the person completing the form can stipulate a type. This is normally not required.

I.3) Main activity –
Main activities are selected from the following list –

1 Production, transport and distribution of gas and heat,
2 Electricity,
3 Exploration and extraction of gas and oil,
4 Exploration and extraction of coal and other solid fuels,
5 Water,
6 Postal services,
7 Railway services,
8 Urban railway, tramway, trolleybus or bus services,
9 Port-related activities,
10 Airport-related activities
11 Or the catchall option named 'Other' where the person completing the form can stipulate the main activity.

I.4) Contract award on behalf of other contracting entities –
This is completed by choosing the yes or no option on the form.

If Yes is chosen then Annex A is used to provide details of the contracting entities. This is used where a single contract is shared between a number of different departments.

Section II: Object of the contract

II.1) Description
This should contain the Title and the Contract Type filled in in completing sections II.1.1 and II.1.2

II.1.1) Title attributed to the contract by the contracting entity:-
This is completed by entering a short title to identify the project.

II.1.2) Type of contract and location of works, place of delivery or of performance –
The person completing the form has to choose between Supplies, Services or Works.

Works categories are split into:-

1 Execution;
2 Design and execution; and
3 Realisation.

In addition the means of work is specified, corresponding to the requirements of the contracting authorities.

The supplies section allows choice of the following options: -

1 Purchase,
2 Lease,
3 Rental,
4 Hire purchase and
5 A combination of these.

The final section is services. This needs to be read in conjunction with Annex C1 provided on the OJEU website which provides codes for the different service categories.

The person completing the advertisement should select the category or categories that apply to the services required.

The Main site or location of works, place of delivery or of performance is the next thing that is required followed by the NUTS code. The NUTS code identifies where the procurement is from. UK NUTS codes start with UK and then an identifier of the location. A full list of NUTS codes is found on the SIMAP website (SIMAP, 2015).

II.1.3) Information about a public contract, a framework agreement or a dynamic purchasing system (DPS) –
This is a choice between the three options. Put the dot in the relevant option.

II.1.4) Information on framework agreement –
This section is only completed if the Framework Agreement option in section II.1.3 is chosen. It provides a choice between a Framework agreement with several operators or a single operator. It also allows the maximum number of participants to the framework agreement to be specified. The next section of II.1.4. allows specification of the duration of the framework agreement in years or months. This is normally 3–5 years. The final part of the section again is reliant in the framework agreement option being chosen. It allows the estimated total value of purchases for the entire duration of the framework agreement or estimated overall value to be entered.

II.1.5) Short description of the contract or purchase(s) –
A summary of the issues involved in the contract are entered in this section. This provides a short description of the works.

II.1.6) Common procurement vocabulary (CPV) –
This identifies the type of work to be carried out. BIP (2012) provides a CPV Code Search on its website. The CPV codes should be provided for the main object and any additional objects.

II.1.7) Information about Government Procurement Agreement (GPA) –
To complete this it requires a yes or no choice.

II.1.8) Information about lots –
This section will become more important due to the change in the EU Procurement directives. It allows the project to be divided into smaller lots which is being promoted in the new directives. Three choices are available to the person completing the advertisement. These are:-

1 one lot only;
2 one or more lots; or
3 all lots.

Description of the Lots is added at the end of the section if Lots are selected in the initial yes or no option.

II.1.9) Information about variants –
This section requires a yes or no response.

II.2) Quantity or scope of the contract

II.2.1) Total quantity or scope –
This allows entries which describe the estimated total quantity and value to be entered.

II.2.2) Information about options –
This allows the specification of a contract with an option to extend it. This can be specified by choosing yes. A description needs to be included. An example of this is a 2 year contract with an option to extend for a further 2 years.

II.2.3) Information about renewals –
This allows the compiler to specify whether or not there will be a renewal and if there is the number of possible renewals.

II.3) Duration of the contract or time limit for completion –
Duration in months or days and start and end dates are specified here.

Section III: Legal, economic, financial and technical information

III.1) Conditions relating to the contract –
This section allows various conditions to be specified. It needs to specify construction related conditions here, such as NEC and the Z clauses.

III.1.1) Deposits and guarantees required –
This should include insurance requirements for public bodies. Any sections relating to deposits and guarantees in the tender documents needs to be replicated here. One which should be stipulated in most cases is the right to require a parent company guarantee or suitable performance bond in relation to the works required.

III.1.2) Main financing conditions and payment arrangements and/or reference to the relevant provisions governing them:
Similar to III.1.1. any sections of the tender documents needs to be replicated here.

III.1.3) Legal form to be taken by the group of economic operators to whom the contract is to be awarded:
Normally for Public Sector contracts joint and several liability should be specified here. All other relevant insurances required for the contract are also specified including Professional Indemnity Insurance where necessary.

III.1.4) Other particular conditions:
The performance of the contract is subject to particular conditions: In construction contracts yes should be selected. The applicable Conditions of Contract should be specified again here. An example would be the New Engineering Contract NEC3 Conditions of Contract amended as per Z Clauses included in the Tender Documents. This should be specified. Stating 'as per Tender documents' or as per 'Prequalification Questionnaire' is not acceptable.

III.2) Conditions for participation –
Again a replication of the information in the Prequalification or tender is required here.

III.2.1) Personal situation of economic operators, including requirements relating to enrolment on professional or trade registers. –
APUC (2010) states 'The regulations require that the information is requested in the Contract Notice or in the invitation to tender, but it is common practice for the PQQ to be referred to in the Contract Notice; this is not 100% compliant'. It is therefore advisable to replicate the information in the contract notice or tender documentation in this section.

III.2.2) Economic and financial ability –
This is similar to section III.2.1.

III.2.3) Technical capacity –
Reference should be made to the Public Contacts Regulations requirements in this section.

III.2.4) Information about reserved contracts –
This is not normally required. However, there are two options to choose from 'The contract is restricted to sheltered workshops' or 'The execution of the contract is restricted to the framework of sheltered employment programmes' if required.

III.3) Conditions specific to services contracts

III.3.1) Information about a particular profession –
The response required is yes or no. It allows restriction to a certain profession. If this is required then reference to the relevant law, regulation or administrative provision relating to this must be supplied. It is sometimes used to stipulate a profession such as an Architect or Engineer if design elements from these professions are specifically required.

III.3.2) Staff responsible for the execution of the service –
If required choosing yes in this section forces the supplier to get Legal persons to indicate the names and professional qualifications of the staff responsible for the execution of the service.

Section IV: Procedure

IV.1) Type of procedure

IV.1.1) Type of procedure –
This allows the compiler to select the various types of procedure described in the Public Contracts Regulations. Restricted and Open Procedures are the most common. PF2 schemes and other innovative procedures form the remaining options (TSO, 2015). The compiler can also choose the accelerated procedures due to electronic procurement as options.

IV.1.2) Limitations on the number of operators who will be invited to tender or to participate (restricted and negotiated procedures, competitive dialogue) –
This section allows the compiler to select the number of economic operators that will be allowed to proceed to the second stage of the tender process. This provides the opportunity to stipulate the minimum and maximum numbers of economic operators.

IV.1.3) Reduction of the number of operators during the negotiation or dialogue –
This is left blank unless the negotiation or competitive dialogue options are chosen. This is to allow the compiler to stipulate by a choice of yes or no, whether recourse to staged procedure to gradually reduce the number of solutions to be discussed or tenders to be negotiated is going to be used to determine award of the contract.

IV.2) Award criteria

IV.2.1) Award criteria –
Initially in this section a choice is provided as to whether the tender is to be awarded on Lowest Price or Most Economically Advantageous Tender (MEAT). If MEAT is chosen this section is vital to the OJEU advertisement

after a Northern Ireland court case changed European Law in regard to disclosure of criteria and weightings. McLaughlin and Harvey successfully challenged the Department of Finance and Personnel over the lack of transparency in relation to how the PQQ was assessed and marked (Craven and Olsen-Welsh, 2008). McLaughlin and Harvey argued that under the requirement for transparency in public procurement that both award criteria and the weightings should be disclosed during the advertisement stage. CourtsNI (2008) stipulated in the decision that the bidders should be made aware of each of the elements or sub elements which affect their preparation of the bid. From 2008 on, the criteria, sub-criteria and weighting of each element need to be published beforehand. These are stipulated in the order of most important criteria first.

IV.2.2) Information about electronic auction –
This relates to stipulating whether or not an electronic auction is to be used. The terms and mechanism that the Auction uses need to be stated. Details of electronic auctions and their *modus operandi* are included later in the chapter.

IV.3) Administrative information –
All the remaining sections IV.3.1 to IV.3.7 in this section are self-explanatory and the information headings for it are included for completeness.

IV.3.1) File reference number attributed by the contracting entity-
The file reference number needs added here.

IV.3.2) Previous publication(s) concerning the same contract-
This section needs added if the tender is a retender

IV.3.3) Conditions for obtaining specifications and additional documents –
Any special conditions are noted here.

IV.3.4) Time limit for receipt of tenders or requests to participate –
If restricted procedure is being used the time for return of the PQQ is specified. In the Open procedure the time to respond to the tender is specified.

IV.3.5) Language(s) in which tenders or requests to participate may be drawn up-
The language is specified here.

IV.3.6) Minimum time frame during which the tenderer must maintain the tender-
The timeframe for keeping the prices and quality statements in the tender is specified here. This should include enough time to allow assessment.

IV.3.7) Conditions for opening of tenders-
The tender opening procedure is specified here.

Section VI: Complementary information –

VI.1) Information about recurrence –
This is used in construction for recurring maintenance contracts

VI.2) Information about European Union funds-
Any European funds included in the contract needs to be specified here. This is especially important in Highways Schemes.

VI.3) Additional information:
Any additional information can be specified here.

VI.4) Procedures for appeal –
Care must be taken in completing this section of the advertisement to ensure that it does not contradict elements of the Conditions of Contract. The appeal procedures that are stipulated there should be replicated here unaltered.

VI.4.1) Body responsible for appeal procedures
This should match that specified in the Conditions of Contract for Construction.

VI.4.2) Lodging of appeals
Where appeals should be lodged should be specified here.

VI.4.3) Service from which information about the lodging of appeals may be obtained
If there is additional information about appeals at a certain location it needs to be specified here.

VI.5) Date of dispatch of this notice.
This is normally automated and the date that the advertisement is posted.

4 Drivers for electronic procurement and building information modelling in the construction industry

Robert Eadie and Srinath Perera

4.1 Background

Chapter 3 sets out the process of e-procurement and Chapter 5 within this book will set out the background for building information modelling (BIM). Toktaş-Palut *et al.* (2014) indicate that Eadie *et al.* (2007) were the first to investigate whether the construction procurement drivers and barriers were similar to those in other industries. They initially reviewed e-procurement in construction from a worldwide, European, and UK perspective. Eadie *et al.* (2007) compared e-procurement in construction initially with detailed studies on e-procurement in the goods and services sectors published in the United States and Australia. This was followed by a larger study into the subject in 2010 (Eadie *et al.*, 2010a,b). The deadline for the implementation of building information modelling (BIM) on government contracts is 1st April 2016 (Efficiency and Reform Group, 2011) and this chapter compares and contrasts the drivers and barriers to BIM with those of e-procurement. Furthermore it looks at the interoperability between the two systems. The comparison of the drivers and barriers for both systems shows that similar barriers still remain to those found in the Eadie *et al.* (2010a,b) studies and provides a current update on the issues identified.

4.2 Data collection and analysis

In order to determine the most important factors in adopting e-procurement and BIM, and in collecting data from dispersed organisations across the UK, a questionnaire approach was adopted for both e-procurement and building information modelling. Data was collected through structured questionnaires using the Limesurvey software which has a PHP frontend to a MYSQL database for survey management and analysis purposes. The drivers for each of the surveys were ranked using mean rank analysis and the relative importance index (RII) formula to establish the respondent's ranking on each of the drivers.

RII is defined by the following formulae:

$$\text{Relative importance index (RII)} = \frac{\Sigma w}{A \times N} (0 \leq \text{index} \leq 1)$$

Where:

W is the weighting given to each element by the respondents. This will be between 1 and 5, where 1 is the least significant impact and 5 is the most significant impact;

A is the highest weight; and

N is the total number of respondents.

4.3 The drivers for construction e-Procurement

e-Procurement is now used worldwide. An investigation of literature into the subject shows that it has been investigated in many countries. This is due to the perceived benefits it brings which act as drivers for its implementation. The literature shows that benefits have been identified in the USA (Gebauer *et al.*, 1998, Davila *et al.*, 2003, Gunasekaran *et al.*, 2009a), Europe (Panayiotou *et al.*, 2003 in Greece), Asia (Gunasekaran and Ngai, 2008 in Hong Kong), Australasia (Hawking *et al.*, 2004; Nguyen, 2013) and the UK (Eadie *et al.*, 2007; Gunasekaran *et al.*, 2009b; Eadie *et al.*, 2010a,b) among others. While most studies remained general for procurement Eadie *et al.* (2010a) produced a collated set of twenty-one drivers for construction e-procurement. These indicated that the drivers and barriers performed differently for the construction industry. The rankings and literature related to these are summarised in Table 4.1. For the drivers in Table 4.1, Eadie *et al.* (2007, 2010a,b) and Toktaş-Palut *et al.* (2014) are deemed to be included for all ranked drivers. Extra drivers that have been identified since Eadie *et al.* (2010a,b) are added to the bottom of Table 4.1. However, even though Eei *et al.* (2012) and Toktaş-Palut *et al.* (2014) were not specifically for construction organisations the drivers identified are included for completeness in Table 4.1.

4.4 The drivers for building information modelling

In a similar way to e-procurement, BIM is now used worldwide. Initially used earlier in the process to e-procurement. BIM can be used through the lifecycle, from concept through to facilities management and demolition. Therefore the drivers for it are much wider than those for e-procurement. However, due to its perceived benefits to all aspects of the design and construction process, the drivers have some similarity to those for e-procurement. The rankings and literature related to these are summarised in Table 4.2 with the most important driver ranked first (1) and the least important driver (18).

Table 4.1 Drivers to e-procurement identified from literature

Drivers in rank order	Referenced in	Band	Private sector rank	Public sector rank	Construction industry rank
Process, Transaction and Administration Cost Savings	Knudsen (2003), Minahan and Degan (2001), Martin (2009), Davila et al. (2003), Egbu et al. (2004), Panayiotou et al. (2003), Hawking et al. (2004), Raghavan and Prabhu (2004), Kong et al. (2001) and Yu et al. (2015)	Cost	1	1	1
Convenience of archiving completed work	Eadie et al. (2010a)	General	2	1	2
Increased Quality through increased accuracy (Elimination of errors through Computer use)	Eadie et al. (2010a), Alvarez-Rodriguez et al. (2014), Williams and Lynnes (2002) and Minahan and Degan (2001)	Quality	2	6	3
Shortened Internal and External Communication Cycle times	Yu et al. (2015), Knudsen (2003) and Kalakota et al. (2001)	Time	6	5	4
Increased Quality through increased efficiency	McIntosh and Sloan (2001), Ribeiro and Henriques (2001) and Martin (2009)	Quality	4	7	5
Shortened Overall Procurement Cycle Times	Yu et al. (2015), Minahan and Degan (2001) and Kalakota et al. (2001)	Time	9	3	6
Increased Quality through Improved Communication	Hawking et al. (2004)	Quality	5	10	7
Strategic Cost Savings	Knudsen (2003)	Cost	13	3	8
Service/Material/Product Cost Savings	Minahan and Degan (2001) and Martin (2009)	Cost	8	10	9
Reduction in Evaluation Time	Yu et al. (2015), Panayiotou et al. (2003) and Martin (2009)	Time	10	7	10
Develops the Technical Skills, knowledge and expertise of procurement staff	Eadie et al. (2010a)	General	14	16	11

(*Continued*)

Drivers in rank order	Referenced in	Band	Private sector rank	Public sector rank	Construction industry rank
Increasing Profit Margins	McIntosh and Sloan (2001), Wong and Sloan (2004) and Ribeiro and Henriques (2001)	Cost	7	19	12
Increased Quality through Benchmarking (Market Intelligence)	Hawking et al. (2004)	Quality	17	9	12
Reduction in purchasing order fulfilment time—Contract Completion	Davila et al. (2003)	Time	11	13	14
Enhanced Inventory Management	Hawking et al. (2004) and Martin (2009)	General	17	14	15
Reduction in time through greater transparency (Less objections)	Panayiotou et al. (2003)	Time	16	12	16
Reduction in time through increased visibility	Yu et al. (2015) and Eadie et al. (2010a)	Time	14	18	17
Increased Quality through increased visibility in the supply chain	Kalakota et al. (2001), Minahan and Degan (2001) and Hawking et al. (2004)	Quality	19	17	18
Increased Quality through increased competition	Kalakota et al. (2001)	Time	20	14	19
Gaining Competitive Advantage	Wong and Sloan (2004)	General	11	20	20

Extra Drivers investigated since 2010

		Comments
Compliance with laws and regulations	Eei *et al.* (2012)	While Europe promoted e-procurement through the European Directives it did not enforce it. However, other countries such as Hong Kong and Malaysia did. This driver is specific to those countries that made e-procurement a legal necessity
Decentralization of procurement power	Eei *et al.* (2012)	This driver is only mentioned as a benefit to SME's in a Table in Eei *et al.* (2012). However, it is not described. Subsequently Toktaş-Palut *et al.* (2014) mention it but do not rank it.
Minimization of process errors	Eei *et al.* (2012)	This is a similar driver to 'Increased Quality through increased efficiency' included in the Eadie *et al.* (2010a,b) studies. While the driver above is much more inclusive it included the effectiveness and efficiency of the process.
On-line and real-time reporting	Eei *et al.* (2012)	This driver relates mainly to the e-advertising and e-auction element of e-procurement discussed in the previous chapter. The e-tendering and e-award element of e-procurement is not affected by this as it is a closed bid system.

Table 4.2 Summary of the drivers for BIM in rank order

BIM drivers	References	Respondents using BIM		Respondents not using BIM	
		RII	Rank	RII	Rank
Clash Detection	BIMhub (2012), Bentley (2012) and Leite et al. (2009)	0.811	1	0.818	4
Cost Savings through Reduced Re-work	Woo (2007), Bazjanac (2005) and Samuelson and Björk (2010)	0.756	2	0.764	10
Improve Design Quality	Emmitt (2007), Woo (2007), Bazjanac (2005) and Samuelson and Björk (2010)	0.744	3	0.691	16
Accurate Construction Sequencing	BIMhub (2012), Bentley (2012), Eastman et al. (2011), Azhar et al. (2008) and Azhar (2011)	0.733	4	0.800	6
Improve Built Output Quality	IDC (2009) and Samuelson and Björk (2010)	0.722	5	0.764	9
Desire for Innovation	Li et al. (2008) and TRADA (2012)	0.689	6	0.691	17
Competitive Pressure	Coates et al. (2010), Bew and Underwood (2009) and Liu et al. (2010)	0.667	7	0.909	2
Government Pressure	Efficiency and Reform Group (2011), buildingSMART Australasia (2012), The BIM Industry Working Group (2011), Fitzpatrick (2012) and Arayici et al. (2011)	0.656	8	0.927	1
Improve Capacity to Provide WLV to Client	Azhar et al. (2011), Grillo and Jardim-Goncalves (2010) and Azhar et al. (2008)	0.656	9	0.727	12
Streamline Design Activities	Azhar et al. (2008, BIMhub (2012), Bentley (2012) and Eastman et al. (2011)	0.644	10	0.673	18
Time Savings	Azhar et al. (2008), Azhar (2011), BIMhub (2012), Eastman et al. (2011) and Faniran et al. (2001)	0.644	11	0.800	7

Cost Savings through Reduced RFI's	Azhar et al. (2008), Barlish and Sullivan (2012) and Deutsch (2011).	0.622	12	0.691	14
Improve Communication to Operatives	Sacks et al. (2009) and Tutt et al. (2011)	0.622	12	0.782	8
Facilitate increased Pre-Fabrication	Eastman et al. (2011) and Nawari (2012)	0.611	14	0.709	13
Client Pressure	Coates et al. (2010), Bew and Underwood (2009), Samuelson and Björk (2010) and Liu et al. (2010)	0.600	15	0.909	2
Automation of Schedule/Register Generation	Azhar (2011), BIMhub (2012) and Bentley (2012)	0.589	16	0.745	11
Design H & S into the Construction Process	Kiviniemi et al. (2011)	0.556	17	0.818	5
Facilitate Facilities Management Activities	Lewis et al. (2010)	0.556	18	0.691	14

Source: After Eadie et al. (2013b).

4.5 Discussion on drivers

The following section will describe and compare the drivers for e-procurement and BIM in detail. Each of these have deadlines over the next period. In order to meet these deadlines the impetus given by adopting the important drivers must be harnessed. In order to identify which of these most emphasis should be placed upon they require the ranking provided in Table 4.2. Over fifty years have passed since the iron triangle of cost, scope/quality and time started to be used as a measure of project success (Atkinson, 1999). Additional factors such as sustainability (Ebbesen and Hope, 2013) and internal environment (Bryde and Robinson, 2007) have been added since then. For the purposes of this chapter, costs for e-procurement and BIM relate to: (1) the process being carried out such as exchange of documents through e-procurement and with BIM through a common data environment. (2) Transactions through electronic collaboration and contract formation and (3) administration cost savings where the reduction in administration through use of the processes is achieved. Secondly scope/quality is defined for e-procurement as the element of quality that is brought to the project through electronic means. A reduction of error can be achieved by the use of both e-procurement and BIM. Time is the length of the process and time reductions can be achieved through using both processes. Sustainability can be defined as implementing the triple bottom-line in construction (Dyllick and Hockerts, 2002). For the purposes of this chapter, it is seen as producing a balance of the social, economic and environmental issues involved in sustainable development. Both electronic systems can deliver this as described later in the chapter. Finally the internal environment relates to the processes, habitus and facilities elements within an organisation (Bryde and Robinson, 2007). The drivers for both technologies have been segregated into these sections for discussion.

4.5.1 Cost

4.5.1.1 e-Procurement cost reductions

It can be seen from Table 4.1 that for the construction industry, *Process, Transaction and Administration Cost Savings* was the e-procurement driver ranked highest by both the public and private sectors. The findings of Eadie *et al.* (2010a,b) verify that the findings of a much earlier study in the US by Gebauer *et al.* (1998) also apply to construction. Gebauer *et al.* (1998) stated that cost and time were the two most important measures of success during a procurement process. Erridge *et al.* (2001) carried out work highlighting this in the UK. Further work in Australia in 2004 (Hawking *et al.*, 2004) suggested the cost element was the more important of the two issues with between 75% and 80% of their sample in Australia in support. Recent work has also highlighted cost savings in South Africa (Laryea and Ibem, 2014), India (Panda and Sahu, 2015), Bangladesh (Simu, 2012), and China (Wang et al., 2014). Through this substantial body of literature a variety of reasons for cost savings are proposed across a number of industries (Ribeiro and Henriques, 2001; McIntosh and Sloan, 2001; Hawking *et al.*, 2004). Rankin *et al.* (2006) finds cost saving achieved in the construction industry in Canada. Some such as Sone (2011) suggests e-procurement generally reduces costs, others such as Chen *et al.* (2011) suggests costs are reduced due to an increase in efficiency in the business processes involved in e-procurement. Estimates of the amount of cost savings vary, however, little evidence is provided in support of large cost savings with the exception of a few case studies. Auriol (2006) reports case study examples of a 20% reduction in Brazil, 20% in Mexico and 22% in Romania. In the papers these are cited from, a breakdown of how these figures were arrived at is not fully provided. Strategic Cost Savings, where procurement strategy is aligned with corporate strategy and entry levels to new markets is achieved through e-procurement, were only ranked in eighth place out of twenty overall showing that the process savings ranked first overall are more significant than the strategic ones. The strategy remains the same but as e-tendering has widened the possible range of input from organisations by provision of access a smaller increase can be achieved. Additionally, on the process side, Rankin *et al.* (2006) acknowledges that electronic systems of procurement result in less paperwork, producing a more streamlined operation and reducing administration costs. Albano and Sparro (2010) have suggested that using e-procurement in the public sector to centralize procurement can reduce public spending by making the whole system more compliant. No geographical or time limits exist as a result and the public sector becomes more efficient through economies of scale. Kalakota *et al.* (2001) acknowledges this stating e-procurement 'allows procurement activities 24 hours a day, 7 days a week, 365 days a year'. Wong and Sloan (2004) show the results are gaining competitive advantage, reducing procurement costs, and increased profitability. Three further studies (Ribeiro and Henriques, 2001; McIntosh and Sloan, 2001; Wong and Sloan, 2004) all point to e-procurement use as a means of gaining competitive advantage through increasing profit margins. McIntosh and Sloan (2001) and Ribeiro and Henriques (2001) show that financial gains are accompanied by quality improvement achieved through efficiency.

4.5.1.2 *Building information modelling cost reductions*

It can be seen from Table 4.2 that similar cost reductions to e-procurement can be achieved through the BIM process. Those surveyed who were using BIM ranked it the second highest driver for its adoption McGraw Hill (2009) show that two-thirds of BIM users have made a positive return on investment from BIM implementation. The BIM Industry Working Group (2011) identify financial savings for all stages of the project identified in the RIBA Plan of Work 2013 (RIBA, 2013) including the procurement process. Grilo and Jardim-Goncalves (2011) prove that BIM use can mean contractors also benefit from more accurate costs and additional specifications. The interoperability between different BIM and procurement software packages, according to Furneaux and Kivvits (2008) could result in cost savings of up to two thirds of the overall $15.8 billion per year spent in Australia. Yan and Damian (2008) suggest that fully worked up drawings can be produced at 'half time at half cost'. These can be used for both the tender and construction process. Cost savings were ranked in second place for BIM in the construction sector indicating its high importance in relation to BIM adoption.

4.5.1.2.1 COSTS FROM IMPLEMENTATION OF BOTH SYSTEMS

Grilo and Jardim-Goncalves (2011) show that the introduction of both systems will bring further benefits. However, they further point out that the interoperability factor becomes a factor as e-platforms need to allow communication of BIM with traditional e-procurement. Cost savings from BIM use can be achieved at the following stages: e-tendering, e-ordering, e-invoicing and e-catalogues through the use of product and process models.

4.5.2 *Quality*

4.5.2.1 *e-Procurement increased quality*

The third ranked driver overall in the construction industry was '*Increased Quality through increased accuracy (Elimination of errors through Computer use)*'. The Private Sector ranked it much higher than the public sector that ranked it in sixth place. Current authors such as Alvarez-Rodríguez *et al.* (2014) support early work such as Williams and Lynnes (2002) and Minahan and Degan (2001) in suggesting that use of e-procurement removes error-prone manual data entry as data entry to the e-award section of the process is seamless from the e-tendering section. Elements of the tender are linked to the electronic system showing when they are uploaded and asking for confirmation as to why they are missing. Lindskog and Mercier-Laurent (2014) show that for public sector projects Artificial Intelligence (AI) techniques can be used to assist in identification of the elements of knowledge necessary to successfully process purchasing. In addition to electronic processes the public sector also has the rigorous Gateway process and other internal processes that the private sector does not have. This means that reliance on electronic systems is not as important in the public sector resulting in the driver counting as less important. However, Spalde *et al.* (2006) still indicate that many public sector organisations still use e-procurement to increase accuracy of produced procurement documentation. Tindsley and Stephenson (2008) show that mistakes can be removed from contract documentation through implementing e-procurement. Principally this is as a result of systems highlighting missing sections, removal of the re-entry of data such as rates and other associated information from construction tender documentation numerous times. This efficient way of data management provides increased quality and time savings during the assessment and contract management stages of the process. Previous early work by McIntosh and Sloan (2001), and Ribeiro and Henriques (2001) identify it as a driver, but do not rank it. The increased quality due to increased efficiency is ranked in fifth place overall indicates its importance. In an investigation of quality of e-procurement, Hawking *et al.* (2004) divided market intelligence from the decisions made on the information. Eadie *et al.* (2007) combined these into a single driver for construction. The lack of a geographical barrier to those bidding is highlighted by Kalakota *et al.* (2001) who show that as there is an increase in numbers tendering resulting in increased competition, this has a further outcome of increased quality. Hawking *et al.* (2004) further indicate that e-procurement allows easier benchmarking resulting in greater market intelligence and subsequent quality. Minahan and Degan (2001) state that quality increases due to visibility of the supply chain. However, research by Eadie *et al.* (2010a,b) has identified visibility as a driver of quality and ranked it for the construction industry (Table 4.1), but it is not as important as some other quality drivers such as design quality and increased accuracy.

4.5.2.2 *Building information modelling quality improvements*

BIM can produce better designs through clash detection, ranked in first place. The accuracy in using BIM is ranked in fourth place in Eadie *et al.* (2013a). Eastman *et al.* (2011) show that walk-through visualisations provided through BIM models assist clients in refining the brief and reduce 'preference' changes. These changes can be indicated in real time on the screen (Azhar *et al.*, 2008; Bentley, 2012). Therefore at tender stage the ideas and concept design more accurately reflects the client's desires. This produces a quality model where clients can accurately assess what it is exactly that they are purchasing in 3D, knowing how it looks, feels and operates. As the BIM model can output quantities it is relatively easy to check that the build meets the available budget while still satisfying client needs and needing little by way of re-work (Emmitt, 2007; Woo, 2007).

4.5.2.3 *Quality improvements from implementation of both systems*

It can be seen from the above paragraphs that the quality can be improved in the design and tendering process by implementing both e-procurement and BIM together. The implementation of BIM in the early stages of the design process improves the design and therefore the accuracy of the tendering process. The accuracy of the automated quantity take-off and the process quality improvements of an e-tendering system further maximises the positive impact on quality.

4.5.3 **Reduction in time**

4.5.3.1 *e-Procurement time reductions*

Time reductions through the use of e-procurement can be achieved for new build projects (Alshawi and Ingirige, 2003) and maintenance, repair and operation (MRO) projects (Yu *et al.,* 2015). Eom *et al.* (2015) show how pro-curement can filter down into the supply chain to promote time savings in relation to the subcontracting process. They noted time savings in relation to all elements of the process strategy, bidding, evaluation and management. Yu *et al.* (2015) support time savings in sourcing goods in the procurement of maintenance, repair and operating (MRO) materials. This builds on earlier work by Kalakota *et al.* (2001) which emphasises time to market, producing product quality through competition, coping with customer uncertainty and reducing costs in a shorter space of time. Rankin *et al.* (2006) determines a reduction in overall procurement cycle time in construction projects in Canada. In the public sector where legal challenges can substantially stall a project, Panayiotou *et al.* (2003) state that electronic systems of procurement mean an overall reduction in time through less objections due to the greater transparency that this type of system can bring. The speed of electronic bid evaluation is a further benefit highlighted by Panayiotou *et al.* (2003).

4.5.3.2 Building information modelling time reductions

Time is a vitally important project parameter to the success of a construc-tion project, (Walker, 2007). Timing of activities is vital within a project and there is often an iterative process of planning and re-planning with the associated cost- and time-forecasts throughout the design and construction phases of a project. BIM assists with this as the impact of changes can be quickly costed. All members of the design team can meet online in the com-mon data environment, an example of such is A-Site where online discus-sions on design changes and associated costs can take place in real time. Azhar *et al.* (2008) stated that as a result BIM can produce time savings of up to 80% on estimation of costs due to client changes.

4.5.3.3 Time savings from implementation of both systems

The design change process and associated cost estimates are impacted by the implementation of both systems. This is a result of the on-line collaboration through BIM, resulting in quicker decision making during the design pro-cess and the associated cost estimation with automated production and up-dating of registers and schedules, alongside the use of e-procurement to get revised prices from subcontractors, can reduce time from days in duration to hours for changes (Faniran *et al.*, 2001; Eastman *et al.*, 2011).

4.5.4 Sustainability

4.5.4.1 e-Procurement and sustainability

Removal of expensive storage space and holding documents in electronic form has added to the sustainability of e-procurement. The second highest ranked driver for e-procurement in construction overall, and joint top driver for Public Sector clients is '*Convenience of archiving completed work*'. Huq *et al.* (2010) indicated that each tender is unique. They suggest a way to use Six-Sigma to archive tender bids electronically. Huq *et al.* (2010) suggest this was a difficult job as normally the tenders are not accessible to other staff as they are stored on the personal drives of those who prepare the tenders. The use of e-procurement solves this issue as submitted documentation can be archived easily in electronic form. Xerox (2007) showed that in a case study of Islington Council when all its documents were scanned and electronically stored that it enabled disposal of 250,000 pages, equivalent to 12 m^2 of storage space costing £2,400 annually. The National Archives (2012) Clause 3.2 specifies that the length of time before disposal of contract documents for local government is six years after they become non-current and twelve years for contracts under seal. Contract documents are retained indefinitely by the Scottish Office (Gray, 2006). This means substantial savings can be made through space savings.

A further driver investigated in this regard is the sustainability of the e-procurement process itself. Eadie *et al.* (2010a, 2010b) indicated that the implementation of e-procurement resulted in the reduction of procurement staff making the system leaner and less expensive. Egbu *et al.* (2004) showed that as much as an 80% reduction in staff on a multi-million pound project could be achieved at a steel supplier through use of an e-procurement system. Egbu *et al.* (2004) is by no means unsupported as Kong *et al.* (2001) and Davila *et al.* (2003) also suggest that e-procurement means less staff. This means that either more work can be carried out by fewer staff or the reallocation of staff to other duties can result in the company being more productive over a greater period of time.

4.5.4.2 *BIM and sustainability*

A number of authors such as Nawari (2012) and Eastman *et al.* (2011), have suggested that BIM allows prefabrication and off-site construction to a greater degree than traditional design methods. Eastman *et al.* (2011) indicate that BIM provides manufacturers of building components, models which contain comprehensive information, reducing information requests and improving output quality. They further show that on a large healthcare project in the USA, a BIM model allowed 100% pre-fabrication for mechanical systems, resulting in zero clashes during installation. Olofsson and Eastman (2008) show that as a result 20–30% savings were achieved in labour. Nawari (2012) identifies BIM as resulting in the following benefits: speed, economy, sustainability, and safety.

4.5.4.3 *Sustainability from implementation of both systems*

From the section above it can be seen that the implementation of both systems increases the sustainability of both the design, procurement, construction and facilities management aspects of the construction process. From a BIM perspective the sustainability aspect addresses the environmental issues by contextualising the building through visualisation. e-Procurement also addresses the environmental aspects through propagation of the specification and other sections of the contract documents electronically. BIM also addresses the social needs through brief realisation with e-procurement also addressing these issues by passing on elements of the contract such as the social clauses electronically. The economic issues are addressed by both e-procurement and BIM through the interoperability between both systems in automating take-off for prices.

4.5.5 Internal environment

4.5.5.1 e-Procurement improving communication

Communication and collaboration throughout a project is vital to its success. Within e-procurement Hawking et al. (2004) investigated three communication drivers: 'Improving visibility in supply chain management', 'Improving visibility in customer demand' and 'Increased compliance'. Rankin et al. (2006) highlighted the accuracy of each data transaction as a driver for e-procurement. Hawking *et al.* (2004) comment that improved communication results in a product that the client wants and improves its quality. 'Shortened Internal and External Communication Cycle times' is ranked in construction as the fourth highest driver overall (Eadie *et al.*, 2010b) with the private sector ranking it sixth and the public sector ranking it fifth. Gunasekaran *et al.* (2009b) purport that e-procurement improves coordination and collaboration, therefore resulting in cost savings and reduced procurement cycle times. A high ranking of third in Davila *et al.* (2003) for '*Purchasing Cycle time*' for the goods and services industries in the USA further supports the assertion that communication is improved through e-procurement.

4.5.5.2 BIM improving communication

The BIM Industry Working Group (2011) suggest that one of a number of significant organisational impacts through BIM implementation is the collaboration or communication aspect of BIM for all stages in construction of a project. Arayici *et al.* (2011) suggest that after BIM implementation that organisational boundaries are no longer as restriction as stakeholder collaboration improves the performance of the project organisation during the design and construction process. Kymmell (2008) suggests BIM produces a whole new paradigm in construction collaboration improving it immeasurably. To achieve this BIM cannot be just treated as a standalone software tool. Howard and Björk (2008) show it needs to be incorporated into the fabric of each business process and managed holistically to get the best value return. In addition to communication throughout the design process BIM offers contractors an additional way to communicate with subcontractors and their directly employed labour. The use of animated construction sequences being displayed on screen means 4D BIM has the ability to communicate operation sequences to site based staff (Sacks *et al.*, 2009). This ability to show even unskilled operatives pictorially the sequence of operations can only improve the process on site. This is not impacted by the language barrier that confronts many construction sites today (Tutt *et al.*, 2011). While the subcontractor and on-site operative issue was only ranked 8th in importance in Eadie *et al.* (2013b) the overall importance of collaboration during the design phase was ranked in first position in Eadie *et al.* (2013a).

4.5.5.3 *Improving communication from implementation of both systems*

Similar to the results of the foregoing sections, it can be seen that the implementation of both systems improve communication within a construction project. The importance of this aspect of can be seen in the relative rankings – first in Eadie *et al.* (2013a) for BIM and fourth in Eadie *et al.* (2010b) for e-procurement.

4.6 Conclusions

This chapter examined the ranked drivers for BIM and e-procurement within the construction industry. They were examined from the aspect of five project management headings: Cost, Quality, Time, Sustainability and Internal Environment. From examination of these aspects of the construction project it showed that the implementation of BIM and e-procurement could provide a win-win solution to improving these facets of a construction project.

The European Commission has provided an incentive for e-procurement adoption by making it mandatory by 2016 (European Commission, 2015). A similar incentive has been adopted for BIM in the UK with April 2016 the date for mandatory adoption for all government projects (Efficiency and Reform Group, 2011). These apply to Construction Works, in addition to Goods and Services procurement. Therefore both the European Union and the UK Government hope to harness the positives that adoption will bring.

This chapter further discussed the overall ranking of drivers to e-Procurement and BIM in construction. *Process, transaction* and *administration cost savings* were the e-procurement drivers ranked highest by both the public and private sectors. The cost savings due to reduced rework element were ranked in second place for BIM adoption. It was revealed that the cost savings element from adoption of e-procurement and BIM has already been widely documented. As the economy exits recession and the European Union still struggles with ongoing economy issues it is important for the construction industry to adopt both these systems to achieve efficiency throughout the complete construction process. The process of clash detection within BIM has reduced the amount of rework due to poor designs. These savings can result in savings accrued that are a high proportion of the overall scheme costs. The use of interoperability between BIM and e-procurement has resulted in a more efficient, accurate and automated take-off provided the level of detail in the BIM model is high resulting in less chance of items being missed from the Bill of Quantities.

Furthermore savings in time and improved quality are further by-products of adoption of these systems. Sustainability is achieved through the elimination of storage of large tender documentation with the use of e-procurement. In relation to BIM sustainability is increased throughout the complete design process. With natural resources running out and the need to plan for future generations it becomes imperative to adopt systems which assist in the movement towards these goals. Adoption of building information modelling and e-procurement is a step in the right direction.

This chapter provides a useful insight into BIM and e-procurement adoption in the construction industry. It provides a detailed account of all drivers for construction and shows with the targets that the acceleration in e-procurement and BIM adoption levels in the construction industry that many benefits can be accrued.

References

Albano, G. L. and Sparro, M. (2010). Flexible strategies for centralized public procurement. *Review of Economics and Institutions*, 1(2), Fall 2010, Article 4, 1–32.

Alshawi, M., and Ingirige, B. (2003). Web-enabled project management: an emerging paradigm in construction. *Automation in Construction*, 12(4), 349–364.

Alvarez-Rodríguez, J., Labra-Gayo, J., and Ordóñez de Pablos, P. (2014). New trends on e-procurement applying semantic technologies: current status and future challenges. *Computers in Industry*, 65(5), 800–820.

Arayici, Y., Coates, P., Koskela, L., Kagioglou, M., Usher, C. and O'Reilly, K. (2011). Technology adoption in the BIM implementation for lean architectural practice. *Automation in Construction*, 20(2), 189–195.

Atkinson, R. (1999). Project management: cost, time and quality, two best guesses and a phenomenon, it's time to accept other success criteria. *International Journal of Project Management*, 17(6), 337–342.

Auriol, E. (2006). Corruption in procurement and public purchase. *International Journal of Industrial Organization*, 24(5), 867–885.

Azhar, S. (2011). Building information modelling (BIM): trends, benefits, risks and challenges for the AEC industry. *Leadership and Management in Engineering*, 11(3), 241–252.

Azhar, S., Carlton, W., Olsen, D. and Ahmad, I. (2011). Building information modeling for sustainable design and LEED® rating analysis. *Automation in Construction*, 20(2), 217–224.

Azhar, S., Hein, M. and Sketo, B. (2008). building information modeling (BIM): Benefits, Risks and Challenges, Retrieved December 2016, from http://ascpro.ascweb.org/chair/paper/CPGT182002008.pdf.

Barlish, K. and Sullivan, K. (2012). How to measure the benefits of BIM—a case study approach. *Automation in Construction*, 24(1), 149–159.

Bazjanac, V. (2005). Model based cost and energy performance estimation during schematic design. In *Proceedings of CIB W78, 22nd International Conference on Information Technology in Construction*, Dresden (eds. Scherer, R., Katranuschkov, P., and Schapke, S.), Institute for Construction Informatics, Technische Universität Dresden, 677–688.

Bentley. (2012). About BIM. Retrieved March 2015 from http://www.bentley.com/en-US/Solutions/Buildings/About+BIM.htm.

Bew, M., and Underwood, J. (2009). Delivering BIM to the UK Market. Handbook of Research on Building Information Modeling and Construction Informatics Concepts and Technologies, IGI-Global, 30–64. Retrieved March 2015 from http://www.igi-global.com/book/handbook-research-building-information-modeling/37234.

BIMhub (2012). Sophisticated BIM is now available to everyone, with great benefits and high ROI, Retrieved December, 2016, from http://www.bimhub.com/level-bim/.

BIM Industry working Group. (2011). A report for the Government Construction Client Group Building Information Modelling (BIM) Working Party Strategy Paper. Retrieved March 2015 from http://www.bimtaskgroup.org/wp-content/uploads/2012/03/BIS-BIM-strategy-Report.pdf.

Bryde, D. and Robinson, L. (2007). The relationship between total quality management and the focus of project management practices. *The TQM Magazine*, 19(1), 50–61.

buildingSMART Australasia (2012). *National Building Information Modelling Initiative Report*. Retrieved December, 2016, from http://buildingsmart.org.au/advocacy/the-national-bim-initiative-nbi/#.WGPqno_XIw0.

Chen, S., Ruikar, K., Carrillo, P., Khosrowshahi, F. and Underwood, J. (2011). *Construct IT for Business, e-Business in the Construction Industry*, Loughborough University, UK.

Coates, P., Arayici, Y., Koskela, L. and Usher, C. (2010). The changing perception in the artefacts used in the design practice through BIM adoption. In *Proceedings of CIB 2010*, 10–13 May, University of Salford, Salford, UK.

Davila, A., Gupta, M., and Palmer, R. (2003). Moving procurement systems to the Internet: the adoption and use of e-procurement technology models, *European Management Journal*, 21(1), 11–23.

Deutsch, R. (2011). *BIM and Integrated Design*, 1st ed. John Wiley & Sons, Hoboken, NJ.

Dyllick, T. and Hockerts, K. (2002). Beyond the business case for corporate sustainability. *Business Strategy and the Environment*, 11(2), 130–141.

Eadie, R., Perera, S., Heaney, G., and Carlisle, J. (2007). Drivers and barriers to public sector e-procurement within Northern Ireland's construction industry. *Journal of Information Technology in Construction*, 12, 103–120. Retrieved 28 January 2012 from http://www.itcon.org/cgi-bin/works/Show?2007_6.

Eadie, R., Perera, S., and Heaney, G. (2010a). Identification of e-procurement drivers and barriers for UK construction organisations and ranking of these from the perspective of quantity surveyors. *Journal of Information Technology in Construction*, 15, 23–43. Retrieved March 2015 from http://www.itcon.org/cgi-bin/works/Show?2010_2.

Eadie, R., Perera, S., and Heaney, G. (2010b). A cross-discipline comparison of rankings of e-procurement drivers and barriers for UK construction organisations. *Journal of Information Technology in Construction*, 15, 217–233. Retrieved March 2015 from http://www.itcon.org/cgi-bin/works/Show?2010_17.

Eadie, R., Browne, M., Odeyinka, H., McKeown, C., and McNiff, S. (2013a), BIM implementation throughout the UK construction project lifecycle: an analysis. *Automation in Construction*, 36, 145–151.

Eadie, R., Odeyinka, H., Browne, M., McKeown, C. and Yohanis, M. (2013b). An analysis of the drivers for adopting building information modelling. *Journal of Information Technology in Construction*, 18, 338–352.

Eastman, C., Teicholz, P., Sacks, R. and Liston, K. (2011). *BIM Handbook: A Guide to Building Information Modelling*, 2nd ed. John Wiley & Sons, Hoboken, NJ.

Ebbesen, J. and Hope, A. (2013). Re-imagining the iron triangle: embedding sustainability into project constraints. *PM World Journal*, 2(3). Retrieved March 2015 from http://pmworldjournal.net/article/re-imagining-the-iron-triangle-embedding-sustainability-into-project-constraints/.

Eei, K., Husain, W., and Mustaffa, N. (2012). Survey on benefits and barriers of e-procurement: Malaysian SMEs perspective. *International Journal on Advanced Science Engineering Information Technology*, 2(6), 14–19.

Efficiency and Reform Group. (2011). *Government Construction Strategy*. Cabinet Office, London.

Egbu, C., Vines, M. and Tookey, J. (2004). The Role of Knowledge Management in E-Procurement Initiatives for Construction Organisations, Paper presented at ARCOM Twentieth Annual Conference 2004, 1–3 September, Heriot Watt University, (Khosrowshami, F. Ed.), Vol. 1, Arcom, University of Reading, Reading, 661–671.

Emmitt, S. (2007). *Design Management for Architects*, 1st ed. Blackwell Publishing, Oxford, UK.

Eom, S., Kim, S. and Jang W. (2015). Paradigm shift in main contractor-subcontractor partnerships with an e-procurement framework. *KSCE Journal of Civil Engineering*, 19(7), 1951–1961, doi:10.1007/s12205-015-0179-5.

Erridge, A., Fee, R. and McIlroy, J. (2001). *Best Practice Procurement: Public and Private Sector Perspectives*, Gover Publishing Company, Burlington, VT, USA.

European Commission. (2015). Delivering savings for Europe: moving to full e-procurement for all public purchases by 2016. Retrieved March 2016 from http://europa.eu/rapid/press-release_IP-12-389_en.htm.

Faniran, O., Love, P., Treloar, G. and Anumba, C. (2001). Methodological issues in design-construction integration. *Logistics Information Management*, 14(5/6), 421–426.

Fitzpatrick, T. (2012). MOJ demands level 2 BIM by 2013. Retrieved March 2015 from http://www.cnplus.co.uk/news/moj-demands-level-2-bim-by-2013/8627140.article.

Furneaux, C. and Kivvits, R. (2008). BIM—Implications for Government. CRC for Construction Innovation, Brisbane. Retrieved March 2015 from http://eprints.qut.edu.au/26997/.

Gebauer, J., Beam, C. and Segev. A. (1998). Impact of the Internet on purchasing practices. *Acquisitions Review Quarterly*, 5(2), 167–184.

Gray, P. (2006). The Management, Retention and Disposal of Administrative Records, Scottish Executive Health Department, Directorate of Primary Care and Community care. Retrieved March 2015 from http://www.sehd.scot.nhs.uk/mels/HDL2006_28.pdf.

Grilo, A. and Jardim-Goncalves, R. (2010). Value proposition on interoperability of BIM and collaborative working environments. *Automation in Construction*, 19(5), 522–530.

Grilo, A. and Jardim-Goncalves, R. (2011). Challenging electronic procurement in the AEC sector: a BIM-based integrated perspective. *Automation in Construction*, 20(2), 107–114.

Gunasekaran, A., McGaughey, R., Ngaic, E. and Rai, B. (2009a). E-Procurement adoption in the Southcoast SMEs. *International Journal of Production Economics*, 122(1), 161–175.

Gunasekaran, A., McGaughey, R., Ngaic, E. and Rai, B. (2009b). Impact of e-procurement: experiences from implementation in the UK public sector. *Journal of Purchasing & Supply Management*, 13(4), 294–303.

Gunasekaran, A. and Ngai, E. (2008). Adoption of e-procurement in Hong Kong: an empirical research, *International Journal of Production Economics*, Vol. 113 (1). 159–175.

Hawking, P., Stein, A., Wyld, D. and Forster, S. (2004). E-procurement: is the ugly duckling actually a swan down under? *Asia Pacific Journal of Marketing and Logistics*, 16(1), 1–26.

Howard, R., and Björk, B. (2008). Building information modelling—experts' views on standardisation and industry deployment. *Advanced Engineering Informatics*, 22(2), 271–280.

Huq, Z., Seyed-Mahmoud, A., Lotfollah, N., and Saeedreza, H. (2010). Employee and customer involvement: the driving force for Six-Sigma implementation. *Journal of Applied Business & Economics*, 11(1), 105–123.

IDC. (2009). *Westfield Uses Building Information Modelling to Reduce Time and Eliminate Rework across the Property Development Supply Chain.* International Data Corporation, Framingham, MA.

Kalakota, R., Tapscott, D. and Robinson, M. (2001). *e-Business 2.0: Roadmap for Success*, 2nd ed. Addison-Wesley Publishing Company, Boston MA.

Kiviniemi, M., Sulankivi, K., Kahkohnen, K., Makela, T. and Merivirta, M. (2011). BIM-based Safety Management and Communication for Building Construction. VTT Technical Research Centre of Finland, Vuorimiehentie, Finland.

Knudsen, D. (2003). Aligning corporate strategy, procurement strategy and e-procurement tools. *International Journal of Physical Distribution & Logistics Management*, 33(8), 720–734.

Kong, C., Li, H., and Love P. (2001). An ecommerce system for construction material procurement. *Construction Innovation*, 1(1), 4354.

Kymmell, W., (2008). *Building Information Modelling: Planning and Managing Construction Projects with 4D CAD and Simulations.* McGraw Hill, New York.

Laryea, S. and Ibem, E. (2014), Barriers and prospects of e-procurement in the South African construction industry. In *Proceedings of the 7th Annual Quantity Surveying Research Conference*, Tshwane University of Technology, Pretoria, South Africa, 21–23 September 2014, 1–12.

Leite, F., Akinci, B., and Garrett, J. (2009). Identification of data items needed for automatic clash detection in MEP design coordination. In *Proceedings of Construction Research Congress 2009: Building a Sustainable Future* (eds. Ariaratnam, S. and Rojas, E.), Seattle, Washington, 5–7 April 2009, 416–425.

Lewis, A., Riley, D. and Elmualim, A. (2010). Defining high performance buildings for operations and maintenance. *International Journal of Facility Management*, 1(2), 1–16.

Li, X., Aouad, G., Li, Q., and Fu, C. (2008). An nD modeling enabled collaborative construction supply chain information system. In *Proceedings of the Second International Symposium on Intelligent Information Technology Application, 2008. IITA '08.* 20–22 December, Vol. 3, 171–175. Retrieved March 2015 from http://ieeexplore.ieee.org/stamp/stamp.jsp?tp=&arnumber=4739981.

Lindskog, H. and Mercier-Laurent, E. (2014). Knowledge management applied to electronic public procurement. *IFIP Advances in Information and Communication Technology*, 422, 95–111.

Liu, R., Issa, R. and Olbina. S. (2010). Factors influencing the adoption of building information modeling in the AEC industry. In *Proceedings of the International Conference on Computing in Civil and Building Engineering* (ed. Tizani, W.), University of Nottingham, Nottingham. Retrieved March 2015 from http://www.engineering.nottingham.ac.uk/icccbe/proceedings/pdf/pf70.pdf.

Martin, J. (2009). 2009 BCIS eTendering Survey Report. Retrieved 28 January 2012 from http://www.bcis.co.uk/downloads/2009_BCIS_eTendering_Survey_Report_pdf_2_pdf

McGraw Hill (2009), The Business Value of BIM: Getting Building Information Modeling to the Bottom Line, Retrieved December, 2016, from http://www.slideshare.net/JohnathanE/mcgrawhill-the-business-value-of-bim.

McIntosh, G. and Sloan, B. (2001). The potential impact of electronic procurement and global sourcing within the UK construction industry. In *Proceedings of Arcom 17th Annual Conference 2001* (ed. Akintoye, A.), University of Salford, Salford, September, 231–239.

Minahan, T. and Degan, G. (2001). Best Practices in e-Procurement, The Abridged Report. Boston: Aberdeen Group. Retrieved March 2015 from http://www.inkoopportal.com/inkoopportal/download/common/e-procurement_1_.pdf.

National Archives. (2012). Records Management: Retention Scheduling 5. Contractual Records. Retrieved March 2015 from http://www.nationalarchives.gov.uk/documents/information-management/sched_contractual.pdf.

Nawari, N. (2012). BIM standard in off-site construction. *Journal of Architectural Engineering*, 18(2), 107–113.

Nguyen, H. (2013). Critical factors in e-business adoption: Evidence from Australian transport and logistics companies. *International Journal of Production Economics*, 146 (1), 300–312.

Olofsson, T. and Eastman, C. (2008). Benefits and lessons learned of implementing building virtual design and construction (VDC) technologies for coordination of mechanical, electrical, and plumbing (MEP) systems on a large healthcare project, *Journal of Information Technology in Construction (ITcon)*, 13(1), 324–342.

Panayiotou, N., Sotiris, G. and Tatsiopoulos, I. (2003). An e-procurement system for governmental purchasing. *International Journal of Production Economics*, 90(1), 79–102.

Panda, P. and Sahu, G. (2015). Electronic government procurement implementation in India: a cross sectional study. *International Journal of Business Information Systems*, 18(1), 1–25.

Raghavan, N. and Prabhu, M. (2004). Object-oriented design of a distributed agent-based framework for e-Procurement. *Production Planning & Control*, 15(7), 731–741.

Rankin, J., Chen, Y. and Christian, A. (2006). E-procurement in the Atlantic Canadian AEC industry. *Journal of Information Technology in Construction*, 11, Special Issue e-Commerce in construction, 75–87. Retrieved March 2015 from http://www.itcon.org/cgi-bin/works/Show?2006_6.

RIBA. (2013). RIBA Plan of Work 2013. Retrieved March 2015 from http://www.ribaplanofwork.com/.

Ribeiro, F. and Henriques, P. (2001). How knowledge can improve e business in construction. *Proceedings of 2nd International Postgraduate Research Conference in the Built and Human Environment*, University of Salford, Blackwell Publishing, Salford UK.

Sacks, R., Treckmann, M. and Rozenfeld, O. (2009). Visualization of work flow to support lean construction. *Journal of Construction Engineering and Management*, 135(12), 1307–1315.

Samuelson, O. and Björk, B. (2010). Adoption processes for EDM, EDI and BIM technologies in the construction industry. Retrieved March 2015 from https://helda.helsinki.fi/handle/10227/779.

Simu, T. (2012). E-government for good governance in Bangladesh: trade and commerce perspective. *SUST Journal of Social Sciences*, 18(4), 28–39.

Sone, J. (2011). E-Governance in Central Texas: Patterns of e-Gov Adoption in Smaller Cities. Applied Research Projects, Texas State University-San Marcos. Paper 381. Retrieved March 2015 from http://ecommons.txstate.edu/arp/381.

Spalde, C., Sullivan, F., de Lusignan, S. and Madeley, J. (2006). e-Prescribing, Efficiency, Quality: Lessons from the Computerization of UK Family Practice, Journal of the American Medical Informatics Association, 13(5). 470–475.

Tindsley, G. and Stephenson, P. (2008). E-tendering process within construction: a UK perspective. *Tsinghua Science & Technology*, 13(1), 273–278.

Toktaş-Palut, P., Baylava, E., Teomanb, S., and Altunbeyc, M. (2014). The impact of barriers and benefits of e-procurement on its adoption decision: an empirical analysis. *International Journal of Production Economics*, 158(2014), 77–90.

TRADA. (2012). *Construction Briefings: Building Information Modelling.* Timber Research and Development Association, High Wycombe, UK.

Tutt, D., Dainty, A., Gibb, A. and Pink, S. (2011). *Migrant Construction Workers and Health & Safety Communication*, Construction Industry Training Board, King's Lynn, UK.

Walker, A. (2007). *Project Management in Construction*, 5th ed. Blackwell Publishing, Oxford, UK.

Wang, Y., Xi, C., Zhang, S., Yu, D., Zhang, W. and Li, Y. (2014). A combination of extended fuzzy AHP and fuzzy GRA for government E-tendering in hybrid fuzzy environment. *The Scientific World Journal*, 2014, 1–11.

Williams, C. and Lynnes, R. (2002). e-Business without the commerce. In *Proceedings of the Water Environment Federation, WEF/AWWA Joint Management 2002*, Water Environment Federation, Vol. 6, 298–303.

Wong, C. and Sloan, B. (2004). Use of ICT for e-procurement in the UK construction industry: a survey of SMES readiness. In *Proceedings of ARCOM 20th Annual Conference 2004* (ed. Khosrowshami, F.), 1–3 September, Heriot Watt University, Vol. 1, Arcom, University of Reading, Reading, UK, 620–628.

Woo J. H. (2007). BIM (Building Information Modeling) and Pedagogical Challenges, *Proceedings of the 43rd Annual Conference by Associated Schools of Construction* (eds. Sulbaran, T. and Cummings, G.), Northern Arizona University,

Flagstaff USA, 12–14 April. Retrieved March 2015, from http://ascpro0.ascweb. org/archives/cd/2007/paper/CEUE169002007.pdf.

XEROX. (2007). Digital Archiving Promotes Sustainability and Improved Service Delivery for the London Borough of Islington. Retrieved March 2015 from https:// docushare.xerox.com/pdf/ds_casestudy_LBI_En-UK.pdf.

Yan, H. and Damian P. (2008). Benefits and Barriers of Building Information Modelling. Retrieved March 2015 from http://homepages.lboro.ac.uk/~cvpd2/ PDFs/294_Benefits%20and%20Barriers%20of%20Building%20Information%20 Modelling.pdf.

Yu, S., Mishra, A., Gopal, A., Slaughter, S., and Mukhopadhyay, T. (2015). E-procurement infusion and operational process impacts in MRO procurement: complementary or substitutive effects? *Production and Operations Management*, 24(7), 1054–1070.

5 Building information modelling and management

David Greenwood

5.1 Introduction

This chapter is the core of a three-chapter offering on building information modelling, sometimes known as building information management, but most easily recognised under the acronym, BIM. The chapter is one of three on BIM with the other two covering, in turn, two BIM application related detailed case studies to demonstrate its uptake, issues and advantages. The aim of the current chapter is to be, as far as possible, a state of the art account with a full coverage of BIM from its origins to current level of development and beyond, to its future potential. This, in itself, brings limitations. The first of these is that with BIM we are dealing with a radical, disruptive and fast-moving phenomenon: this makes the contemporaneity of its reporting most difficult. Until BIM becomes fully established and prevalent, anything published on the topic will almost immediately require updating. Secondly, the uptake of, and progress in BIM has been different in the different construction industries around the world. This account, out of necessity, takes a focus that is predominantly on the UK. It will, to some extent report on developments in BIM elsewhere in the world, nevertheless it should be kept in mind that a thorough global coverage is not intended.

5.2 Historical development, terminology and scope

The term *building information modeling* [note the US spelling: in UK English this is 'Modelling'] was used in an Autodesk 'Building Industry Solutions White Paper' in 2002 (Autodesk, 2002) though the term 'building information model' was in use before that (see, for example, van Nederveen and Tolman, 1992) and earlier references to the concept appear in Eastman *et al.* (1974). It has been suggested (Laserin, 2008) that the term was actually first coined from a merger of the earlier expressions 'Building Product Model' (largely used in the USA) and 'Product Information Model' (which was common at the time in Europe). In the UK, problems of interoperability and model data translation were being addressed as early as the 1990s in projects such as the European Union funded 'COMBINE' research programmes

(Wright *et al.*, 1992) and explorations of n-D modelling involving Aouad and others (Ford *et al.*, 1994).

U.S. commentator and architectural director John Tobin, drawing on the terminology of Bower and Christensen (1995) notes that BIM began as a 'sustaining technology' (i.e. using 3D models to produce construction information more efficiently) but has become a 'disruptive innovation' that has 'created brand new value networks' that will 'change markets and expectations' (Tobin, 2013). Thus BIM, with its origin as a design tool, 'becomes VDC' (Virtual Design and Construction) with its adoption by contractors. Not only this, but 'smart owners' ['clients' in the U.K.] then enter the picture as they 'realize how these information-rich 3D models could be useful as an active decision-making tool during construction, and then used as stores of information for facility operations purposes' (Tobin, 2013). The process Tobin has described has caused some to question the applicability of the term *Building Information Modelling*: and variants such as *Building Information Management, Virtual Design and Construction* and *Digital Design and Construction* have appeared. The advantage of the last two descriptors is that they remove the early misconception that BIM was only applicable to *building* – a misconception that prompted commentators to ask questions such as 'is BIM relevant to civil engineering?' The question has, of course, been answered in the affirmative. As Corke (2012) writing in New Civil Engineer states, 'the core principles and workflows associated with BIM apply equally to all infrastructure projects, including roads, railways, bridges, dams and water works.' There are publications specifically aimed at civil engineers (e.g. Barnes and Davies, 2014), asset managers (Shetty *et al.*, 2013; BSI, 2014) and the formation of sector-specific groups such as 'BIM 4...' groups (http://www.bimtaskgroup.org/bim4-steering-group/) which include Clients, Retail, Water, Health, FM (facilities management), and Housing. Although it is arguable that BIM (taken literally as *building information modelling*) is far from ideal as a representation of the concepts it has come to embody, the fact remains that it is now commonly-accepted and all-pervasive.

The reader will encounter many definitions of BIM as well as a number of substitute names for the concepts that it represents. Examples of BIM definitions include, from the (United States) National Institute of Building Sciences (2007)

> Building Information Modeling (BIM) is a digital representation of physical and functional characteristics of a facility.

5.3 Drivers for BIM adoption – 'push' and 'pull' reasons

Within the 'Construction Strategy' published by UK Government (Cabinet Office, 2011) was a requirement for 'collaborative 3D BIM (with all project and asset information, documentation and data being electronic)' by 2016.

The twin key objectives of the strategy were 'cost reduction in the construction and operation of the built environment' and the 'implementation of existing and emerging government policy in relation to sustainability and carbon'. The report highlights BIM as an important route to meeting these objectives. This was influenced by the so-called 'BIM Strategy Report (BIM Working Party, 2011) that recommended the adoption (by the government) of a 'Push-Pull' strategy: a *push* by the supply side (i.e. the industry itself) and a *pull* from the client side (specifically the public sector client). The government's 'BIM edict' has produced an enormous interest in BIM in the UK as illustrated by the annual (since 2011) NBS National BIM Reports (2011–2015). In NBS's 2011 survey (National Building Specification, 2011) 43% of respondents were unaware of BIM, and this has reduced to 5% in the 2015 survey. There was a corresponding increase in those 'aware of and using BIM': 13% in the 2011 survey, to 48% in 2015 (National Building Specification, 2015). In the absence of client *pull*, the *push* by the industry supply side would naturally require a business case and evidence of a return on investment (ROI). Some of the earliest indications of ROI in BIM came from the United States, where Holness (2006) reports savings on construction costs of between 15% and 40%. More recently a survey of over 1,000 industry participants by McGraw-Hill has pronounced that '63% of BIM users are experiencing a positive perceived ROI on their overall investment in BIM' (McGraw-Hill, 2012: p. 39) with the most common range being between 10% and 25%. Evidence from the UK is relatively scarce and generally unquantified, though case studies are emerging on the Constructing Excellence website.[1] In all cases the evidence suggests that experienced users derive an increased ROI, or, to put this conversely, BIM adoption may show an initial productivity loss, followed by considerable gains.

5.4 Applications and uses of BIM

From its initial manifestation as an enhanced design tool BIM technology has been developed and extended into a wide range of functional applications. For example, Bryde *et al.* (2013) map the potential uses of BIM against the 'knowledge areas' specified by the Project Management Institute's (PMI) Project Management Body of Knowledge (PMI, 2008) and others have similarly examined the potential contributions of BIM. What follows is an attempt at providing a fairly comprehensive (but not exhaustive) list of how BIM-based applications can inform, facilitate and improve the operations involved in the design, construction and management of the built environment.

5.4.1 3D design and visualization

The earliest and most basic application of BIM is in design – architectural, structural, services, and so on. The integrated nature of BIM and its capability in information re-use complements the iterative nature of the building design process. Furthermore, automatic code and regulation compliance checking can be incorporated (see, e.g. Malsane *et al.*, 2015). Links are available to structural and environmental analysis software. Visualised renderings of 3D designs (to a range of realism) are useful communicators of design intent, particularly to 'lay' stakeholders, including clients. 3D renderings can be easily generated in-house with little additional effort. In addition, BIM 'viewing software' enables constructors to view and manipulate 3D design outputs.

5.4.2 Design coordination, clash detection and change management

Applications have evolved to enable co-ordinators (lead designers, main contractors, etc.) to resolve design clashes or problematic interfaces between different construction systems. The use of BIM has been shown not only to reduce design clashes and subsequent requests for information (RFIs) and variations/change orders, but also, where changes do occur, BIM facilitates their management as a change to any part of the model can be coordinated with all other parts.

5.4.3 Off-site fabrication

'Shop drawings' can be readily generated and designs can be integrated into the production software used by some manufacturers. Leading examples are structural steel, precast concrete and ductwork, where manufacturers and fabricators have for many years used digital 'design for manufacture' applications. There is now a real prospect of integrating these systems with the design-procure-construct process of the individual construction project.

5.4.4 Construction sequencing ('4D BIM')

4D planning involves linking a time schedule to a 3D-model to improve construction planning techniques. Schedules can be generated by interrogating the design model(s) to identify activities, calculate durations (using automated quantity extraction), impose installation logic, schedule resource requirements and visualise the time/space relationships of the delivery process: the overall aim being to improve communication between project team members, through informative animations of the build process.

5.4.5 Estimating, cost management and procurement ('5D BIM')

Applications are available that cater for the procedures currently undertaken by quantity surveyors and estimators in the bidding, procurement and cost management and project accounting functions. This includes 'time-cost-value' analysis techniques such as earned value management (see e.g. Barlish and Sullivan, 2012: p. 153). Work is also underway to integrate BIM applications with enterprise resource planning (ERP) systems at the business level of the organisation (see e.g. Babič *et al.*, 2010) to inform their sales, purchasing and logistics functions.

5.4.6 BIM for Sustainability ('6D BIM')

The potential of this application relates to sustainability targets for a building, allowing information such as energy use, resource efficiency and other aspects of sustainability from a materials and management point of view to be better analysed, managed and understood (see e.g. Hamza and Horne, 2007; Azhar and Brown, 2009; Nour *et al.*, 2012). The object-modelled data in the BIM model can accommodate information such as embodied carbon, including that created by the process of construction. Work is underway to enable the automated performance of BREEAM[2] and LEED[3] tracking and assessment (see e.g. Azhar *et al.*, 2011).

5.4.7 Whole-life and facilities management ('7D BIM')

7D BIM relates to Life Cycle Costs, and represents the management of facilities or assets. This could be delivered to the client or end-user in the form of an 'asset tagged', 'as-built' BIM model at handover, or in a more specific format (e.g. as COBie[4] information) and may be populated with appropriate component and product information, operation manuals, warranty data, and so on. Information based on BIM can thus be re-used for driving efficiencies in the management, renovation, space planning, and maintenance of facilities. Recognising the potential for such capabilities, and seeing BIM's value in efficiency gains in OPEX as well as CAPEX, the UK Government has linked BIM with its 'government soft landings', the objective of which is 'to ensure that value is achieved in the operational lifecycle of an asset' (BIM Task Group, 2013). Mention should also be made of BIM's potential role in the retrospective modelling of existing buildings and other structures through point-cloud capture using laser scanning (see e.g. Volk *et al.*, 2014).

5.5 Technology, process and people

The familiar trio of *technology, process* and *people* is a combination based on work such as that of Davis (1993) on 'technology acceptance' (i.e. its acceptance by *people*) and that of Brynjolfsson and Hitt (2000), Bower and Christensen (1995), and David (1990) on the impact of technology on workflow and process in organisations. In turn, this echoes earlier work on socio-technical systems by the likes of Emery and Trist (1969) that examines how people and organizations respond to technology, particularly information technology. It is probably true to say that most of the aforementioned work revolves around what happens when one of the three intervenes to cause a change in a relationship between the other two (e.g. the intervention of new technology between people and process, and so on).

5.5.1 BIM as technology

First and foremost (or at least, initially), BIM is 'a modelling technology' (Eastman *et al.*, 2011: p. 16). The model incorporates and reconciles information from different professionals (architect, structural engineer, etc.) initially in the form of graphical object data, in the form of size, shape, location and other visible characteristics. Grilo and Jardim-Goncalves (2010) refer to this as a 'surface model': however, in what they describe as a 'solid' or 'smart' model BIM can incorporate characteristics other than geometry, such as material composition, design life or other supplier information in object-based, parametric components. Such information can then be re-utilised to produce allied applications such as detailed building performance analyses, 4D scheduling and 5D modelling for cost management. The extensive range of uses to which the model could potentially be put is revealed in the term '*n*D modelling' (see Marshall-Ponting and Aouad, 2005) and some are considered later here in a little more detail.

The modelling takes place within a variety of different native software platforms, usually proprietary and commercially-driven. Eastman's original 'BIM Handbook' (Eastman, 2008) lists over 70 different software companies with hundreds of different software packages. These have developed to suit the functional needs of their current users (architects, structural engineers, services engineers, constructors, and so on). Consequently, they differ structurally and semantically (Lockley *et al.*, 2013). In the 2014 NBS National BIM Report 25% of BIM users felt that 'information models only work in the software they were made on'. (NBS, 2014: p. 14). This in itself is a limiting factor in reaching fully-collaborative BIM, and is what is referred to as the issue of interoperability or, more specifically 'semantic data interoperability' (Yang and Zhang, 2006).

5.5.2 Interoperability

Interoperability is defined by the International Alliance for Interoperability (now 'buildingSMART') as 'an environment in which computer programs can share and exchange data automatically, regardless of the type of software or where the data may be residing' (Fischer and Kam, 2002: p. 14).

As noted above and articulated by Cerovsek, achieving 'inter-operability between multiple models and multiple tools that are used in the whole product lifecycle' is a key challenge (Cerovsek, 2011: p. 224). The ambition to achieve the automatic and efficient use and re-use of information throughout the design, delivery and operation of a built asset is hindered by the current existence of a range of commercially available, 'native' BIM software platforms, each with its different functionalities, naming conventions and classification. The problem is particularly acute when it comes to the authoring and use of standard BIM object libraries (Howard and Bjork, 2008).

To overcome this situation, the secure and reliable exchange of data is essential and this has led to work by various national and international bodies, including the International Organization for Standardization (ISO), producing 'requirements for the exchange of building element shape, property, and spatial configuration information between application systems' (Grilo and Jardim-Goncalves, 2010: p. 525). The tangible result, for AEC practitioners, is an intermediary format for exchange of data called *Industry Foundation Classes* (IFC). IFC is a 'schema' that defines and provides 'Model View Definitions' to meet the needs of information exchange between different applications. The latest version, IFC4, is registered with ISO as an official International Standard, ISO 16739:2013.

According to NBS (2014: p. 17) 'IFC, as a platform-neutral, open file format, allows models to be shared among the design team, irrespective of software choices'. In its 2014 survey, NBS found that 45% of BIM-using respondents used IFC (an increase of 6% from the previous year). A detailed review of the history, development and current status of IFC is provided in Laakso and Kiviniemi (2012).

Recognising the importance of IFC standards, the promoters of the many proprietary BIM software platforms aim to ensure that their products support them fully: in other words, are, through IFC, compatible with one another. The extent of this compatibility has been questioned (see, e.g. Lockley *et al.*, 2013), however, most products are undergoing continual improvement by their authors and are offered for testing and certification by buildingSMART, the organisation that champions the use of IFC. As a result, full support of IFC, and thus, interoperability is now a realistic prospect.

5.5.3 BIM as process

In their 'BIM handbook' Eastman and his co-authors comment that, as well as being a 'technology' BIM has also come to represent an 'associated *set of processes* to produce, communicate and analyse building models' (Eastman *et al.*, 2011: p. 16: italics added for emphasis). Thus, in their foreword to the same work, the authors note that BIM adoption requires not just a change in technology but a 'process change' (Eastman *et al.*, 2011). It is precisely this aspect of BIM that accounts for its identification as a 'disruptive innovation'. Reflecting this, the 2014 NBS National BIM Survey found that 92% of respondents who were BIM users agreed that 'adopting BIM requires changes in our workflow, practices and procedures' (National Building Specification, 2014: p. 21).

The extent of these changes depends on the extent to which BIM is adopted. A well-accepted measure of BIM adoption is provided by the 'BIM Maturity Diagram' developed by Mervyn Richards and Mark Bew in 2008, and subsequently adopted by the UK Government's BIM

Task Group (see http://www.bimtaskgroup.org/). Consequently, in the UK, the model has become an industry reference point for measuring levels of BIM maturity. A similar representation was produced by Succar (2009). In the Bew-Richards model, there are four maturity levels (from 0 to 3). Level 0 corresponds to Succar's 'Pre-BIM' stage (i.e. where designers work using manual methods or CAD). Bew-Richards' Level 1 sees the introduction of 3d modelling, though this is isolated (sometimes referred to as 'lonely BIM') with no collaboration between disciplines. This corresponds with Succar's 'BIM Stage 1: Object Modelling'. The essence of Level 2 in the Bew-Richards model is that there is some degree of model collaboration. Typically, this would involve 'federating' individual models to work in a 'common data environment'. In Succar's model this is described as 'BIM Stage 2: Model-based collaboration'. BIM Stage 3, In Succar's model, is 'network-based integration' and in the Bew-Richards scheme Level 3 represents the full collaboration of all disciplines in a single, shared project model.

5.5.4 *Integrated project delivery*

A feature of both the Bew-Richards and Succar (2009) representations is that increased BIM maturity is characterised by increasing collaboration: in order to exploit BIM technology to its utmost, requires a 'collaborative environment'. Succar explicitly has 'Integrated Project Delivery' as the 'long-term goal of BIM implementation'. The term *Integrated Project Delivery* (IPD) was coined in the construction industry of the United States, and, according to the American Institute of Architects (AIA) is 'a project delivery approach that integrates people, systems, business structures and practices into a process that collaboratively harnesses the talents and insights of all participants to reduce waste and optimize efficiency through all phases of design, fabrication and construction.' (AIA California Council, 2007: p. 1).

5.5.5 *Innovative approaches to project procurement strategies*

The AIA definition of IPD includes its descriptive elements but does not specify how it should be achieved or what form of project governance structure or procurement strategy might enable it. However the UK Government Cabinet Office has proposed three 'new models of procurement' that would best correspond to 'high levels of supply chain integration, innovation, and good working relationships between client and industry [and] will lead to a significant change in the costs and risks of construction projects' (Cabinet Office, 2014: p. 7).

5.5.5.1 *Cost-led procurement*

The basis of this, the most conventional of the three 'new models' is that two or three integrated framework supply teams develop their bids in competition. The successful proposal is then selected on affordability and quality criteria.

5.5.5.2 *Two-stage open book*

The aim of this approach is to enable the early engagement of the supply-side whilst deferring the commercial commitment of the client. In the first stage, contractor-consultant teams compete on the basis of a development fee and an appropriate set of qualitative elements. In the second stage, the successful team transparently develop the project proposal to the client's cost bench-mark, with risks being addressed during Stage 2, but before commencement.

5.5.5.3 *Integrated project insurance*

In this, the most innovative of the three 'new models', there is an initial competition between integrated project teams, one of which is selected on criteria of qualitative elements and a 'fee declaration' and then develops an acceptable design solution. A single joint-names project insurance policy is executed to cover risks associated with delivery of the project; this in-cludes traditional construction-related insurances together with an element of cover for cost overrun. The latter may be subject to an 'excess' that is underwritten by a 'pain-share' agreement between the parties (client and delivery team).

In the future, Level 3 BIM will 'raise significant legal, contractual and in-surance issues' (Golden, 2015) that will undoubtedly have some impact upon procurement. Currently, however, new procurement approaches are not es-sential for achieving Level 2 BIM on a project and traditional arrangements may suffice, providing that the parties collaborate sufficiently. However, to achieve Level 3 BIM, new procurement models would probably be needed.

5.5.6 *Problems with collaboration*

The increased collaboration that is required for more effective BIM ex-ploitation brings a variety of accompanying challenges. In terms of Level 1 ('lonely') BIM, where BIM users operate in isolation, these are few: BIM users can easily settle for a 'business as usual' attitude. But for Level 2 (model-based collaboration) to work to its best effect it is necessary to set rules, conventions and ways of working to cope with the individuality of the different participants. This comprises both technical and non-technical challenges. Technical issues revolve around data exchange and interopera-bility, as discussed earlier. But when BIM becomes collaborative there are also questions about *what* information is to be expected, *when, from whom, to whom*, and in *what form* or level of detail and information. Collectively known in the USA as levels of definition, the concepts of levels of detail and levels of information represent a recognised ascending order of devel-opment in models. Without guidance, there is the prospect of chaos, and, as Gu and London (2010) have recognised, the very problem for which BIM might offer a solution (i.e. the construction industry's fragmentation) would be a major factor in inhibiting its implementation. Consequently, the UK Government, as an important industry client (and, as will be discussed later, a major influence in the uptake of BIM in the UK) has taken a lead in creating standardised solutions to some of these problems.

5.5.7 *Documentary requirements for information*

A number of standard protocols for BIM process have been developed in the United States, most notably the AIA's 'Digital Practice Documents' (AIA, 2012) and these were normally adapted to the specific needs of individual projects. It is from these that the central requirements for a BIM-enabled project were categorised. Some examples are listed below in order of their normal appearance in a project. For a more complete list, with fuller definitions the reader is referred to UK process standards such as PAS 1192-2:2013 (BSI, 2013) and PAS 1192-3:2014 (BSI, 2014). Also, note that the *NBS Digital Toolkit* (see later coverage of this) is a digital application that is designed specifically to address many of the requirements of these documents.

5.5.7.1 *Employer's information requirements (EIR)*

The EIR sets out, in terms of content and form, a client's requirements for the delivery of information by its project supply chain.

5.5.7.2 *BIM execution plan (BEP) or project execution plan (PEP)*

A BEP or PEP is prepared to demonstrate how the EIR will be delivered. It communicates to the client how the information modelling will be implemented and presented.

5.5.7.3 *Master information delivery plan (MIDP)*

The MIDP is, according to the definition in PAS 1192-2:2013 (BSI, 2013) the 'primary plan for when project information is to be prepared, by whom and using what protocols and procedures, incorporating all relevant task information delivery plans'. The role of the MIDP is encompassed in such BIM management tools as the *Digital Toolkit* (see below).

5.5.7.4 *Project information model (PIM)*

The PIM is, according to PAS 1192-2:2013 (BSI, 2013) the information model 'developed during the design and construction phase' of a project, consisting of documentation, non-graphical information and graphical information defining the delivered project. For the purpose of managing, maintaining and operating the asset this is eventually superseded by the asset information model (AIM).

5.5.7.5 *Common data environment (CDE)*

A CDE, as defined by PAS 1192-2:2013 (BSI, 2013) is 'a single source of information … for multi-disciplinary teams in a managed process'. This requires the integration or federation of information from a variety of sources and in a variety of native software platforms. Most of the existing providers of file-sharing collaborative extranets are creating products to fulfil the same function with BIM databases, rather than files. Currently, there are other barriers to all project parties working in a CDE; these will be examined later.

5.5.8 **Standardised process solutions**

From 2011 onwards the UK Government has commissioned 'standardised solutions' for working digitally in the built environment. These take the form of eight components, in the form of official standards, implementation tools and guides that comprise the 'rules of engagement' for Level 2 BIM. These are available at, or *via* www.bimtaskgroup.org and are introduced in the following section.

5.5.8.1 *PAS 1192-2:2013 (BSI, 2013) specification for information management for the capital/delivery phase of assets using building information modelling*

This document builds upon the existing BS 1192:2007 (Collaborative production of architectural, engineering and construction information information) in order to specify what is required for delivering projects in Level 2 BIM. The PAS also describes how models evolve through increasing levels of development (in both graphical design detail and information content). The *level of model detail* is a description of graphical content of models at each of the stages from client's brief requirements and establishing site constraints to the stage where the model is equipped to facilitate the operation and maintenance of the asset. The *level of model information* is a description of non-graphical content of models at each of the stages.

5.5.8.2 *PAS 1192-3:2014 (BSI, 2014) specification for information management for the operational phase of assets using building information modelling*

PAS 1192-3 extends the project information delivery cycle into extended to the operating phase of the built asset's life cycle. It specifies information requirements from the viewpoint of the operational phase of a constructed asset or group of assets. According to PAS 1192-3 itself, this includes 'data and geometry describing the asset(s) and the spaces and items associated with it, data about the performance of the asset(s), supporting information about the asset(s) such as specifications, operation and maintenance manuals, and health and safety information.'

5.5.8.3 *BS 1192-4 collaborative production of information.*
Part 4: fulfilling employers information exchange requirements
using COBie

The existing British Standard – BS 1192:2007 (Collaborative production of architectural, engineering and construction information) has been revised (as BS1192-4:2014) to encompass the handling of information using COBie. COBie is a data schema presented in the form of a spreadsheet which serves as a standardised index of information about new and existing assets throughout their lifecycle. The format permits open exchange of the relevant data; thus users are able to access COBie files (and create them automatically from models) with little or no software investment cost or operating knowledge. In the schema, data are classified spatially (into *Facility – Floor – Space – Zone*) and physically (into *Type – Component – System – Assembly – Connection*). In the case of a new facility, the COBie data file 'grows' as the project progresses, with a series of increasingly comprehensive *data drops* as the project advances through its stages. The file can be retained in its spreadsheet form or imported into decision-making tools such as an FM database.

5.5.8.4 *PAS 1192-5:2015 (BSI, 2015) Specification for security-*
minded building information modelling, digital built environment and
smart asset management

The document aims to provide stakeholders (particularly building and asset owners) with protocols and controls to ensure the appropriate security of their data whilst they are collaborating using digital information.

5.5.8.5 *CIC BIM protocol and associated publications*

The Protocol, produced by the Construction Industry Council is a document to supplement the contractual agreements of parties working in a BIM environment. It is expected that, in the longer term, construction contract drafting bodies will amend their documents to cater for projects at Level 2, and eventually Level 3 BIM. CIC has also produced a guide to explain how BIM relates to Professional Indemnity (PI) insurance, particularly applicable if the party insured is acting in an information management role. The scope of this role is described in the *Outline Scope of Services for the role of Information Management* which was also published by the CIC in 2013.

5.5.8.6 *Government soft landings (GSL)*

GSL is a protocol that specifies the (gradual, where appropriate) handover of an asset (and related structured information) to enable owners and/or asset managers to make best use right from its handover. A more detailed description is available online at http://www.bimtaskgroup.org/gsl-summary/. The relationship between GSL and structured information underpins its association with BIM.

5.5.8.7 BIM toolkit and digital plan of work/digital toolkit

Following a project funded by Innovate UK, a consortium led by NBS has produced a digital framework that is able to be customised for individual projects or project types to provide a data delivery template. Set against the eight stages of the RIBA's revised Plan of Work (RIBA, 2013) the Toolkit gives the ability to specify and verify the delivery to the client of required levels of geometric and other data and documentation. As well as a specifier of information deliverables, it is seen as an enabler for collaboration between all stakeholders and at every stage. It defines and allocates responsibility and facilitates the verification of information delivery.

5.5.8.8 Unified BIM classification system

Standard classification systems are an essential component of information management and sharing data. By providing the taxonomy, or rules that govern a common language, they enable the search and retrieval of information, and its integration and aggregation. This is the very foundation of working in a common data environment. The current classification systems that relate to construction (for example, Uniclass and NRM) have been developed at different times by different organisations and, therefore, do not entirely align. For the full implementation of Level 2 BIM this needs to be rectified and reflected in a comprehensive new classification system compliant with the revised ISO 12006 standard for Building construction – Organization of information about construction works.'

5.6 Barriers to BIM adoption

In an earlier part of the chapter, some of the non-technical (organisational) problems with operating collaboratively in Level 2 BIM were identified. Technical issues over data exchange and interoperability were also discussed earlier. As shown in the previous sections, guidance has been forthcoming and continues to emerge from government sources, professional bodies and from solution providers, such as software developers. Some of the complex and unresolved legal and contractual challenges (including issues regarding insurance) to working in increasing levels of BIM are dealt with elsewhere in this book. Matters such as computer failure and data security will have increased importance. Finally, as shown in studies, such as those by van der Smagt (2000) and Dossick and Neff (2010) 'human factors' (such as leadership, capability, education, organisational culture, team working) play a leading part in the likely success of BIM-enabled construction operations.

5.7 Conclusions and prospects for the future

BIM has been in gestation for more than twenty years. Before 2000, and under other names, it was the province of researchers in the USA and Europe. After that, it began to take hold in the construction industry of the USA, and in the UK, following the government's 'construction strategy' and 'BIM mandate'. Such 'pushes to adoption' are now becoming more common around the world, as governments follow the example of the UK and other early adopters from the demand side of the industry. In terms of its advantages to the 'supply side', BIM has a variety of applications, exemplified by the 'dimensions' (from 3-D to n-D) by which the different applications are characterised. It is becoming clear that BIM is not simply 'new technology' but requires us to understand and respond to the effects that the technology will have on process and people. In contrast with technology, these elements are often slow to change, and it remains to be seen how quickly, for how many, and to what extent BIM will become part of the normal way of working within the industry.

So much for the current goal: what has been called 'Level 2 BIM'. In terms of the 'near future' we have the prospect of Level 3 BIM. This aims at the use, on construction projects, of a fully collaborative single real-time model in which software interoperability, IT infrastructure, and contractual and legal obstacles have been resolved. In October 2014 a 'strategic plan' for Level 3 BIM, entitled 'Digital Built Britain' was published by the UK Government and previewed 'the next stage in the digital revolution ... [in]...the way we plan, build, maintain and use our social and economic infrastructure'. The next steps are envisaged as the creation of new, international open data standards; new contractual frameworks for BIM-enabled projects; co-operative cultural environments; appropriate training in BIM techniques particularly for the public sector; domestic and international growth and jobs in technology and construction. The predicted technological adoptions that flow from this digital revolution will include *automated digital decisions* (i.e. those that do not require human input) and *predictive digital* (i.e. solutions that are automatically predicted based upon digital information).

In September 2014 the Construction Industry Council's BIM2050 produced a future-gazing report on 'Our Digital Future' (CIC BIM2050 Group, 2014). The following is a selection of their summarised findings that are most directly applicable to the present topic.

Cyber security: An interesting conflict is highlighted, namely that between the movement towards free and open information, connectivity and collaboration, on the one hand, and the need to secure that information; the current strategy for which is identified as 'to throttle access'.

Interoperability for smart cities: Recognising the advent of 'intelligent infrastructure' in future smart cities, the Group highlights the need for organisations to 'review the interoperability of the products that their supply chain installs on to projects'.

Nano-second procurement and performance: Organisations are advised to 'accelerate their digitisation of business management and enterprise resource planning systems' to approach the reaction speeds that are currently in existence in the stock market.

Constructing in space: The future demands of 'extra-terrestrial construction' will necessitate ever more radical approaches in the industry.

Robotics and autonomous systems: A future that requires a minimum 'field population' of construction operatives but an increase in those managing 'people–plant interfaces in complex operations' would require organisations to 'consider automation and design for manufacture strategies as early as possible in the asset lifecycle'.

Demountable organisations: Overall, future business operations are predicted to show a 'shift from employers owning employees to entrepreneurs trading talent as a commodity'.

It is appropriate to finish this chapter with an extract from the Executive Summary of the 'Digital Built Britain' publication. 'Building Information Modelling (BIM) is changing the UK construction industry – a vitally important sector that employs more than three million people and in 2010 delivered £107 billion to the UK economy. Over the next decade this technology will combine with the Internet of things (providing sensors and other information), advanced data analytics and the digital economy to enable us to plan new infrastructure more effectively, build it at lower cost and operate and maintain it more efficiently. Above all, it will enable citizens to make better use of the infrastructure we already have.' (H.M. Government, 2014: p. 5).

Notes

1 http://constructingexcellence.org.uk/bim/.
2 Building Research Establishment Environmental Assessment Methodology – a method of assessing, rating and certifying the sustainability of buildings, developed by the Building Research Establishment in the UK.
3 Leadership in Energy and Environmental Design (the US equivalent of BREEAM).
4 COBie stands for Construction Operations Building Information Exchange. The concept was developed by the US Corps of Engineers, but a UK version (COBie UK 2012) has been produced.

References

AIA California Council (2007) *Integrated Project Delivery: A Working Definition (Version 2)*. McGraw-Hill Construction and The American Institute of Architects, California Council, Sacramento, CA.

Autodesk (2002) Building Information Modeling. Autodesk Building Industry Solutions. Available at: http://www.laiserin.com/features/bim/autodesk_bim.pdf. Accessed 10 May 2015.

Azhar, S. and Brown, J. (2009) 'BIM for sustainability analyses'. *International Journal of Construction Education and Research*, 5 (4): pp. 276–292.

Azhar, S., Carlton, W.A., Olsen, D., and Ahmad, I. (2011) 'Building information modeling for sustainable design and LEED® rating analysis'. *Automation in Construction*, 20 (2): pp. 217–224.

Babič, N.C., Podbreznik, P., and Rebolj, D. (2010) 'Integrating resource production and construction using BIM'. *Automation in Construction*, 19: pp. 539–543.

Barlish, K. and Sullivan, K. (2012) 'How to measure the benefits of BIM—a case study approach'. *Automation in Construction,* 24: pp. 149–159.

Barnes, P. and Davies, N. (2014) *BIM in Principle and in Practice*. Institution of Civil Engineers, ICE Publishing, London.

BIM Task Group (2013) Government Soft Landings. Available at: http://www.bimtaskgroup.org/gsl/. Accessed 03 June 2015.

BIM Working Party (2011) BIM: Management for Value, Cost and Carbon Improvement: A Report for the Government Construction Client Group. Available at: http://www.bimtaskgroup.org/wp-content/uploads/2012/03/BIS-BIM-strategy-Report.pdf. Accessed 03 June 2015.

Bower, J.L. and Christensen, C.M. (1995) 'Disruptive technologies: catching the wave'. *Harvard Business Review*, 73 (1): pp. 506–520.

Bryde, D., Broquetas, M. and Volm, J.M. (2013) 'The project benefits of Building Information Modelling (BIM)'. *International Journal of Project Management*, 31 (7): pp. 971–980.

Brynjolfsson, E. and Hitt, L. M. (2000). Beyond computation: Information technology, organizational transformation and business performance. The Journal of Economic Perspectives, 14(4), 23–48.

BSI. (2013) *PAS 1192-2 Specification for information management for the capital/ delivery phase of construction projects using building information modelling.* BSI Standards Limited, London.

BSI. (2014) *PAS 1192-3 Specification for information management for the operational phase of assets using building information modelling (BIM).* BSI Standards Limited, London.

BSI. (2015). PAS 1192-5:2015 (BSI, 2015) *Specification for security-minded building information modelling, digital built environment and smart asset management.* BSI Standards Limited, London.

Cabinet Office (2011) Government Construction Strategy. Available at: https://www.gov.uk/government/uploads/system/uploads/attachment_data/file/61152/Government-Construction-Strategy_0.pdf. Accessed 04 June 2015.

Cabinet Office (2014) New Models of Construction Procurement. Available at: https://www.gov.uk/government/publications/new-models-of-construction-procurement-introduction. Accessed 06 June 2015.

Cerovsek, T. (2011) 'A review and outlook for a Building Information Model: a multi-standpoint framework for technological development'. *Advanced Engineering Informatics*, 25 (2): pp. 224–244.

CIC BIM2050 Group (2014) Built Environment 2050: A Report on Our Digital Future. Construction Industry Council BIM2050 Group. Available at: http://www.bimtaskgroup.org/wp-content/uploads/2014/09/2050-Report.pdf. Accessed 31 May 2015.

Corke, G. (2012) BIM: Constructing a Virtual World. New Civil Engineer. Available at: http://www.nce.co.uk/bim-constructing-a-virtual-world/8640433.article. Accessed 31 May 2015.

David, P.A. (1990) 'The dynamo and the computer: a historical perspective on the modern productivity paradox'. *American Economic Review Papers and Proceedings*, 80 (2): pp. 355–361.

Davis, F.D. (1993) 'User acceptance of information technology: system characteristics, user perceptions, and behavioral impacts'. *International Journal of Man Machine Studies*, 38: pp. 475–487.

Dossick, C. S., and Neff, G. (2011). 'Messy talk and clean technology: communication, problem-solving and collaboration using Building Information Modelling'. *The Engineering Project Organization Journal*, 1(2), 83–93.

Eastman, C., Fisher, D., Lafue, G., Lividini, J., Stoker, D., and Yessios, C. (1974) *An Outline of the Building Description System*. Institute of Physical Planning, Carnegie-Mellon University, Pittsburgh, PA.

Eastman, C., Teicholz, P., Sacks, R., and Liston, K. (2008) *BIM Handbook: A Guide to Building Information Modeling for Owners, Managers, Designers, Engineers, and Contractors*. John Wiley & Sons, Hoboken, NJ.

Eastman, C., Teicholz, P., Sacks, R., and Liston, K. (2011) *BIM Handbook: A Guide to Building Information Modeling for Owners, Managers, Designers, Engineers and Contractors*. (2nd ed.). John Wiley and Sons, Hoboken, NJ.

Emery, F.E. and Trist, E.L. (1969) 'Socio-technical systems'. In: Emery, F.E. (ed.) *Systems Thinking: Selected readings*. Penguin Books, London.

Fischer, M. and Kam, C. (2002) PM4D Final Report, CIFE Technical Report Number 143. Available at: http://cic.vtt.fi/vera/Documents/PM4D_Final_Report.pdf. Accessed 31 May 2013.

Ford, S., Aouad, G., Brandon, P., Brown, F., Child, T., Cooper, G., Kirkham, J., Oxman, R., and Young, B. (1994) 'The object oriented modelling of building design concepts'. *Building and Environment*, 29 (4): pp. 411–419.

Golden, A. (2015) Where Does BIM Leave JCT and NEC3 Contracts? Chartered Institute of Building. Available at: http://www.bimplus.co.uk/management/where-does-bim-leave-jct-and-nec3-contracts/. Accessed 08 June 2013.

Grilo, A. and Jardim-Goncalves, R. (2010) 'Value proposition on interoperability of BIM and collaborative working environments'. *Automation in Construction*, 19: pp. 522–530.

Gu, N. and London, K. (2010) 'Understanding and facilitating BIM adoption in the AEC industry'. *Automation in Construction*, 19 (8): pp. 988–999.

Hamza, N. and Horne, M. (2007) 'Building Information Modelling: empowering energy conscious design'. In: *3rd International Conference of the Arab Society for Computer Aided Architectural Design (ASCAAD)*, 26–28 November, Alexandria, Egypt.

H.M. Government (2014) Digital Built Britain: Level 3 Building Information Modelling Strategic Plan. Available at: http://www.digital-built-britain.com/DigitalBuiltBritainLevel3BuildingInformationModellingStrategicPlan.pdf. Accessed 30 May 2015.

Holness, GVR. (2006) 'Building Information Modeling'. *ASHRAE Journal*, 48 (8): pp. 38–40, 42, 44–46.

Howard, R. and Bjork, B-C. (2008) 'Building information modelling—experts' views on standardisation and industry deployment'. *Advanced Engineering Informatics*, 22: pp. 271–280.

Laakso, M. and Kiviniemi, A. (2012) 'The IFC standard—a review of history, development, and standardization'. *Journal of Information Technology in Construction (ITcon)*, 17: pp. 134–161.

Laserin, J. (2008) *BIM Handbook: A guide to Building Information Modeling for Owners, Managers, Designers, Engineers and Contractors.* In: Eastman, C., Teicholz, P., Sachs, R., and Liston, C. (eds.). John Wiley & Sons, Hoboken, NJ.

Lockley, S., Greenwood, D., Matthews, J., and Benghi, C. (2013) 'Constraints in authoring BIM components for optimal data reuse and interoperability: results of some initial tests'. *International Journal of 3-D Information Modeling (IJ3DIM)*, 2 (1): pp. 29–44.

Malsane, S., Matthews, J., Lockley, S., Love, PED., and Greenwood, DJ. (2015) 'Development of an object model for automated compliance checking'. *Automation in Construction*, 49 (A): pp. 51–58.

Marshall-Ponting, A.J. and Aouad, G. (2005) 'An nD modelling approach to improve communication processes for construction'. *Automation in Construction*, 14 (3): pp. 311–321.

McGraw-Hill Construction (2012). Business Value of BIM for Construction in North America. Available at: http://static-dc.autodesk.net/content/dam/autodesk/www/campaigns/BTT-RU/MHC-Business-Value-of-BIM-in-North-America-2007-2012-SMR.pdf. Accessed 04 June 2015.

National Building Specification (2011) NBS National BIM Report 2011. Available at: http://www.thenbs.com/topics/bim/reports/index.asp. Accessed 04 June 2015.

National Building Specification (2014) NBS National BIM Report 2014. Available at: http://www.thenbs.com/topics/bim/articles/nbs-national-bim-report-2014.asp. Accessed 31 May 2015.

National Building Specification (2015) NBS National BIM Report 2015. Available at: http://www.thenbs.com/topics/bim/reports/index.asp. Accessed 04 June 2015.

National Institute of Building Sciences (2007) National BIM Standard-United States. Available at: https://www.nationalbimstandard.org/faqs. Accessed 26 December 2016.

Nour, M., Hosny, O., and Elhakeem, A. (2012) 'A BIM based energy and lifecycle cost analysis/optimization approach'. *International Journal of Engineering Research and Applications*, 2 (6): pp. 411–418.

PMI (2008). *A Guide to the Project Management Body of Knowledge (PMBOK® Guide).* (4th ed.). Project Management Institute, Newton Square, PA.

RIBA (2013). The RIBA Plan of Work. Available at: https://www.ribaplanofwork.com/.Accessed 26 December 2016.

Shetty, N., Hayes, A., Pocock, D., and Watts, J. (2013) Building Information Modelling and Asset Management. Institute of Asset Management. Available at: https://theiam.org/knowledge/Building-Information-Modelling. Accessed 31 May 2015.

Succar, B. (2009) 'Building information modelling framework: a research and delivery foundation for industry stakeholders'. *Automation in Construction*, 18 (3): pp. 357–375.

Tobin, J. (2013) 'BIM becomes VDC: A Case Study in Disruption. Building Design and Construction Network. Available at: http://www.bdcnetwork.com/bim-becomes-vdc. Accessed 31 May 15.

van der Smagt, T. (2000) 'Enhancing virtual teams: social relations v. communication technology'. *Industrial Management & Data Systems*, 100 (4): pp. 148–156.

van Nederveen, G.A. and Tolman, F.P. (1992) 'Modelling multiple views on buildings'. *Automation in Construction*, 1 (3): pp. 215–224.

Volk, R., Stengel, J., and Schultmann, F. (2014) 'Building Information Modeling (BIM) for existing buildings—literature review and future needs'. *Automation in Construction*, 38: pp. 109–127.

Wright, A.J., Lockley, S.R., and Wiltshire, A.J. (1992) 'Sharing data between application programs in building design: product models and object-oriented programming'. *Building and Environment*, 27 (2): pp. 163–171.

Yang, Q.Z. and Zhang, Y. (2006) 'Semantic interoperability in building design: methods and tools'. *Computer-Aided Design*, 38: pp. 1099–1112.

6 BIM in practise – industry case studies

Tristan Gerrish, Kirti Ruikar, Malcolm Cook, Mark Johnson, and Mark Phillip

6.1 Introduction

The Architecture, Engineering and Construction (AEC) industry is facing a mass change in the methods and technologies used to produce construction projects (Becerik-Gerber and Kensek, 2010). Nowhere else in the AEC industry has BIM impacted so many processes than in the engineering design of a building, with research by Yalcinkaya and Singh (2015) showing significant trends in its research throughout the industry on the implications of BIM adoption and application. Depending on the definition of BIM to which you subscribe, how it is used and the effects resultant from its implementation can change significantly across different projects. If considered as primarily a process (the path through which the project develops and information is created and exchanged), then how that path is followed is governed by the interactions between contributors to that project. If BIM is considered to be a new technological paradigm, you may see the tools developed to harness the information being created as a great benefit to the ongoing management of that built asset. Generally, both are correct interpretations of what BIM is and means for industry, but the practical implementation of the changes imposed by it as a process and a technology often vary significantly between projects. Given the diversity of projects throughout the AEC industry there is no exemplar BIM project, as each project is sufficiently different to warrant its own interpretation of how BIM should be used, and encounter different benefits and challenges to and from this use as a result.

The adoption of BIM as a common development tool, technology or process means the way in which projects are developed, and the information generated therein is becoming a far more holistic environment where input from the contributing disciplines must be managed more effectively. Here, three engineering design projects are examined, at different stages of a multidisciplinary engineering design consultancy's adoption of BIM, showcasing the challenges experienced during the early adoption of BIM, and the solutions developed as a result. These challenges continue to appear as innovative methods of using information and require innovative management of that information. Each case study describes the project and its uses of

BIM in place of conventional design and construction processes, and then looks at the challenges and solutions developed as part of the BIM adoption throughout these projects. A key issue identified throughout each case study is the need for the entire AEC industry to be ready to use BIM to its fullest potential, for a project is only as successful as its least capable contributor.

6.1.1 The organisation

The point of view from which this section is written is from that of Buro-Happold Engineering, a multidisciplinary engineering design consultancy, with specialisms in structural, building services, sustainability, civil, geo-technical and infrastructure engineering. The roles primarily concerned with BIM adoption currently are the primary structural and building services disciplines that form the core part of BuroHappold expertise. While these are the main areas where BuroHappold is currently undertaking a whole business change to implement BIM, its impact on all aspects of engineering design development mean that wholescale change needs to be made in how we manage our projects and collaborate both internally and externally to make use of BIM effectively. As such, quantifying what we already can do is essential to avoid duplication of efforts and to optimise our BIM adoption strategy. As BIM has been around for some time now, it may be considered to be well within the 'trough of disillusionment' on the Gartner Hype Cycle (Gartner, 2014) (Figure 6.1) as many business are struggling to implement it successfully in the conventional workflow. For context, BuroHappold is part of the engineering industry as a whole trying to implement an ever-evolving tool/technologies/processes to suit its needs and existing methods. It could therefore be seen as part of the 'Early Majority' in Rogers's technology adoption lifecycle (Rogers, 2003). This is echoed in the sentiments of the industry being that the initial hype surrounding the potential for BIM did not account for the difficulties in implementing it (Gardiner, 2014). However, the recurring National BIM Report issued by the National Building Specification (NBS) each year for the UK shows continuing uptake in response to the perceived benefit from its implementation (Figure 6.2).

The role that BuroHappold undertakes is that between the architect and contractors for the building, providing a preliminary and developed design around which details can be added as the project progresses. These services mainly take place during the Royal Institute of British Architects (RIBA, 2013) Preparation and Brief, Concept Design, Developed Design and Technical Design stages of the buildings development. The specialisms Buro-Happold contribute toward each have challenges associated with how they handle information developed in a BIM environment, or the transfer of information from their traditional formats (often CAD or another proprietary formats) into a BIM environment.

Where necessary, a project calls for involvement at stages where these roles aren't usually found; but as is often experienced, taking on these roles

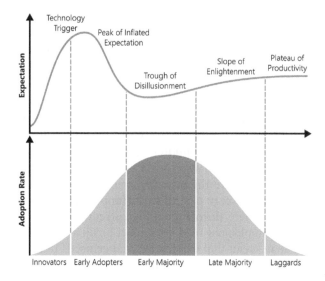

Figure 6.1 The hype cycle (Gartner, 2014) and adoption lifecycle (Rogers, 2003) showing coincidental adoption phases.

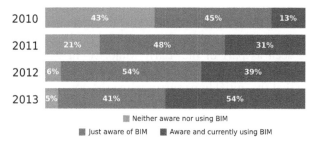

Figure 6.2 Ongoing uptake and awareness of BIM in UK AEC industry (NBS, 2011, 2012, 2013, 2014).

have shown benefits such as the experience of working across a design project from its definition to handover and the opportunity to influence design in all stages. In particular, it has been found that early stage involvement in the project by contributors that will be basing engineering decisions on actions taken at these early stages can reduce issues by pre-empting them before they arise. A key driving factor for this is early stage agreement about the BIM tools, process and techniques being used on the project. As is demonstrated by the case studies, early stage agreement about how to manage design using BIM benefits the project greatly and helps avoid potentially costly issues in interoperability and progress management.

6.1.2 *The case studies*

Here, three case study projects are presented, all within the past seven years, but each utilising BIM at varying stages of BuroHappold's ongoing adoption process and for different purposes. Each project uses BIM to varying extents also, given the clients brief and how the key designers at its initiation set out the project. Seeing how projects are evolving and making use of new and novel innovations in design development can help us understand where improvements might be made in the future and where successes can be replicated, something that BuroHappold regularly reviews to ascertain where further effort needs to be made to improve capabilities.

The majority of activities described within the case studies are from the perspective of the Building Services engineer, designing the environmental conditioning systems throughout the building to create a comfortable environment for its inhabitants. This role is one currently undergoing significant upheaval during adoption of BIM tools, processes and technologies. In particular, the use of large datasets to quantify performance metrics and automate aspects of design is an area where most effort is being placed, but the challenges inherent to parametric design are especially apparent across the Mechanical, Electrical and Plumbing (MEP) disciplines. The majority of BIM authoring tools are based around the needs of architects and those providing objects linked to parametric data. When considering the ongoing conditioning and operations of a building these parameters often change and evolve throughout that buildings operation – meaning modelling of such data becomes that much more complex. These case studies demonstrate how this challenge is being addressed, following three projects over the past 7 years at different stages of BIM adoption in the MEP organisation. The three major disciplines covered in these case studies are shown in Figure 6.3 alongside their periods of involvement in the projects.

In each of the case studies, the project is given a brief overview, detailing its location, client, purpose and key challenges faced in the design of that project. The design process is described (or followed in the most recent case study) indicating where BIM was used to enhance traditional project development activities. Where issues were encountered in the implementation of

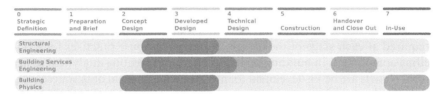

Figure 6.3 Roles and design involvement undertaken by the organisation referenced here.

these, contributing factors and actions undertaken to move past or avoid these issues, these are recounted in detail to give a comprehensive and accurate account of how BIM is being used in industry currently to effectively further its engineering designs.

	Case Study 1 (CS1)	*Case Study 2 (CS2)*	*Case Study 3 (CS3)*
Description	A flagship high performance office building, aiming for the highest BREEAM accreditation	A multi-function university teaching facility and laboratory	A multi-function university teaching facility undergoing rapid development following a redesign
Current RIBA 2014 Plan of Work Stage	7 (In-Use)	5 (Construction)	3 (Developed Design)
Start and Finish Dates	January 2009– March 2013	January 2013– September 2015	November 2013– September 2016
Level of BIM Use (according to BIM Task Group (BIM Task Group, 2011))	1	2	2

6.1.3 Roles represented in each project

As is experienced in each project, the number of contributing stakeholders means managing interactions throughout the process can be complex. In projects undertaken by BuroHappold, those contributing to the design are overseen by a project lead (representing that discipline e.g. structures, building services, etc.) on that particular project or aspect of that project. In cases where multiple disciplines are working on the same project (e.g. CS1), these are often coordinated further by a project lead that ensures each discipline is aware of the others developments and coordinates these efforts. It is now customary for each project, be it multidisciplinary or single discipline, to have one of the contributing team managing its use of BIM tools. This person, often called a BIM Manager, is there to help create the BIM Execution Plan for that project and set forth the conventions for modelling, interoperability processes, model sharing conventions and Level of Detail (Figure 6.4) required at each stage of the design. Standard in-house practises for modelling components in building services design and creating analytical structural models are being developed to interface with external data sources and models, slowly moving conventional workflow to a BIM enhanced design development environment.

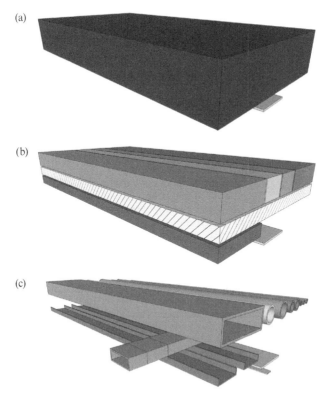

Figure 6.4 (a–c) Increasing levels of detail in object modelling.

6.2 Case study 1 – high performance office

CS1 constitutes the flagship building of an £800 m, 20-acre mixed-use redevelopment in Manchester (Table 6.1). Now used as a commercial office headquarters, the building brings several of this enterprises core businesses into a more sustainable building over 14 floors. Initial development began in 2008 with its programme completed at occupation in 2013 when the building was handed to the client. While the building is now occupied, commissioning of its numerous systems to ensure it meets the strict energy performance criteria imposed upon it by its designers is ongoing, with optimisation of its building services and facilities maintenance activities being conducted continuously. This continuous commissioning has demonstrated the potential to improve a buildings overall energy performance by between 10% and 20% (Neumann and Jacob, 2010). The potential for BIM within the operational stage of a buildings lifecycle is great, and support for the Facilities Management (FM) teams in doing can be facilitated through management of project information in a BIM environment (International Facility Management Association, 2013; Volk *et al.*, 2014).

Table 6.1 Case study 1 project datasheet

Criteria	Description
Location	Manchester, UK
Completion Date	March 2013
Construction Cost	£105 million (WHR and Jones Lang LaSalle, 2013)
Floor Area	30439 m^2
Function/Purpose	High-rise Office
Servicing Strategy	Combined Heat and Power providing heat and electricity from rape seed oil, supplemented by gas boilers. Extensive heat recovery from roof mounted and Earth tube air handling units. Heat recovery from server rooms and double skin façade. Cooling towers, water cooled chillers and absorption chiller, supplying chilled beams throughout occupied spaces. CO_2 regulated air supply from underfloor plenum.
Current Status	Completed/Occupied
Use of BIM	3D BIM was used extensively to coordinate the structural engineering, architecture and building services layouts, as well as the basis for energy modelling. BIM was also used to develop fly-through films and animations to communicate design ideas to the client and within the construction team.

Initially, the clients brief aimed for a flagship building to achieve the highest sustainability accreditation according to the BREEAM Offices category in which it falls. Following careful simulation of proposed design concepts, and optimisation of layout, systems and operations the building was given BREEAM 'Outstanding' accreditation. At the time, all buildings reaching this level were required to undergo in-use accreditation to maintain good environmental practises; however, since then this requirement has been withdrawn. Nevertheless, the occupants are still keen to optimise their use of the building and are currently undertaking the assessment to meet their long-term environmental management plan.

6.2.1 Project description

The time in which this building was designed spanned between 2008 and 2011, with construction following until 2013 when the building was occupied. At design commencement BIM was a relatively new term to the construction industry, with the extent of its use being the application of element coordination and the sharing of 3D geometry between contributing designers across several disciplines. At this early stage of BIM adoption, its primary uses were fairly limited to clash detection across modelling packages and the visualisation of finished designs for client approval. These functions, while useful, do not represent the full extent to which BIM is used today; however, this represents an early stage of the incremental adoption of its various possible uses. Additionally, these features are those that initially brought about the identification of the benefits BIM could bring to the industry as a whole.

BuroHappold took part in the project as Structural, Fire, and Sustainability engineers (amongst other related areas), taking the architects ideas and creating the superstructure and conditioning systems to support such a design. Initial concepts developed by the architect in conjunction with the client were made available at Stage 2 (Concept Design) in order for the building physics team to create preliminary energy performance models (using simplified space uses and layouts to indicate potential whole building energy consumption/issues). These 'mass models' resulted in the 'beehive' shape being selected for its performance characteristics. At this stage the architects then continued with their designs and completed an Autodesk Revit model of the building for coordination with the Structures and MEP teams.

The tools available to the stakeholders contributing to the design dictate the tools being used here. Where possible, all contributors use a common development environment (Autodesk Revit, or Graphisoft ArchiCAD are the current market leaders) in order to reduce issues with interoperability when sharing models across design teams. This process is further enhanced by subscription to a common data development methodology, which specifies *how* the project should be modelled and *what* should be modelled in *which* format. As this project was an early example of BIM adoption for the organisation, the role of the BIM Manager was yet to be defined and it was the responsibility of each of the project teams to ensure interoperability and data accessibility. This project also predated the availability of cloud based model servers meaning collaboration and coordination required fixed handover points and careful management. The central model was hosted by the architects and periodically shared with the engineers for the addition of structural details, changes to the layout reacting to the clients, servicing or structural requirements.

BIM was used throughout this design process primarily as a coordination tool. The architects and structural engineers both used Autodesk Revit, meaning the Autodesk Navisworks tool was suitable for clash detection using the export to .nwc format utility from Autodesk Revit. This enabled the highlighting of misplaced structural members or locations where architectural layout needed to be rethought to accommodate essential structural members. Analytical structural models were created using the same Autodesk Revit model developed by the structural design team, supported by manual calculations to support output from these models.

Not within the remit of BuroHappold's activities described here, the sequencing of installation was simulated at the time of the design of structural elements, to ensure that precast floor slabs could be installed through the steel superstructure prior to their specification. While not the most advanced use of BIM, this formed a precursor to 4D scheduling or construction sequencing, utilising the structural layout and manual manipulation of objects around these forms to test how slabs could be brought into the structure at various points during times of assembly. In addition to sequencing the fabrication process, this also informed the location of plant around the site to reduce unnecessary cost in specifying more plant equipment than required.

Mechanical engineers used tools conventionally used to model building services layouts such as AutoCAD for schematics and layouts imposed upon .dwf's exported from the architectural Revit model. For the creations of whole building schematics, 3D modelling of complex service runs used the CAD-Duct plugin to AutoCAD (which was also used for clash detection using Autodesk Navisworks) and various external calculation tools for pipe-run sizing, thermal load calculations and service supply sizing. It is important to know that the modelling of mechanical service runs at Stage 3 does not always require discrete coordination of intersections and riser joints; however, given the complexity of this buildings conditioning system efforts were made to avoid issues here at this stage prior to problems encountered further in the design process. Use of Revit for building services is something that is still undergoing development today, given the complexity and range of disciplines this field encompasses. As such, proprietary tools for the design of complex systems are used in favour of BIM authoring tools that are yet to meet their capabilities.

A similar issues is faced by building energy performance simulation, where the method of modelling geometries in an accurate, or simplified manner to support detailed simulation are fairly strict. In conjunction with this difference in modelling accuracy, performance analysis for the optimisation of systems and plant equipment sizing was completed almost entirely separate to the BIM environment prepared by the architects, due to the lack of interoperable modelling formats allowing paces to be shared between the two modelling environments. For energy performance analysis to be undertaken, the model must be prepared in such a way that it can be exported to a suitable format (gbXML or IFC) to be read by the energy simulation tool.

6.2.2 Issues encountered and lessons learnt

The issues encountered throughout the development of CS1 have been categorized under Technological or Process based, indicating the primary means by which these issues were overcome or have yet to be addressed in a conventional BIM design development process. Both use of technologies and processes need to change in order to enable effective BIM implementation.

6.2.2.1 Process-based issues

Coordinating several distinct disciplines, each with their own contingent design aspects can be a daunting task, but through basic use of BIM tools the design of CS1 was made more efficient than it would have been otherwise.

Coordination between the structural engineer and architect is especially indicative of where this coordination becomes useful, enabling the movement of structural members and layout according the requirements of the other party. When the mechanical engineer begins to specify their requirements (spaces for risers, requests for cellular beams allowing service pass through and specification of coffered floor slabs to house servicing equipment), coordination becomes even more complex. At the time, technologies such as BIM servers were not available, meaning regular touchdown meetings to plan changes prior to assembling each constituent model were necessary.

6.2.2.2 *Technological issues*

Early servicing coordination was undertaken in conjunction with the structural framing at complex intersections of the building geometry. The majority of the building is repeated sections of long, thin floor plates; however, at each of the three core sections the structural frame transitioned to accommodate the lifts, risers and, double height spaces. These required careful planning of service interface with the structural frame, resulting in the specification of cellular beams through which conduits and pipework could pass, and customised precast concrete floor plates. The ability to visualise these interfaces at a detailed design stage prevented issues later in construction with the fitting of these systems.

Repeated structural joints across the whole building were tagged in Revit, generating schedules for later use in design and manufacture. This played a key part in the value engineering of the project and made the transition from designer to fabricator much simpler. Where previously, the number and type of joints was extracted manually (count each joint and list its details in a spreadsheet), this information was automatically generated to enable optimisation during design. The handover of this information to the fabricator gave information such as number of joints, joint type and strength specification, which would have taken far longer to compile manually for a building of this size.

Further issues that were encountered on the other case studies included problems with file interoperability between different software (mainly between architectural models and proprietary software models for specific tasks such as CFD analysis, performance simulation and MEP systems calculations). These are common to most projects using BIM to any extent and are discussed in CS2.

6.3 Case study 2 – multi-function teaching facility and laboratory space

CS2 is a new, purpose built teaching, office and laboratory facility due for completion in time for the 2015/16 academic year (Table 6.2). Following expansion of student and staff numbers, and the need for state-of-the-art

Table 6.2 Case study 2 project datasheet

Criteria	Description
Location	York, UK
Completion Date	September 2015 (expected)
Whole Project Cost	£12 million (University of York, 2014)
Floor Area	4000 m^2
Function/Purpose	Multi-Function University Teaching Facilities
Servicing Strategy	Heat from campus district heating, Free cooling from adjacent lake and supplementary chillers, Mixed mode ventilation
Current Status	RIBA Stage 5 (Construction)
Use of BIM	Extensively used by architects (ArchiCAD IFC-COBie links) to create package ready for handover to facilities
	Coordination using IFCs between tool, remodelling required for performance analysis due to interoperability issues.
	Asset tagging with parametric performance data for single model source of performance data (also linked to Room Data Sheets for conventional access of performance specifications)

facilities for the teaching department, this development aims to create a space suitable for interdisciplinary groups collaborating within a research environment and informally within communal spaces. BuroHappold took on responsibilities of mechanical and electrical engineering, fire engineering, building physics simulation, acoustics and assessment of the buildings BREEAM rating.

Throughout this project, the modelling of component building services was essential given the amount of equipment present in the building (primarily in laboratory spaces requiring fume cupboards, various gas supplies and with the potential for free cooling from the adjacent lake). This project demonstrates the advancements made in BIM adoption by the organisation since its previous project, where coordination of design specifications has begun to move into a BIM environment alongside some aspects still within their traditional development frameworks. A good example of this is in the modelling and storage of extensible parameters to objects and spaces in the Revit model. Here, the benefits of using BIM are demonstrated through the creation of a whole building performance strategy and creation of a model based library of spaces, their conditioning requirements and objects providing these conditions.

6.3.1 Project description

This buildings design took place during the Early Majority/Slope of Enlightenment phases shown in Figure 6.1. At this point, the engineering designers were familiar with the basic concepts of BIM and its potential for design improvements and optimisation. This doesn't mean that it was fully integrated into every aspect of engineering, but where possible this project made use of BIM for the management of large amounts of design data. The roles undertaken by the design engineers on this project were primarily mechanical, electrical and sustainability based, encompassing the design of the buildings systems for environmental conditioning, communications, services and thermal performance simulation.

The model from which the engineering design progressed was developed by the architects who then shared it at a stage where further changes to layout and form are minimal. Again, a model server was not used as the determination of information sharing specifications was not made. Until a document is available detailing in depth the amount of information that can be shared, the means via which that information is shared and whose responsibility it then becomes, means use of external project serves are unlikely to become convention.

At this point the model is used for modelling pipe runs, conduits, outlet placements and 3D coordination of services in complex intersections. Proprietary tools were used in conjunction with Revit to calculate systems performance characteristics. 3D representation of local and main plant equipment was achieved using libraries of components developed in-house and modified to include calculated parameters. This bespoke modification of standard objects is now common practise in projects where objects need to be modelled in conjunction with their specification. These in-house objects require modification to meet project requirements, so that supplementary tools developed on a project-by-project basis can be accommodated. These could then be used to output detailed schedules indicating system capacity. The model to which these systems are based around is linked to the mechanical services model, requiring periodic updates to ensure clashes do not occur, nor changes impacting the capacity of a mechanical service to serve its purpose has not been hindered.

At the start of the design process, simulations were built to test form, function and layout – enabling the optimisation of these aspects at Stage 2 prior to developing design. This function is yet to be integrated into the BIM process; however, for preliminary analysis several tools are emerging that allow for some indication of building performance from 3D models and basic user specified criteria. BuroHappold uses detailed thermal analysis to determine likely building loads, specify plant equipment to meet these loads, and have encountered several issues in the use of BIM models for this purpose. At later stages of design, these preliminary models are updated to include more representative data regarding the buildings form and function (once fabric has been confirmed and space use has been finalised ready for services installation). The purpose of this stage of modelling is to show compliance with building regulations.

CS2 was the first test of a method of using BIM for the storage of environmental performance criteria by the organisation. Based on the Level of Development framework (AIA and AGC, 2013), building performance criteria were assigned to the progressive levels of project development, indicating where these needed to be specified for use by the building performance modeller (Figure 6.5). This framework was used to ensure information generated by one party (the engineer) was available to others (the energy performance modeller) to ensure an accurate representation of the building for modelling.

Information used to create these models is commonly stored alongside the BIM model in a Room Data Sheet, but in practise this separates some of the design and analysis so that care must be taken to ensure up to date information in used where necessary. This project in particular used Room Data Sheets, but linked these to spaces in the Autodesk Revit model using a plugin designed to interface Excel with object parameters ensuring up-to-date

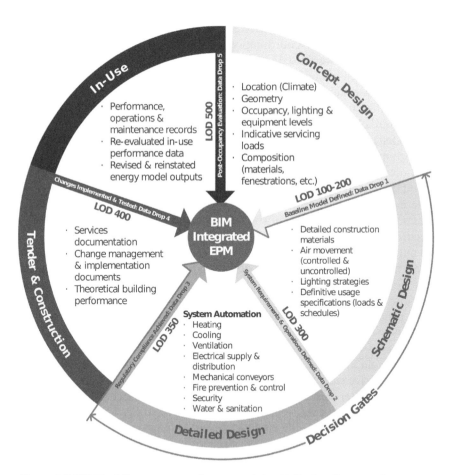

Figure 6.5 BIM building energy performance data attribution framework.
Source: Created by T. Gerrish.

information stored in the Revit model semi-automatically populated the room data sheet for use in other engineering activities. These still required minimal user interaction to specify what variables should be shared in order to allow this data to be accessed by the two software platforms.

In addition to the extensive work completed in linking building performance modelling with the BIM tools at BuroHappold's disposal, conventional use of BIM for clash detection, sequencing and value engineering also took place – now a common feature of BIM enabled projects across the organisation.

6.3.2 Issues encountered and lessons learnt

As the organisation continues to adopt BIM technologies and processes into its conventional design framework, the issues encountered continue to evolve, raising new opportunities for data use and problems to be overcome.

6.3.3 Process based issues

Fragmentation of disparate datasets is an issue BIM is attempting to solve, creating a reference point for linked data attributed to modelled projects, spaces and objects. For this system to work properly, all users of the BIM environment need to know what is expected of them and how best to use the tools available to them. Often, the person specifying a particular piece of equipment does not know where in the object parameters certain values should be attributed, leaving exported schedules from the model to be poorly populated and not representative of the building as a whole. Training is required for all contributors on a BIM based project to ensure this does not cause a problem and that data generated is collated and useful for other contributing designers.

The role of the BIM Manager has become a common part of project development teams now, whose job it is to ensure contributing designers and engineers can work together and apart using the relevant data available to them. The UK governments guidance in this area has been bolstered by the introduction of PAS 1192-2 (BSI, 2013), aiming to give guidance in project delivery and specify the actions that should be undertaken to facilitate and effective project delivery using BIM. Supported by this guidance, and in line with the rest of industry the organisation has developed its own BIM execution plans and modelling conventions to enable these actions to be undertaken, and work cooperatively with other outside stakeholders in the design and engineering processes.

The use of a central model within which all changes to the building's design can be made and agreed upon by contributing designers is still some way off. The act of reconciling all changes to a model currently requires regular handover periods where each discipline can check their current plans against other designer's changes to ensure clashes can be avoided. This is particularly true in the design of mechanical services, which rely on a fixed layout from which calculations can be based. The adoption of BIM by architects has enabled the automatic update of custom sheets (showing elevations and plans) from which proprietary building services tools base their detailed models. The exchange of these general arrangements at fixed points means changes by the architect can be made at any point up to handover (including minutes beforehand), and work done in designing service run layouts for these designs can be quickly superseded by new layouts or changes to building geometries. Thankfully, these issues rarely cause any big problems due to coordination between designers, but until a fully coordinated BIM environment that can handle several contributors changing aspects concurrently these issues must be accounted for procedurally.

6.3.4 Technological issues

A common issue encountered on projects where BIM is used is the potential for misuse of the models to give indications for potential building performance. Early stage analysis tools such as the IES VE-Ware plugin to Revit and Sketchup can be useful in offering comparative performance metrics for different designs; however, in the detailed design of buildings and their systems an experienced modeller using a dynamic simulation tool is required to interpret and take action from the output of these simulations. Tools are being developed to enable greater interoperability between BIM authoring and energy performance simulation packages, and there is currently the potential to export geometry from BIM to these tools via IFC and gbXML formats. The method with which the BIM model was created has a great effect on the quality of these geometries, often requiring the geometries to be built again in the native energy performance simulation tool to avoid issues in the simulation of that building.

A problem experienced all too often by the energy modeller is in the quality of BIM obtained from contributing designers. While the modelling conventions used by architects are suitable for planning and layout, energy models use geometry intersections as nodes for shape bounding, meaning the way in which the model is built has some bearing on how that model will be interpreted by another software package. For example, a simple room may be built in Revit using external face snaps to define the external wall finish within a specified boundary, but within that boundary the wall centreline may be used for ease of model building. An energy simulation tool doesn't differentiate between a face snap and a midpoint snap meaning the room may no longer be enclosed and therefore unsuitable for simulation purposes.

In other projects the organisation has taken part in, links between Autodesk Revit and energy performance simulation package TAS have been developed through which the buildings geometry (controlled by the organisation) can be shared between the two tools. This tool relies on interim steps interpreting and modifying the Revit geometry to suit thermal simulation (simplifying joins between rooms, removing extraneous detail that detrimentally impacts simulation time), and enables two way communication between the two environments, but is only suitable in a small range of cases.

6.4 Case study 3 – university teaching facility

The plans for a new building to house a college of social sciences originally aimed to have a finished construction ready for occupation by the new academic year in 2016 (Table 6.3). In late 2014 due to unforeseen circumstances the Developed Design required significant changes to accommodate new design challenges. At that time the majority of the building's design had been finalised, with only detailed specification of mechanical systems left before progression to detailed design and construction. These late stage changes in

Table 6.3 Case study 3 project datasheet

Criteria	Description
Location	Lincoln, UK
Completion Date	September 2016 (proposed)
Construction Cost	~£19 million
Floor Area	~6000 m^2
Function/Purpose	Multi-Function University Teaching Facilities
Servicing Strategy	Gas-fired condensing boilers, Air cooled chiller, Mixed mode ventilation
Awards/Ratings	N/A
Current Status	RIBA Stage 2 (Concept Design)
Use of BIM	Originally used to coordinate MEP service runs and architectural/structural layouts (incorporating an existing building onsite). Now used to store space-describing data to enable the quick redesign of its layout to accommodate changing client requirements and restrictions. Performance modelling has already taken place so this data can be included in the BIM model almost directly.
	While the architect used BIM from the Concept design stage, contract documents specify very little in terms of BIM requirements *'All CAD drawings are to be drawn in an industry standard format that is fully compatible with other consultants in the design team. Where drawings are produced in a 3D model or BIM (building information model) format all members of the design team will need to work in this format. 'As-built' drawings are to be supplied in Adobe 'pdf' format.'* This means without requiring BIM, participants are using it to their own advantages without a clear definition of what is required in these formats. A good example of Lonely-BIM in a project requiring careful coordination.

conjunction with the shortened timeframe required significant effort by the designers to move design from Concept through to Technical Design in a matter of months, aided by use of BIM to coordinate this process.

Fast paced design development is somewhere the use of BIM excels. This is not yet experienced across industry as a major benefit, but is listed by 19% of contractors as one of the top three benefits of BIM adoption in the 2014 Business Value of BIM report (McGraw Hill Construction, 2014). In a project with a short turnaround, this is key to ensuring a fast and effective design process, especially when reducing the number of clashes and foreseeing potential problems requiring additional time later in the design process.

6.4.1 Project description

As with the previously described projects, the process this design followed was that of Architect-led BIM-based model creation in Revit, about which other disciplines made their own supporting designs for coordination at fixed points in time. In early 2015 the redesign was approaching the Stage 2 issue (the design has been specified to the level of detail required for the concept design to be fulfilled) where the majority of architectural changes had been made and finalised. Prior to this stage, the design had already moved towards Stage 3 completion, so key design criteria such as space requirements, likely loading per spaces (determined using occupancy density and use) had been predetermined. This shortened the determination of form through reducing the time taken to model and simulate options.

Upon deciding on the new form the architects created the Stage 2 general arrangement and the model around which further design developments can be made. The practise of using a central model means changes made to the design of various aspects of the building need to be reconciled against this model if its author makes any further changes. This project made use of a model server from which the most recent model could be taken; however, this required the author (the architect) to re-upload the most recent model for access by the other designers, meaning the most recent model may not be the one hosted there.

The standards for modelling had been defined for the previous iteration of this project, meaning templates for the structural model were available as were the components suitable for reuse. Structural modelling of buildings using standard components can be achieved far more efficiently than mechanical systems modelling for building services. The finite number of structural components reduces the number of potential variations and the mostly fixed grid layouts of the building means changing a certain aspect of the structure can be achieved relatively easily. The aspects describing these structural elements are also attached to the objects describing the superstructure in a similar manner to that of the CS1 project, using BIMs extensible parameters to store object take-offs for cost estimation and value engineering created automatically from the model.

During the short Stage 2 process, the structural engineer found that frequently updated architectural models meant the creation of a definitive structural layout was made difficult; however, the grid layout meant that those changes that were made could be accommodated through the manipulation of geometry (snapping beams and columns to snap to the newly created points). In doing so, the originally created structural concept could be edited without loss of the additional information added (such as indicative sizing, joint types and strength) and changed with fairly minimal effort at the handover of architectural changes.

Prior to embarking on a second iteration of design, the information generated through the prior iteration described spatial requirements, layout adjacencies and likely usage scenarios feeding into the building energy performance model. While the output from these simulations (for plant sizing and comfort evaluation) will not be usable for the second design iteration, its inputs form templates about which to model the new design and shorten simulation time. This information is not stored in a BIM environment, but as with CS2, the ability to export and manipulate BIM created information (zones, areas, use etc.) is used extensively throughout the design process.

Modelling of the building services layout also encounters the same problems as structural layouts, requiring a base model to which the design can be attributed. These aspects of engineering design tend to begin once the layout has been finalised given the number of systems reliant on the central models geometry to ensure they are given the appropriate clearance and assignment to the correct space. A small change in the central models geometry could mean a number of building services need to be rerouted to accommodate that change and subsequently the calculation of those service runs capacity could change resulting in an over or under provision of conditioning services to that space. Changes to a building between its initial design and operational stage have been found to impact the overall building energy performance and contribute to the 'performance gap' (Bordass *et al.*, 2004).

6.4.2 Issues encountered and lessons learnt

At this early stage in project development many of the potential issues related to use of BIM for design are yet to be experienced. Once again, the main issues of coordinating multiple discipline contributions toward a single or federated model are present, but at this stage most of these are process based – preparing the right information in the right manner for use further along the design process.

6.4.3 Process-based issues

At an early design development stage the majority of design decisions are still within the hands of the architects – specifying the layout and composition of the building prior to engineers providing the technical knowledge to ensure this layout works. Sharing of full models between disciplines at this stage is fairly uncommon, instead favouring export of simple layouts (general arrangements) for mark-up and comment by other designers using Revit's commenting functionality. Sadly this fragmented development recreates an issue encountered across the industry where silo development of particular aspects of a building's design, often due to the physical limitations of concurrently developing a design, happens. However, given the early stage of development this project is in, the creation of a BIM Execution Plan is yet to be made and closer collaboration has been identified as a key part of delivering the project in such a short timeframe.

Both the structural and building services engineers working on this project made note of the changing role of technician and engineer. Traditionally, the engineer contributes the technical design and justification behind the choices being made while the technician models these designs and contributes where necessary using modelling and calculation tools to supplement engineering knowledge. Much of this design work is moving to an entirely model based format where the technicians are often best placed to adopt new functionality within BIM tools due to their familiarity with these tools. It was recognised that engineers need to understand how best to model and record design information while the role of technician becomes more that of an information coordinator who takes on responsibility for BIM execution plans, data access agreements between designers and revision control.

6.4.4 Technological issues

Coordination of varying design disciplines and constituent inputs into the overall building model continues to be the major problem encountered between the architectural and engineering design disciplines. While this project is at an early stage, these issues are more likely to be encountered (as changes to the design are more likely to occur at this stage), but the difficulties in surmounting them are one the entire industry is struggling with. Use of a decentralized project server could provide a means through which these concurrent designs could be handled, but given the complex nature of design calculations in disparate disciplines, making changes to a single aspect could potentially impact many other systems and change the entire design.

In time it is likely that these changes could be made in one area; for example a partition could be moved resulting in a different space thermal characteristic, meaning changes to the HVAC system in that zone and therefore the overall HVAC requirements for the project. This would carry across to other disciplines requiring review this change by those designing in those disciplines; however, for now distinct 'lines in the sand' need to be specified in the design schedule about which changes can be coordinated manually.

These 'lines in the sand' are information exchanges set out at distinct points of project delivery for the coordination of design between disciplines, to ensure issues are met at an early stage and the information is being developed at an appropriate rate for project progression. Usually described in the world of BIM as COBie Data Drops (BIM Task Group, 2012), these are meant to help populate a programme of assets and their related information for handover to the buildings operator upon completion, but also serve as a benchmark within the design process measuring the level of progress. In actuality, information exchanges to happen more frequently to make use of the number of changes BIM has enabled through its rapid design progression.

An interim stage of this coordination is through automatic comparison of models with previous iterations, essentially checking for clashes against a previous model. Several BIM server tools are available, but experiences of the organisation in using these has been problematic given the sizes of files being worked on and shared. In particular, these systems operate as 'software as a service' meaning collaboration with another designer halfway around the world can be held up by problems in transmitting information where connectivity is limited and transfer of large model files can often be unsuccessful. In this project a model server is not used in lieu of a cloud based platform for model sharing, resulting in more effective information exchange, but requiring subscription to a regular schedule of update and communication between the stakeholders to coordinate fragmented development.

6.5 Case study synthesis and future direction

Bringing each of the case studies presented here together, several themes can be observed, indicating what challenges are yet to be addressed, and where industry implementation of BIM is heading. Using the categorization of issues encountered as applied to the three case studies described here, the issues commonly encountered throughout a construction design programme (following the RIBA (RIBA, 2013) development framework) are given below, alongside where this may change in the coming years.

6.5.1 Preparation and brief

At such an early stage of design development, the processes to be followed throughout a project must be well defined and agreed upon by the contributing parties. These standards set the framework about which information is going to be developed, shared and applied to the various engineering aspects of the design to ensure the building meets its expected performance. Whilst adherence to standards for information exchange is not enforced in the UK on non-public funded projects, the guidance set out in documents such as PAS 1192 (BSI, 2013) give clear instruction as to the necessary procedures for consistent information exchange in a suitable format. As experienced in CS3, this is not yet applied to non-public funded projects meaning the procedures followed often do not enable efficient information exchange and management of design expectations across different disciplines in different designing organisations.

Setting the standard for information integrity and form at such an early stage can really push the project to be of higher quality (Eadie *et al.*, 2013). These rules mean procedures to interact with information between disciplines throughout the design process can be applied with the knowledge that that information is held in a certain standard. While standardisation is often seen as a limitation to innovative design, it is essential when management information across platforms, stakeholders and projects.

6.5.2 Concept design

At the Concept Design Stage the responsibilities for information and design developed are well established, moving into design options, where the tools used to develop these designs become the forefront issue. The technology available currently doesn't fully meet the needs of the user – instead the existing and more developed proprietary technologies used by distinct disciplines tend to be much better than those capabilities integrated into BIM environments. For example, the models created by the architect in CS2 suit coordination of HVAC services, but are not suitable for energy performance simulation due to their complexity and geometry modelling conventions. Methods of translating data between different tools are developed ad-hoc to enable interoperability, but given the proprietary nature of tool development, wholescale open data translation is not probable.

However, new methods of supporting currently dis-integrated information generating platforms and moving data between tools are becoming available. For example, BuroHappold now uses Autodesk Dynamo (2016) to enable custom exchange of information from outside the Revit environment to enable links between simulation and analysis tools with the Revit platform. Support for concurrent conceptual design support is something the industry is now exploring, to allow multiple disciplines feed into a common conceptual model in which the most effective design outcome can be generated.

6.5.3 Developed and technical design

Once the form of the building has been decided, the detailed design begins, determining the suitable equipment, operational strategies and likely performance levels of the building. Here BIM is yet to fully integrate into the design process, in part due to the lack of capability in modelling complex HVAC systems and in moving data between commonly used HVAC modelling tools, and partly through lack of investment in using the capabilities BIM authoring tools do currently have to model such information. As HVAC systems are intrinsic to the layout and specification of spaces within a building, and are dependent on the form of that building to remain mostly fixed throughout their design, their current modelling in tools without BIM interoperable features means reliance on less dynamic methods of data exchange (e.g. development in 2D CAD using PDF mark-up to make changes before redrawing in 2D CAD).

The issues described in this part of the design process are the most difficult to address, given their link between Technology – where tools lack the necessary features to enable successful BIM use, and Process – where the procedures necessary for BIM to be used successfully are yet to be developed. While not applicable to all projects, these challenges are commonly experienced across all projects BuroHappold undertakes. As we make incremental developments to improve in this area, the potential for integration between disparate systems is enormous.

The majority of design decisions are based on well-established methodologies for determining response to input. For example, the required heating capacity fort a space depends on its gains and losses which are quantifiable. The size and utilisation of that space is the metric about which a value can be calculated; but given the number of inputs this is not yet automated. With a federated model comprising separate discipline specific models; throughout which information is accessible it is foreseeable that such calculations could be made automatically upon slight changes to the central model without requiring manual input (Building Futures, 2003).

6.6 Conclusion

Since the term BIM was originally coined by van Nederveen and Tolman (1992), (though only catching on 10 years later (Autodesk, 2002)), what it means and how it could be used to better manage the design and development of a building project has changed significantly. Rather than seeing it as a new technology, the organisation here has thought of it more as a system of joining up the (still) disparate areas of design development. Early adoption focused primarily on the implementation of enabling technologies and moving workflow from 2D CAD to 3D CAD – but this is not what BIM is. The value of BIM which has been demonstrated in the projects exhibited here is in its holistic approach to the design development of buildings, using information generated by one party in another discipline to produce a more coherent and effective built asset. This value also illustrates issues experienced throughout the AEC industry when reconciling conventional design activities with new methods of designing and sharing design responsibilities. The case studies chosen demonstrate how widespread these challenges are, how they are changing over time and are common to multiple projects demonstrating the scope of those challenges remaining in full-scale BIM implementation.

Now, we are responsible for bringing together those pockets of implementation in small ways across various projects into a cohesive and replicable methodology, and in doing so make BIM the standard process each project can follow. At this point in time there will be challenges in bringing the skills and tools necessary to design in a BIM environment to every member of an organisation, but eventually become more common practise.

The case studies described here indicate that the AEC industry as a whole is generally benefitting from BIM application to a wide range of construction design projects; however, there are still several issues to address before full implementation across all design schemes. While not discussed, BuroHappold is often approached to contribute in some part to a larger scheme where requirements of information developed do not specify use of BIM or coordination with external stakeholder, nor would this be useful; to the scheme. As such, BIM is almost entirely unused in these processes. The organisation might use some methods such as coordinating between internal disciplines and attributing extensible data to modelled objects, but open handover at completion of this contract this data reverts to its conventional form – in 2D drawings and spreadsheets.

Projects where the whole design team understands that coordination and adherence to a standard of modelling and information management is essential is where most benefits are experienced, with the opportunity for innovative use of this data. CS2 is a good example of this, enabling more effective energy performance simulation through better use of data attribution in the BIM model allowing the modeller to extract the relevant data in a more effective way; enabling more simulation and testing in design to optimise the building conditioning systems and therefore the performance of the building upon completion.

If nothing else, these example show that BIM is not just an industry fad, or a buzzword used to peak interest, but a useful and practicable method of managing the vast amount of information developed during the construction process – but only if those involved in that process make use of it in the right way.

References

AIA, AGC, 2013. Level of Development Specification. BIMForum.

Autodesk, 2002. Building Information Modelling. White paper. Available at: http://www.laiserin.com/features/bim/autodesk_bim.pdf. Accessed 9 June 2002.

Autodesk, 2016. Dynamo Studio [WWW Document]. URL http://www.autodesk.com/products/dynamo-studio/overview. Accessed 18 January 2016.

Becerik-Gerber, B., Kensek, K., 2010. Building information modeling in architecture, engineering, and construction: emerging research directions and trends.

Journal of Professional Issues in Engineering Education and Practice 136, 139–147. doi:10.1061/(ASCE)EI.1943-5541.0000023.

BIM Task Group, 2011. A Report for the Government Construction Client Group Building Information Modelling (BIM) Working Party Strategy Paper. Department of Business, Innovation and Skills, London.

Bordass, B., Cohen, R., Field, J., 2004. Energy performance of non-domestic buildings—Closing the credibility gap. Presented at the Building Performance Congress. 3rd International Conference on Improving Energy Efficiency in Commercial Buildings (IEECB'04). The Usable Buildings Trust, Frankfurt, Germany.

BSI, 2013. PAS 1192-2 Specification for information management for the capital/delivery phase of construction projects using building information modelling. BSI Standards Limited, London.

Commission for Architecture and the Built Environment (CABE) and RIBA, 2003. The professionals' choice: the future of the built environment professions. Building Futures. Available at: http://www.buildingfutures.org.uk/assets/downloads/The_Professionals_Choice2003.pdf.

Eadie, R., Browne, M., Odeyinka, H., McKeown, C., McNiff, S., 2013. BIM implementation throughout the UK construction project lifecycle: an analysis. *Automation in Construction* 36, 145–151. doi:10.1016/j.autcon.2013.09.001.

Gardiner, J., 2014. Is BIM what it says on the tin? *Building Magazine*. Available at: http://www.building.co.uk/analysis/is-bim-what-it-says-on-the-tin?/5072149.article. Accessed on 22 December 2016.

Gartner, Inc, 2014. Gartner Hype Cycle. Gartner Research Methodologies. URL http://www.gartner.com/technology/research/methodologies/hype-cycle.jsp. Accessed 4 June 2016.

International Facility Management Association, 2013. *BIM for Facility Managers*. Wiley, Hoboken, NJ.

McGraw Hill Construction, 2014. The Business Value of BIM for Construction in Major Global Markets: how Contractors around the World Are Driving Innovation with Building Information Modelling (SmartMarket Report). New York.

NBS, 2011. National BIM Report (UK) 2011. NBS, London.

NBS, 2012. National BIM Report (UK) 2012. NBS, London.

NBS, 2013. National BIM Report (UK) 2013. NBS, London.

NBS, 2014. National BIM Report (UK) 2014. NBS, London.

Neumann, C., Jacob, D., 2010. Results of the project Building EQ: tools and methods for linking EPBD and continuous commissioning (No. EIE/06/038/SI2 .448300). Fraunhofer Institute for Solar Energy Systems, Freiburg, Germany.

RIBA, 2013. *RIBA Plan of Work 2013 Overview*. RIBA, London.

Rogers, E.M., 2003. *Diffusion of Innovations*, 5th ed. Simon & Schuster International, New York.

University of York, 2014. New Environment Building Construction Works Start [WWW Document]. University of York: environment. URL http://www.york.ac.uk/environment/news-events/news/new-building-august-2014/. Accessed 16 January 2016.

van Nederveen, G.A., Tolman, F.P., 1992. Modelling multiple views on buildings. *Automation in Construction* 1, 215–224. doi:10.1016/0926-5805(92)90014-B.

Volk, R., Stengel, J., Schultmann, F., 2014. Building Information Modeling (BIM) for existing buildings—Literature review and future needs. *Automation in Construction* 38, 109–127. doi:10.1016/j.autcon.2013.10.023.

WHR and Jones Lang LaSalle, 2013. One Angel Square. Available at: http://www. noma-manchester.com/media/1066/one-angel-square.pdf. Accessed 15 January 2016.

Wood, B., 2012. COBie Data Drops. BIM Task Group. Available at http://www. bimtaskgroup.org/wp-content/uploads/2012/03/COBie-data-drops-29.03.12.pdf. Accessed 22 December 2016.

Yalcinkaya, M., Singh, V., 2015. Patterns and trends in Building Information Modeling (BIM) research: a latent semantic analysis. *Automation in Construction* 59, 68–80. doi:10.1016/j.autcon.2015.07.012.

7 Manchester Central Library and Town Hall Extension Project

The BIM journey so far of a public sector client

Jason Underwood, Joanna Chomeniuk, Liam Brady, and David Woodcock

7.1 Introduction

May 2011 witnessed the UK Government's commitment to support a transformation of the construction industry through the launch of their Construction Strategy (2011), which mandates BIM (Level 2) on all public procured projects by 2016. Since its launch the UK construction sector has witnessed a momentum build in the awareness and adoption of BIM. The 2014 NBS report indicates an increase in the awareness and current use of BIM from 13% in 2010 to 54% in 2013 with a decrease in neither being aware of BIM nor using from 43% in 2010 to 5% in 2013. However, the report also identified that one of the key barriers for wider adoption perceived by the industry is the lack of clients demanding BIM, which is considered in part due to limited/no awareness/understanding of and drive for BIM, i.e. clients being uninformed. This chapter aims to explore the BIM journey so far of a local authority public sector client, Manchester City Council (MCC), and to show the transition from being an uninformed client to becoming an informed client through collaborative engagement with their supply chain on the prestige exemplar Manchester Central Library and Town Hall Extension project. The influences of key decisions during this journey and on behaviours that enhance collaboration and cooperation in the understanding of their BIM requirements are discussed. Furthermore, the challenges, drivers and enablers as well as the key roles, e.g. a visionary champion are explored. Two empirical studies with a local university that measured improvements of their BIM maturity and capability, and served as a further driver for MCC in their adoption of BIM are presented. Lessons learned from the project, recognised as an exemplar, and which are being transferred to other projects are shared, and finally recommendations are provided.

Keywords: BIM adoption; Public Sector Client BIM journey; BIM challenges, enablers and drivers, Short term and long term BIM vision; Lessons learned.

7.2 Background

On the 22 March 2014 Manchester Central Library re-opened to the public following 4 years of major renovation, refurbishment, and extension. Manchester Central Library is a Grade II* listed building of national significance which was originally constructed between 1930 and 1934 and officially opened by King George V on 17 July 1934 whereby he declared '... *in the splendid building which I am about to open, the largest library in this country provided by a local authority, the Corporation have ensured for the inhabitants of the city magnificent opportunities for further education and for the pleasant use of leisure*'. Following the passing of the Public Libraries Act 1850, Manchester was the first local authority to provide a public lending and reference library. Designed by E. Vincent Harris, the form of the Central Library building is a Tuscan columned portico attached to a neoclassical rotunda domed structure; inspired by the Pantheon, Rome and libraries in America.

> Central Library is a jewel in Manchester's crown. When people discover it as a resource they can use, they'll be inspired.
>
> (Guy Garvey, Elbow)

In 2008 it was reported that the Central Library needed essential renovation to repair and modernise its facilities, including asbestos and structural integrity problems (Ottewell, 2008). Before the work started only 30% of the building was accessible to the general public, one of the aims of this project was to make 70% of the building available for visitors. The 5 year redesign project of the town hall extension and central library commenced in 2009 and was finally completed in March 2014 at a cost of £48m; having been allocated a budget of £100m for the whole complex.

Manchester Central Library and Town Hall Extension serves as an exemplar flagship project in demonstrating the value that BIM can deliver to the public and a public sector client through the transformation of processes, people (i.e. culture, ways of working, etc.), information, and technology (Brady, 2015). According to Mark Bew, Chairman BIM Task Group, '... *The Town Hall and Central Library project is a shining example of how our industry can collaborate in a complex and commercial world, to learn how to develop better techniques for the future. The learning we see here and across the other Early Adopter projects in the Government Construction Strategy will ensure the UK remains in the vanguard of this exciting and dramatic change to a digital economy*' (Codinhoto *et al.*, 2013).

7.3 Challenges

7.3.1 Project challenges

In relation to the challenges of the project, Central Library and Town Hall Extension are situated in the vibrant city centre of Manchester. Both buildings are of national significance, among the best examples of the architecture of their period; iconic, sophisticated and constructed to high standards. One of the main challenges for the project team was to plan, implement and deliver a strong heritage awareness and respectful refurbishment approach to provide a space to discover, share and learn about Manchester's past, present and to think about the future.

7.3.2 BIM implementation and capability challenges

In terms of BIM and capabilities, a study was conducted by the University of Salford between February and June 2011 with the aim of better understanding the BIM implementation process from MCC's perspective, while also identifying the impacts of adopting BIM-based methods on setting the scope of works for the design, production and FM teams, specifically the Central Library project as a case study (Codinhoto *et al.*, 2011). The study identified a number of barriers including MCC's more advanced use of BIM both within the project and their organisation being considerably impacted by a lack of understanding about systematic approaches for BIM implementation. In addition, there was the lack of an independent BIM coordinator within the client side along with a lack of clarity of understanding how the FM process would benefit from BIM utilisation. Moreover, while participation in BIM-related workshops served to raise BIM awareness, share the vision and get buy-in and ownership from the whole project team, it was perceived that further training should be undertaken by project members to facilitate developing their BIM capabilities.

In addition to the key barriers, a number of inefficiencies were also identified. The design-approval-cost-approval process (target costing approach) caused modelling rework to occur due to a focus on cost reduction and the model being developed to an advanced level before being costed. Furthermore, BIM was not being used in decision-making during the earlier stages of design. Moreover, as is not unusual in the delivery of projects, the accuracy of information between that of the existing building and the actual model was not always up-to-date due to the pressures of the actual project delivery, the relevant protocols and processes being in place for updating the model information, and MCC not completely understanding/defining their information requirements.

7.4 Drivers/enablers

As far back as 2006 Manchester City Council (MCC) had considered using BIM on the Building Schools for the Future programme of works, and it has subsequently been used on a number of capital projects to improve stakeholder engagement and design development. BIM was considered at the start of the Town Hall complex project but was not included in the contractual documentation. However, having won the contract for the re-design, it was the Lead Architect that convinced MCC to adopt BIM on the project following early discussions to explain the benefits that could be realised through BIM to its estate management; thereby facilitating the opportunity to improve their whole approach to asset and facilities management. At the same time, the Central Library project represented the most significant refurbishment that the Lead Architect had undertaken to date. Significantly, as mentioned, BIM was not included in the contractual documentation, but rather the adoption of BIM on this project has been described as 'BIM on a handshake!', i.e. driven by the willingness of the key project members (Brady, 2015). This raises an interesting issue of the level of influence that contractual arrangements can have on either driving or impeding the success of BIM projects in transforming behaviour and culture from one that is traditionally adversarial, in contrast to the willingness of the project team and key stakeholders to engage for the benefit of the project. For example, the experience of a U.S. healthcare client/owner of an Integrated Project Delivery (IPD) contractual approach is '... "relational" contracts try too hard to dictate behaviour and similar outcomes could be achieved through standard contracts with addendums outlining the expectations in relation to collaboration and Lean principles'. According to the American Institute of Architects, the IPD project delivery method is distinguished by a contractual agreement between a minimum of the owner, design professional, and constructor, whereby risk and reward are shared and stakeholder success is dependent on project success (AIA, 2010).

7.4.1 Vison and champion

Key to enabling any successful BIM adoption is establishing a succinct and well-articulated vision, along with effective leadership through a champion to drive the adoption and incremental transformational change required to strive towards realising the vision (AutoDesk, 2012). In the case of MCC, this was particularly important considering BIM was not a contractual requirement.

The MCC BIM champion was the Capital Programme Director who had a vision/aspiration for MCC to be able to use the BIM model and information in a manner that would deliver benefits and value to MCC during the operational life of the Town Hall complex. The MCC BIM champion established very close relationships with leading lights in BIM from across academia, the cabinet office (government BIM Task Group) and industry who were keen to support MCC on this pathfinder project. In addition, he gave the team confidence to explore and experiment, along with providing support when difficulties were encountered. In essence, the MCC BIM champion promoted a collaborative learning environment where the team shared knowledge, ideas and aspirations.

At the core of MCC's vision was not to be driven by technology and therefore led by software vendors. Moreover, similarly not allowing their construction partners dictate how building asset data/information should be presented via their own chosen software platforms and data terminology. MCC believed that such an approach would have led to:

- Incurring costs to manage and pay for additional upkeep of multiple systems.
- High risk of data duplication and inconsistency.
- Data terminology and classification structures not in a common language.
- Data difficult to access for end users.
- Continuous end user training and associated costs on differing software platforms.
- Ongoing software incompatibilities/conflicts with the councils ICT network infrastructure, software version updating and security restrictions.
- Complicated and unsustainable integrated software system links/dependencies.
- Increased confusion for the clients/end users as to where to find information.

In essence, too many 'add-on' systems would have prevented the client/end users knowing exactly where to find accurate and reliable building information, while inconsistent and confusing data terminology would have prevented clients/end users utilising the information effectively or complying with the estate wide data requirements. This would have inevitably led to a loss of confidence in all the various systems and reluctance in using them, which in turn, would have ultimately resulted in the data and information not being updated and maintained; therefore proving redundant, insufficient and impractical to use intelligently for future estate developments and building lifecycle needs and costs.

There is an appreciation and acceptance that these new and innovative digital software systems and web-based platforms are still evolving and to have one central system that can hold rich data and related information in a controlled manor is still yet to be fully achieved. However, with regard to BIM, it is widely accepted that for it to reach its ultimate goal of conception through to building lifecycle and finally into decommissioning and demolishing, there is a need to 'start with the end in mind'. Therefore, from MCC's experience, CAFM/BMS systems need to lead and adapt to ensure they can upload, handle and update this new evolving technology and in differing formats. In addition, software developers need to continue to ensure they can provide open formats that will work on multiple systems and platforms, such as the Industry Foundation Classes (IFC), together with a consistent approach to data naming and terminology classification.

In taking the aforementioned into consideration, MCC developed a short- and long-term vision to key vendors in the early concept stages of BIM4FM to ensure a clear path for data exchange and additional development for model integration was achievable (Figure 7.1).

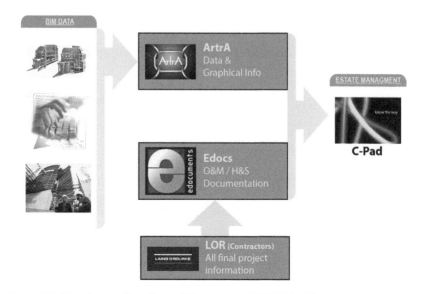

Figure 7.1 Manchester City Council short-term BIM4FM vision.

Although the short-term vision was slightly cumbersome, in that additional software was required to align the mechanical and electrical asset information from the BIM model to the MCC's estate-management software, it helped to prove this vision. Moreover, the vendors were able to demonstrate that the systems could integrate and they were also prepared to work collaboratively in developing this further, which was a significant outcome in itself. Furthermore, the main achievement was the way in which the CAFM developers took the lead as they could see the significant benefits to adopting BIM and data exchange management processes, whilst aligning key data to building lifecycle requirements such as (but not limited to):

- Planned preventative maintenance schedules
- Job helpdesk (and history logs)
- Asset management
- Spares and ancillaries
- Warranties, lifespan and replacement costs
- Energy monitoring
- Health and safety

This meant that MCC's long-term vision, whereby all building data and information is held within one central system, would be more streamlined through the BIM asset information being extracted direct from the model into MCC's estate-management software. Moreover, this would be

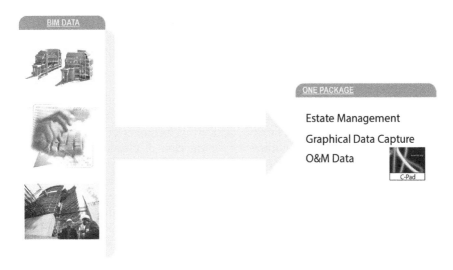

Figure 7.2 Manchester City Council long-term BIM4FM vision.

achievable (Figure 7.2) through ongoing development and collaboration between MCC and the CAFM vendor, with support from leading industry experts, academia and the government BIM Task Group.

7.4.2 'True' collaboration

In its essence, BIM is a collaborative way of working, enabled through new process workflows and underpinned by the digital technologies, to unlock more efficient methods of managing the digital delivery and operation of assets. According to the BIM Task Group (2011), '... other than a digital-tool set you do not actually use BIM, it is way of working, it is what you do, information modelling and management in a team environment, which facilitates collaborative working and shared understanding'. Moreover, according to the NBS National BIM Report 2014, it is clear that the industry are aware that BIM is about real-time collaboration and is not just about CAD drawings, technology/software, etc. (NBS, 2014).

Collaboration has been key to both the success of the project and MCCs adoption of a BIM approach. The project was procured via the North West Construction Hub (NWCH) High Value (2010–2014) Framework, whereby collective knowledge, sharing best practice, partnering behaviour as well as early contractor engagement enabled a 'one team' approach in a true collaborative manner (NWCH, 2015a,b). In addition, a NEC3 Option C contract was agreed in sixteen weeks after selecting the main contractor via a mini competition process, i.e. More for your Money report (NWCH, 2014a). Therefore, the procurement route allowed early contractor involvement in the project. Moreover, this way of working prioritised the client needs, to drive efficiencies and to build long term relationships between project team and stakeholders. In particular, the project team established a co-located office at Manchester City Council to promote the ethos of 'one team' along with giving the MCC team access to BIM expertise from their supply chain and with whom they could share knowledge, ideas and aspirations to develop and deliver the vision. This is in a similar vein to that of the 'BIGRoom' concept utilised in IPD, which serves to enhance 'integrated teams' through physically co-locating members of the design and construction team; thereby allowing for the ease of physical proximity to address questions and solve problems. This affects individuals beginning to perceive themselves more as one team (AIA, 2007). In addition, guidance and support to the project team in making informed decisions and, in turn, building MCC's confidence to the project vision was further enhanced by forming strong engagement beyond the project with the government BIM Task Group and academia with expertise in the area of BIM and Lean.

7.4.3 BIM capability and maturity

A successful BIM implementation should follow several recommended 'steps' and 'actions', which in the main, relate to developing an understanding of the level of capability and maturity in place, identifying the target level of the BIM implementation, including the necessary resources, and the promotion of changes for achieving the desired level of implementation that will have a positive impact (as measured against a baseline). According to Succar (2009, 2010), BIM capability refers to 'the ability to generate BIM deliverables and services' whereas BIM maturity refers to 'the extent, depth, quality, predictability and repeatability of these BIM deliverables and services'.

The understanding of the maturity level of a BIM implementation within a project or organisation is crucial for the successful implementation and development. According to Harmon (2004), 'the basic concept underlying maturity is that mature organisations do things systematically, while immature organisations achieve their outcomes as a result of the heroic efforts of individuals using approaches that they create more or less spontaneously'. A maturity model consists of a certain amount of maturity levels in a sequence that demonstrate the development of the entity, e.g. project, organisation, etc., and for each level a set criterion exists according to the specific characteristics of the entity. Entities progress level by level in an order, progressing from one maturity level to the next as the organisation develops certain capabilities for that particular level. Therefore, the maturity level can be considered as a performance indictor for the entity. A BIM maturity level relates to the positioning of a project or organisational practices in relation to basic, medium or advanced use of integrated (interconnected) information. Measuring the maturity level of a BIM implementation serves two purposes: (1) indicating the state of the art of an organisation in relation to its use of BIM-enabled

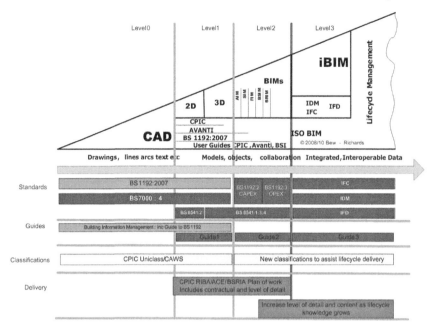

Figure 7.3 B/555 BIM implementation maturity model.
Source: BSI (2011).

set of technologies and group of processes, (2) developing this understanding then provides direction for development and improvement.

An initial BIM capability and maturity assessment conducted as part of the University of Salford study provided a further driver for MCC (Codinhoto *et al.*, 2011). In relation to BIM capability, the integration of information through information modelling was found to be being addressed at the project level but beginning to migrate towards the organisation level. This was assessed in terms of both the NBIMS Exchange Tier Architecture Model (National Institute of Building Science, 2007), which is composed of four levels (tiers), i.e. (1) Information Delivery Manual Activity (IDM), (2) Model view, (3) Derived view, (4) Aggregated view, and the BSI BIM implementation model/evolution ramp (BSI, 2011) (Figure 7.3).

An assessment of the BIM maturity level was conducted according to the NBIMS capability maturity model, which is based on eleven instances to assess the overall maturity, with each having ten levels ranging from basic to more advanced levels of maturity (National Institute of Building Science, 2007):

- *Data richness*: This identifies the completeness of the building information model from initially very few pieces of unrelated data, to the point of it becoming valuable information and ultimately corporate knowledge about a facility.

- *Life cycle views*: These refer to the phases of the project and identifying how many phases are to be covered by the building information model.
- *Roles and disciplines*: Roles refer to the players involved in the business process and how the information flows. Disciplines are often involved in more than one view as either a provider or consumer of information.
- *Business process*: The business process defines how business is accomplished.
- *Change management*: This identifies a methodology used to change business processes that have been developed by an organisation.
- *Delivery method*: Data delivery and access is also critical to success.
- *Timeliness/response*: The closer you can get to accurate, real-time information, the better the quality of the decisions that will be made.
- *Graphical information*: The advent of graphics and 3D images help to develop a common view and a higher level of understanding for all involved especially as standards are applied, whereby information can begin to flow as the provider and receiver must have the same standards in place.
- *Spatial capability*: Understanding where something is in space is significant to many information interfaces and the richness of the information.
- *Information accuracy*: Having a way to ensure that information remains accurate is only possible through some mathematical ground truth capability.
- *Interoperability/IFC support*: An ultimate goal is to ensure interoperability of information of getting accurate information to the party requiring the information facilitated through the use of a standards-based approach.

The initial maturity assessment indicated that MCC was at the early stages of implementation in many of the eleven instances (i.e. scores between two and three out of ten). The initial reaction from MCC to the assessment was one of disappointment and despondency to their 'apparent' low levels of BIM maturity. However, when presented in contrast to that of the wider industry, this served to emphasise that MCC are at a similar level of BIM maturity, i.e. early stages of implementation, but were making significant progress; thereby further building their confidence in their BIM implementation. Figure 7.4 shows an overview of the capability maturity level within the Central Library project (dark grey represents the perceived maturity and light grey the overall industry level of maturity).

7.5 Derived Benefits

7.5.1 Enhancing communication and understanding

Visualisation is currently a key benefit of BIM, specifically in enhancing communication and understanding of the project amongst the project team and stakeholders. NBS (2014) found that 83% of users identified that BIM improves visualisation and similarly in the U.S., McGraw-Hill (2013) suggest that the use of BIM-generated visualisation is expected to grow due

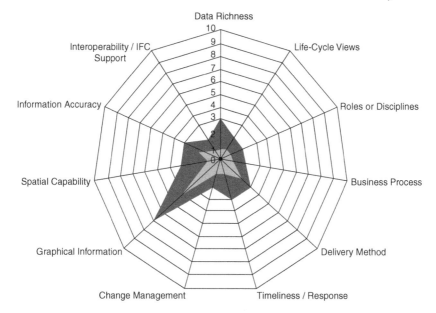

Figure 7.4 MCC initial BIM maturity analysis of the Central Library project.
Source: Codinhoto *et al.* (2011).

to the effectiveness of coordinating project teams and facilitating owners making better and more informed decisions.

In terms of the Central Library project, the use of BIM to communicate the site establishment and boundary proposals to local businesses, neighbours, highways and planning departments by the Lead Contractor, reduced the queries and concerns normally expected on such a large city centre site. In addition, the clarity and understanding of the site access routes and logistics and safety requirements was improved through the use of BIM as a site induction tool. Communication and understanding of the project was further enhanced for the site teams through developing visual method statements using BIM to breakdown complex site works into individual activities. This served to promote a 'right first time' approach and thereby improving workmanship site safety. Moreover, BIM was utilised by the Lead Architect to communicate complex design solutions such as the vertical circulation core and the removal of the four-storey book stacks in the Central Library to key stakeholders (including the project team, councillors, librarians, planning officers, English Heritage), which enabled them to make better informed decisions (Brady, 2015).

7.5.2 Sequencing and programming

The use of 4D, using the model to develop and visualise the sequencing and programming, served to provide key stakeholders with confidence that the Grade II* buildings were being respected, and works were being appropriately considered and planned; thereby giving MCC the confidence in decisions and agreements on design. Furthermore, the design and installation of the Central Library's ground floor sculptured troughed ceiling required a supplier that could use BIM data to manufacture the product. As a result, a local SME gained a commercial advantage over competitors due to their access to BIM technologies and capability to coordinate the extremely complex services, structure, and ceiling finishes (Brady, 2015).

In relation to time and cost savings, £250,000 was achieved on the vertical circulation core, representing 10% of the value. Furthermore, the coordination of complex services across the Town Hall Extension and Central Library saved time and cost due to right-first-time solutions and a reduced number of changes. Time was also saved in consultation with stakeholders, especially English Heritage and conservation specialists. In particular, the use of BIM has enabled all parties to engage with confidence and absolute clarity during design development, planning approval and heritage process. The main contractor's innovative use of the architect's model to develop visual method statements for the vertical circulation core, by breaking down the sequence of works into steps, identifying workmanship and health and safety requirements, has enhanced understanding operational activities and the quality of the end product; whereby all parties had absolute clarity of their individual roles and responsibilities. The use of sequencing animation created the environment to set agendas aside and allowed the translation of 2D drawings into a '*3D common language*' shared by all stakeholders engaged during the pre-construction phase. Moreover, this way of working enabled a constructive dialogue with English Heritage and clear focus on design solutions and benefits (e.g. new vertical core and circulation routes) (Brady, 2015).

> BIM was used during several consultation meetings on Central Library. In addition to drawings and rendered images, using the model 'Live' in presentations gave a unique understanding of the nature of the interventions and gave us greater confidence in our decisions and agreements on design.
> (P. Mason, Manchester City Council Design and Conservation Manager)

In addition, the MCC operational team has been actively involved at all stages of the development of BIM for asset information management (AIM), in particular, focusing on the mechanical and electrical (M&E) data. By working collaboratively with the M&E subcontractor, the two parties were able to ensure that the M&E asset information was eventually uploaded into the BIM model based on the principles of COBie but, altered in accordance with the MCC operational team requirements (Brady, 2015).

7.5.3 *BIM capability and maturity improvements*

The University of Salford conducted a second follow-up study to the initial study with the aim of better understanding the uses and implementation process of BIM for hard and soft FM services (Codinhoto *et al.*, 2013). A comparison with the assessment from the initial study (Codinhoto *et al.*, 2011) identified an overall improvement in all of the eleven instances of the NBIMS capability maturity model with the exception of business process and timeliness/response (Figure 7.5). However, the study notes that the first assessment was conducted during design development at which time it was not clear to what extent BIM was applied to the business.

The second study also assessed the reactive maintenance process in relation to a general comparison with FM practice in the UK (baseline for comparison drawn from anecdotal evidence of industry and academic opinion). Again, the overall results achieved by MCC's reactive maintenance Services could be considered low when considered in isolation and in comparison with design and construction. However, when compared with the industry FM practice baseline, then the achievement in the BIM implementation is significant, especially considering the constraint imposed on the assessment by the technology not currently being ready for full adoption together with very little about the benefits and barriers to the adoption of BIM FM being known (Figure 7.6).

Furthermore, the second study reports that although MCC are at the early stages of their BIM implementation, the progression has been rapid in positively pushing the boundaries of BIM for FM. Furthermore, the willingness of MCC to implement BIM, and their vision in using BIM to improve facilities management has been paramount to their progress. In particular, the attitude of the MCC FM team has been transformed to become more process-oriented and in seeking to identify process inefficiencies

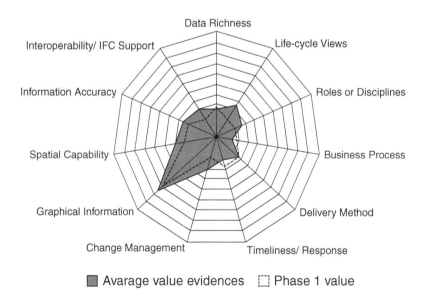

Figure 7.5 MCC comparative BIM maturity analysis of the Central Library project (design and construction).

Source: Codinhoto *et al.* (2013).

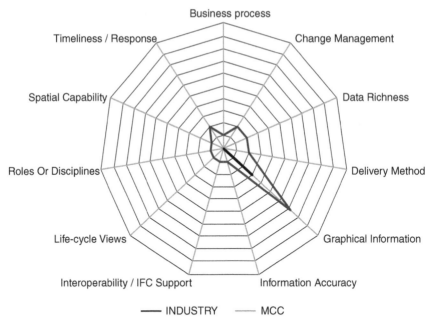

Figure 7.6 MCC comparative BIM maturity analysis of the Central Library project (FM).
Source: Codinhoto *et al.* (2013.

which could be mitigated by BIM. In this manner, the MCC FM team is not just following a trend, but are taking a lead in focusing on promoting continuous improvement by setting new standards for the management of facilities and utilising BIM wherever appropriate; thereby establishing an MCC construction knowledge asset base for improving the future management of this and future projects. Moreover, the MCC FM team have become aware of their advanced position in relation to traditional FM practices, together with the route for redefining the process model for FM in public sector building projects and further BIM implementation.

7.6 Lessons learnt

The biggest lesson learnt from MCC's BIM journey to date through the Central Library project is not to go it alone, while their greatest achievement was developing a collaborative approach team spirit between the client, construction partners, software vendors, academia, and other supportive groups such as the BIM Task Group.

A further key lesson learned is to adopt a very informed client approach with regards to the software vendors and in sharing with them the vision, aspirations and requirements. With regard to the Central Library and Town Hall Extension project, the software vendors have responded very positively to this approach and have not only stepped up to the plate but, at times, have exceeded MCC's expectations.

More specifically, a key lesson is that although the process of uploading the M&E asset information into the BIM model was conducted retrospectively, and thereby proved very time consuming, MCC are using this learning to develop their BIM Employer Information Requirements (EIR), which will then be clearly articulated at the outset of future works.

7.7   Legacy/moving forward

7.7.1   City-wide estates asset management

The transformation to date, derived through the Central Library and Town Hall Extension, has served to build the foundation in developing the basic capability towards ultimately a city-wide estates asset database that supports the entire operational management of the asset, including reactive and planned maintenance, scheduling, purchasing, stock control, replenishment and financial management (Brady, 2015).

7.7.2   Knowledge transfer

In the meantime, the transfer of the current capability is evident through three pilot projects whereby MCC will be testing BIM and data exchange at varying levels with knowledge gained and utilising developments to date within the CAFM system. These three projects are:

7.7.2.1   Town Hall Complex refurbishment

This Town Hall Complex Refurbishment project (THC) introduced MCC to the bigger benefits of BIM and set a basic level of understanding of how BIM and surrounding electronic information could be integrated for facilities and building lifecycle management. The project focused more around mechanical and electrical (M&E) data modelling and exchange and set out the foundation for data transfer from project to the CAFM system.

While this was a path finding exercise, the model is being used effectively by onsite engineers to visually find and understand the installed complex M&E systems. Data from the model is still being fine-tuned and the CAFM is readily awaiting to import this data into the sites asset register and align to the planned preventative maintenance schedule. All the additional electronic information (O&M/H&S/drawings, etc.) is being supplied on a web-based system that will integrate with the CAFM system and is currently being structured to suit the facilities and engineers requirements. A challenge yet to be investigated is how to keep the building information model up to date in a streamlined way. Therefore, until the technology has developed sufficiently to create this streamline function, MCC will have to keep data within an 'end user model' to a minimum, as the cost to change data for a 3D modelled element (for example a sink) could be more than to actually replace the real asset/item (sink) in the first place.

7.7.2.2 Old Town Hall refurbishment

If approved, the refurbishment of the Old Town Hall (OTH) offers MCC a great opportunity to vastly improve on the lessons learned from the THC project and ensure the whole project team take BIM4FM to a whole new level, whilst meeting other areas of BIM use such as 4D, 5D, etc. The ultimate aim is to streamline the process of BIM (plus additional digital information) straight into the CAFM system and assign building lifecycle costs and maintenance requirements from day one of handover.

Phase 1 of the project is predominately electrical works only, but the team are keen to widen the scope to capture a larger need (all within reason). EIR will be developed and shared along with a collaborative approach to naming conventions and data exchange methods. As this project is a refurbishment rather than a new build (as was the THC project), there will be challenges to understand if, and indeed, how some of the existing buildings systems and fabrics will be captured, as this will affect the benefits for the end users in terms of realistically using the BIM for day to day and legacy purposes. However, the data captured will be fully utilised and imported into the CAFM system.

7.7.2.3 Estate wide condition surveys works

Manchester City Council is in the process of carrying out condition surveys on over 170 operational properties within the local authority. There is an aspiration to use point cloud scanning technology and create building information models to capture data and condition surveys for all these sites. However, a realistic and phased approach with more common techniques needs to be applied due to:

- *Costs:* The current financial constraint associated with scanning technology still being relatively new and costly.
- *Risks:* Technology is still developing and could lead to unknown complications down the line or missed opportunities on new and improved functionality of such technology.
- *Not ready:* It is still unclear how BIM can be seamlessly and safely updated internally with little cost, time, up-skilling, and control measures.
- *ICT limitations:* Storage, software versions and security integrity will require a serious overall of the ICT network infrastructure.
- *Timing:* BIM Level 3 protocols, processes, industry knowledge and standard classification structures all still under development.

Therefore, the surveys are being conducted in a traditional method using 2D drawings but with a slight technological advancement in using an app that links straight into the CAFM system. Although the level of data capture for these surveys will be minimal, it is envisaged aligning the dataset standards, were practically possible, to that of BIM and from the structure set up developed from the THC and OTH projects. This will also allow the council to revisit simple data management standards and requirements, including CAD management. In doing so, this will raise staff awareness in the importance of data needs, management and structures, and better prepare MCC and the back office systems for a change in the way that they need to work, and for a full and smoother implementation of BIM and surrounding roles and responsibilities.

Moreover, the Central Library is recognised as an exemplar case, both in the UK and internationally, in demonstrating the importance of protocols and processes along with the attitude of the project team for a successful client BIM implementation. The knowledge transfer is also being extended external beyond Manchester City Council to other major clients such as Buckingham Palace.

7.7.3 Organisational/behavioural legacy

The Central Library project is also an exemplar project of the importance of cultural change in a successful BIM implementation. In terms of organisational/behavioural legacy, the working culture, trust, goodwill and respect that thrived from this project has created a platform for sharing knowledge and experience beyond the team and stakeholders. For example, a BIM Special Interest Group (BIM SIG) was created via the NWCH Framework network to further promote BIM aims, aspirations, ideas and lessons learnt. In addition, one of the main actions of the BIM SIG was to examine the EIR document from the project and to create a proposal EIR document that could be adopted by other schemes procured via the NWCH. The proposed draft is currently being discussed with the National Association of Construction Frameworks (NACF) who took over the stewardship of BIM for local government from Cabinet Office in December 2014. Under the umbrella of NACF, the proposed NWCH EIR draft document can be further examined by East Midlands Property Alliance (EMPA) and NACF with the aim of creating standard EIR that would cover COBie, information exchange and information delivery plan.

7.7.4 Industry standard BIM measurement and KPI model

Moreover, in Q4 of 2014, a small working group was established via the BIM SIG in response to the NWCH, and also in line with a need identified by the BIM Task Group legacy BIM4 community and Constructing Excellence, to focus on establishing a standard industry BIM measurement model for the NWCH framework partners based on an critical evaluation and analysis of existing BIM maturity models and linking this to existing industry Key Performance Indicators (KPIs). This would allow the optimum BIM level that would allow clients to achieve project life cycle benefits to be identified, i.e. the minimum level of BIM that can bring benefits to the project without increasing project cost. The work in progress is focused on real 'paybacks' that could be achieved if CAPEX and OPEX are taken into consideration in the pre-construction phase. Therefore, the aim is to establish the best starting point for a collaborative approach for client and contractor teams to maximise the benefits by assessing existing CAFM systems, agreeing on information that will be required in the future and setting realistic milestones. Another aspect the group is looking at is the dialogue with clients and clients' advisors to achieve better client briefs, more coordinated ways of working, i.e. drawings, plans, decision making, etc. as well as to use the industry collective knowledge to start up skilling project teams. The outputs of this working group are expected to be available in Q3–Q4 2015.

7.8 Awards recognition

As outline previously, the Central Library is recognised as an exemplar project, both in the UK and internationally, for a successful client-led BIM implementation. This recognition is evident through a number of awards (Table 7.1) (NWCH, 2015c).

Mike Gibson, Chair of the Greater Manchester Chamber's Property and Construction Committee and who hosted their Awards event, said 'Manchester Central Library has been selected as 2014 Building of the Year Award winner as it is a high quality reconstruction and refurbishment of one of the city's crown jewels. The attention to detail in preserving the historic character of the building is matched by the fine updating of the Library's facilities, making it a superb asset for the inhabitants of a 21st century city. It's also worth noting that the use of BIM in the reconstruction process has helped establish Manchester as one of the global leaders in the application of this new technology' (NWCH, 2014b).

7.9 MCC's position on their BIM journey

Whilst MCC have achieved a wealth of experience and learning from extremely enthusiastic and engaging project partners and industry experts, it has also highlighted some basic internal housekeeping requirements that need to be addressed before BIM can be truly implemented and integrated into business as usual. This also requires further advances in technological solutions to allow end users to easily manage and update building information models without great cost or overly complex software systems. Finally, in relation to their BIM journey to date, some general recommendations:

Table 7.1 MCC Central Library and Town Hall Complex Project Awards

Awarding body/award	Building/service	Year
EDGE Awards projects must increase participation, while demonstrating innovation and creativity. • Physical category—which honours excellence in library buildings • Digital Category, for the successful application of digital technology in libraries.	Central Library building Interactive Archives + Centre, based at Central Library (winner)	2015 2015
North West Local Authority Building Control	Best Public Service Building and Best Large Commercial Building—Manchester Town Hall Extension and Central Library	2015
Building Awards	Major Project of the Year—Manchester Town Hall Extension and Central Library	2015
Construction News	Project of the Year (over £50m)—Manchester Town Hall Extension and Central Library	2015
Greater Manchester Chamber of Commerce's Property and Construction Awards 2014	BIM Project of the Year—Manchester Central Library	2014
Green Apple Award: the annual international campaign to recognise, reward and promote environmental best practice around the world.	Laing O'Rourke won the Built Environment and Architectural Heritage award for Manchester Town Hall Complex	2014
Forum for the Build Environment	Laing O'Rourke won Public Sector Project of the Year with Manchester Town Hall Complex	2014
RICS The awards recognise property professionals and showcase their talent and ingenuity across a wide range of categories - Building Conservation, Community Benefit, Design through Innovation, Infrastructure, Regeneration, Commercial, Residential and Tourism and Leisure.	Laing O'Rourke won the Building Conservation award for Manchester Town Hall Complex	2014
The North West Regional Construction Awards, hosted by the Centre for Construction, the Constructing Excellence Centre for the North West, celebrate the achievements of the region's best and most innovative construction projects, organisations and individuals.	Laing O'Rourke won Sustainability - Legacy Award and Manchester Project of the Year for Manchester Town Hall Complex	2014
	Integration and Collaborative Working Award, Manchester Town Hall Complex, Transformation Programme, submitted by Laing O'Rourke, Winner	2013
The Considerate Constructors Scheme National Site Awards recognise those sites that have demonstrated best practice in engaging with the workforce and community and consideration of the environment.	Manchester Town Hall Complex Transformation Programme, Laing O'Rourke, Silver Award	2013

- Know your data needs (Strategic needs, i.e. Organisational Information Requirements, Operational/Asset Information Requirements, which need to be articulated in the Employer Information Requirements). Just because the technology can does not mean you should! Be realistic and take a common sense approach when it comes to data requirement and needs. Asking for everything but with no outlined benefit or purpose will just create an unmanageable amount of data and waste time and cost to the project and upkeep of the property management systems. However, on the other hand, do not let this stop aspirational ideas and innovation.
- Small steps, phased approach, learn and share.
- Understand your current CAFM systems data, capabilities and areas where you may wish to expand. This is a good starting point as to what data you may want to include in the EIR and include within the BIM outputs.
- Contractual BIM but with shared risks. Everyone is still learning and understanding how BIM can be used effectively across many disciplines, so establishing open trust and collaboration within the whole team, will allow room for open innovation, development and greater understanding to all involved.
- Do not be led by the multitude of software out there on the market. Take control of how you want the data and models to be viewed and used, and only purchase a new system where current solutions cannot meet your demands and future expectations.
- Do not go it alone. There are many guidelines, classifications, standards and books about implementing BIM now available, but this can be overwhelming to say the least. Speaking to industry experts and sampling real examples can help unravel some of these worrying and over complicated processes, which can help get the creative mind into what your needs may be and also how BIM can assist as a tool to achieve your vision and overall aim.

7.10 Conclusions

The UK Construction Strategy 2011, in an effort to reduce whole life cycle costs and reduce carbon emissions, mandated the use of BIM 'Level 2' on all centrally public procured projects by 2016. The important aspect of this approach concentrates on using data-driven procurement and delivery strategies to achieve project success. What determines achieving or missing the success objectives is largely dependent on effectiveness of processes and capability of stakeholders to communicate a shared understanding. This chapter explored the importance of mutual understanding, visualisation and the sharing of risk and rewards. The journey of each BIM project should start with an organisational maturity analysis in order to empower the client organisation to demand a focus on whole life cycle and not just on the capital expenditure. An informed client is in much stronger position to drive project benefits at a strategic, implementation and organisational level. By focusing on the whole life of built environment assets, the end user and facilities management team can improve existing standards and manage the building more efficiently. It should be an informed client's decision about what 'add-ons' are required to manage data efficiently in each stage of the construction process; pre–construction, construction and following handover. An informed client will work with software vendors to develop tools that provide 'fit for purpose' solutions in achieving the aspirations and vision of clients. The lessons learned and legacy of the above project opened the door to a wider BIM adoption and knowledge transfer between project teams for other MCC clients as well as for other organisations (e.g. Blue Light, Higher Education, etc.). Further research and case studies will allow establishing a best practice guide covering aspects of behaviours, roles and responsibilities, modern procurement routes, sharing risks and rewards, and working in integrated project teams to achieve clients' aspirations and realise their vison.

References

AIA. (2007). Integrated Project Delivery: A Guide, The American Institute of Architects, Washington, DC.

AIA. (2010). Integrated Project Delivery: Case Studies, The American Institute of Architects California Council, Sacramento, CA.

AutoDesk. (2012). White paper: a framework for implementing a BIM business transformation, AutoDesk, Inc.: San Rafael, USA.

BIM Industry Working Group. (2011). A report for the Government Construction Client Group: Building Information Modelling (BIM) working party strategy paper. Retrieved from http://www.bimtaskgroup.org/wp-content/uploads/2012/03/BIS-BIM-strategy-Report.pdf.

Brady, L. (2015). Reaping the Benefits, *RICS Construction Journal*, February/March 2015, RICS, London, 16–19.

British Standard Institute. (2013). Design, Construction and Operation Data and Process Management. [online] BSI. Available at: http://shop.bsigroup.com/upload/construction_downloads/b555_roadmap_june_2013.pdf.

Codinhoto, R., Kiviniemi, A., Kemmer, S. and da Rocha, C. G. (2011). BIM Implementation: Manchester Town Hall Complex—Research Report 1, University of Salford, Salford, UK.

Codinhoto, R., Kiviniemi, A., Kemmer, S., Esiet, U. B., Donato, V. and Tonso, L. G. (2013). BIM-FM: Manchester Town Hall Complex—Research Report 2, University of Salford, Salford, UK.

Harmon, P. (2004). Evaluating an organisation's business process maturity, *Business Process Trends*, 2(3), 1–11.

McGraw-Hill. (2013). The Business Value of BIM in North America: Mulit-Year Trend Analysis and User Ratings (2007–2012), McGraw-Hill Construction, Bedford, MA.

National Institute of Building Science. (2007). 'National Building Information Modeling Standard, version 1, Part 1: Overview, Principles, and Methodologies'.

NBS. (2014). NBS National BIM Report 2014, RIBA Enterprises Ltd., Newcastle-upon-Tyne, UK.

NWCH. (2014a). More for Your Money, North West Construction Hub, http://www.nwconstructionhub.org/mfym.pdf.

NWCH. (2014b). Award Wins to Date in 2014, North West Construction Hub, http://www.nwconstructionhub.org/news/award-wins-to-date-in-2014.

NWCH. (2015a). What is the North West Construction Hub? North West Construction Hub, http://www.nwconstructionhub.org/.

NWCH, (2015b). BIM Focus 2015—Driving back-to-basics best practice Project Delivery ... 21 May 2015, North West Construction Hub, http://www.nwconstructionhub.org/news/bim-focus-2015-driving-back-to-basics-best-practice-project-delivery.

NWCH. (2015c). NWCH Projects Awards List 2015, North West Construction Hub http://www.nwconstructionhub.org/assets/documents/case-studies/Award_update_Apr15_2.pdf.

Ottewell, D. (1 July 2008). '£150m to save Central Library'. *Manchester Evening News (M.E.N. Media)*, Oldham, UK. [Accessed on: 28 May 2016].

Succar, B. (2009). BIM Episode 12: BIM Performance Measurement. BIM Think-Space Blog. Available at: http://www.bimthinkspace.com/2009/09/index.html [Accessed on: 28 May 2016].

Succar, B. (2010). Building Information Modelling Maturity Matrix. In Underwood, J. and Isikdag, U. (Eds.), *Handbook of Research on Building Information Modelling and Construction Informatics: Concepts and Technologies*, Information Science Reference, IGI Publishing, New York.

8 Project extranets and developments in project collaboration

Eric Lou and Anas Bataw

8.1 Introduction

The use of information technology (IT) in the construction industry both in the UK and internationally is rapidly growing. This resurgence has further stepped up due to the rapid internationalization of construction activity involving global teams in practice requiring methods techniques and processes to harness global project collaboration. This growth is primarily led by two factors – the technology push and strategic pull. Technological push has led to the development and implementation of computer hardware and software to improve various functions including that of integrating between design and construction planning systems.

Strategic pull generates the growing awareness of process reengineering, the design for construction and the concurrent engineering concept. This chapter will discuss the state of the art on project extranets that addresses both the push and pull aspects in enabling a collaborative behaviour among team members executing multinational global construction projects.

8.2 Need of technology

Studies from Wirtz *et al.* (2015) and Rodrik (2014) indicated that the use of technology is one of the fastest growing sectors globally, if not the fastest. This growth in technology usage is due to the great need of technology in almost everything around us in our daily life. For instance, computers and cell phones are developed to help different businesses with different opportunities. Social networking is also a big part of many people's everyday life around the world.

The use of technology in the UK built environment industry is rapidly growing. This growth is led by two factors – the technology push and strategic pull (Saxon, 2013). Technological push has led to the development and implementation of computer hardware and software to improve various functions including that of integrating between design and construction planning systems. On the other hand, the strategic pull generates the growing awareness of process reengineering, the design for construction and the concurrent engineering concept. Through the years of IT implementation, there have been severe hitches in total computerised construction – broad spread of stand-alone application packages that do not 'speak' to each other as a result of isolated investment and the inefficiency of data exchange as there are no international standards to follow, among other problems. Information technology in the construction industry is facing the most fundamental issue, the business issues in the organisation. Today, organisations are faced with increasing globalisation of construction products and services, accelerating deregulation of professional roles and organisational systems and the realisation that the industry is facing long term recession over-capacity (Wright *et al.*, 2010)

8.2.1 IT in construction

The construction industry is often hailed as the gauge of the global economy. From a value perspective, the industry contributes an estimated 9.8% towards gross domestic product (GDP) of the European Union (EU) economy (Thompson, 2013). This staggering figure in construction expenditure and project capitals urge leaders in the construction sector to push the urgency and the need of improvements and cost saving steps in procurement and management of new construction, refurbishment and maintenance. However, the industry is often seen as a low-tech industry (Gann and Salter, 2000), with much of the technology being adapted from new technology development or other disciplines to 'fit' into old processes of construction. Information technology (IT) is one of the fastest changing industries in the world. There are new inventions and technologies created every day. Globalisation is making the world a smaller place and the power to communication is endless, projecting a scenario of a 'global village'. The Internet is becoming one of the most powerful media for communication and information exchange (Carty, 2010); statistics show a 528.1% growth in worldwide users from the year 2000–2011, and also show 376.4% growth for Europe and the Asian continent having the most users in the world (Internet World Statistics, 2012).

In this respect, the European Commission's Information and Communication Technology Uptake Working Group report highlighted the importance of IT-based innovation in bringing productivity improvements and competitive advantage to the industry (Goulding and Lou, 2013). This task force also reported a constant decline in labour productivity, which was mainly attributed to the lack of IT related investment since the mid-1990s. This evidence also highlights that higher productivity growth rates were observed in the USA and other world trade partners of Europe through the greater use/integration of IT by all segments of the economy. However, industries have not been in a position to capitalise on the investment in terms of productivity growth. In this context, sustainability, competitiveness and growth of this vital sector of the economy can only be sustained through the pursuit of knowledge and innovation; and the latter of which has historically been driven by rapid developments of IT systems, specifically the ability to capture, store, analyse/manage and exchange data. Therefore, in an increasingly knowledge-based industry such as construction, it is vital to have early access to knowledge-based tools, together with an IT infrastructure that can handle media-rich services; and in order to remain competitive, construction firms will have to fully embrace this technology.

Concerning IT and the construction industry, whilst several success stories (Bryde *et al.*, 2013; Kiviniemi, 2011; Won *et al.*, 2013) can be highlighted over the past decade, these have mainly focused on technical operations such design, planning and estimating. In addition, there has been continued growth in the uptake of extranets. These applications are designed to manage and control project documents among partners, whilst also providing up-to-date information on their progress. However, although these applications can provide 'value' to projects, their actual role in achieving competitive advantage is not overtly documented in seminal reports (e.g. Egan (1998) and Latham Reports (1994), Cabinet Office (2011)). In addition, whilst a 'technology push' approach may bring about 'first comer' advantages to organisations, implementing IT applications to create competitive advantage can only be leveraged by improving businesses processes in line with management objectives, and using IT as the core enabler (Lou *et al.*, 2012) Therefore, in today's economic climate, competitive advantage can be achieved by focusing on issues such as providing high quality services and products with minimal cost, having the flexibility to predict and respond to market needs, or through the efficient management of resources. These can be realised by embracing IT to enable streamlined business processes, which not only make organisations operate efficiently but also allow them to build their knowledge base in order to gain competitive advantage.

Notwithstanding this, investment in IT-based business systems such as building information modelling (BIM), extranets (collaborative environments), enterprise resource planning (ERP) and intelligent systems have not realised their full potential (Alshawi and Goulding, 2008). Furthermore, it is estimated that the worldwide cost of IT failure could account to US$6.2 trillion per year, or an estimated US$500 billion per month (Krigsman, 2009); and similar failure stories are reported elsewhere (Business Wire, 2008; Krigsman, 2010). This situation also pervades the construction industry for example, Salah (2003) identified that 75% of IT investment in business-oriented systems did not meet their intended business objectives. Furthermore, some of these projects were abandoned, significantly redirected, or 'kept alive' despite business integration failures. On reflection, the main attributes of these failures were rarely purely technical by origin, but more often than not related to organisational 'soft issues', which underpin the capability of organisations to successfully absorb IT into its work practices. In this respect, Basu and Jarnagin (2008) noted that business executives did not fully recognise the full functionality and value of technology to the business, nor did IT personnel possess an understanding of the business and its strategic objectives. This was reinforced through a survey of Chief Executives and Directors of construction organisations in the UK, where the results demonstrated a high level of awareness regarding the strategic benefits of IT to achieve innovation and competitive advantage, but a lack of direction on how best to achieve these benefits in their organisations (Alshawi *et al.*, 2008).

Change is needed in the construction industry to create an environment that fosters innovation, clarifies communication and truly integrates design, manufacturing and construction processes (IET, 2014). IT technologies and knowledge-rich environments offer real opportunities and the potential to realise this integration and offer an arrangement in which there is a single contracting entity from which participants provide their expertise. Such a framework promises reduced complexity and seamless integration among multi-disciplined professionals involved in construction projects – architects, structural engineers, civil engineers, mechanical engineers, electrical engineers, building services engineers, project managers, planners, quantity surveyors, land surveyors and so on. A framework of system development is needed to allow sharing of knowledge with organisations and also externally with other organisations to allow merging of skills to deliver the integrated project using the single model for the entire life of the project to meet clients' needs and expectations.

The idea and thought of integrating IT into the construction practice started since the late 1970s. It was not until the early 1980s that computers became more accessible to the public with the innovation of the personal computer. In the year 1980, it was reported that The Liang Group (UK) has been developing and using a wide range in-house computing applications for use within the construction industry where the applications are processed with a carefully develop framework of computer hardware and software which provide a high degree of flexibility and compatibility. The microcomputer was also studied in detail for the application in the construction industry. The construction industry was one of the first industries to explore the computing sector to be integrated into the construction practice (Retik and Langford, 2001). However, the integration process of IT and construction did not 'take off' within the industry over the years.

The construction industry has a multi-party nature, with its project success relaying heavily on timely transfer and availability of information among the parties involved in the project. Therefore, communication is crucial in the construction industry – communication between the multi-discipline construction team members is the core of a successful construction project (Emmitt and Gorse, 2009; Lee and Yu, 2012). The construction industry is also one of the most information-dependent industries, among others, with its diversity of forms of information which include detailed drawings and photos, cost analysis sheets, budget reports, risk analysis charts, contract documents, planning schedules, bills of quantities and many more.

The great advancement in IT and the availability of a wide range of software is changing the industry – the option of borderless communication and changed the world into a 'virtual village' (Dehlin and Olofsson, 2008; Peansupap and Walker, 2006; Virkar, 2015). The Internet, a new member of IT, offers a medium with new opportunities in communication and new methods of project management, tendering and other processes in the construction process – new terms such as e-tendering started to emerge. Thus, creating a new environment, the 'collaborative environment'.

8.2.2 *Integration of IT*

The integration of IT and the construction industry is a major issue in the UK construction industry. In the past, the Department of Trade and Industry (DTI) of the UK government launched various schemes and initiatives to promote integration and use of IT in the construction industry. Steps taken by DTI are such as launching forums, discussions, business initiatives and establishing educational excellence programmes – The IT Construction Forum, Constructing Excellence, Technology Watch, Strategic Forum, Centre for Education in the Built Environment and many more. The government sector is urging the private sector and academia to be involved and play a more active role in transforming the current traditional and fragmented practices to the advanced, modern and more efficient method of construction process – by using and integrating IT in construction organisations. Today, this initiative it taken over by the Department of Business, Innovation and Skills (BIS), whom now launched the Strategic Forum for Construction, Constructing Excellence, BIM Task Group and many others. In order to define what IT integration in construction projects is, there are several key dimensions that reflect the level of IT integration achieved by the organisation and projects. These key dimensions will be discussed and elaborated – information sharing and reuse, project modelling, level of e-business implementation, levels of team integration, design process and information management.

Information sharing and reuse is the starting point based on traditional paper-based information. Information sharing by paper is limited – time consuming for sharing, costly to transmit to other parties and difficult to reuse. The electronic paper overcomes this transmission problem. However, if the receiving system cannot read or interpret the information, problems will arise. With the current establishment of information exchange standards, information becomes more standardise and information reuse becomes greater. Higher levels of information exchange require sophisticated systems to manage and control of information flow. There are various tools to further improve this process – interoperability between systems to ensure all IT systems 'speak' the same language' Other benefits include better efficiency in data interchange for greater speed and fewer errors. There are also improved sharing of information to enable the project team to work more efficiency with up-to-date information resulting in less waste and less duplication of work within the team. The speed of design will increase, leading to faster procurement processes, more completeness at tendering stage and more accurate tenders. The improved sharing or information with the speed of communication enables designs to be 'right first time' and provide a more accurate model for the project. This also feeds through to facilities management where significant benefits can be accrued from more accurate as-build information.

8.2.2.1 Drivers of integration

Organisations in the construction industry have incorporated the use of computers and computing technology in business processes. The main aim and driver of computing usage is to incorporate better efficiency, speed, accuracy and effectiveness in everyday business processes and management (Eadie *et al.*, 2010). However, the integration of IT and the construction industry is slowly accelerating as more and more organisations realise the power of the technology and the massive benefits reaped out of it. There are many drivers in the integration of IT and construction within the organisation and within the project, as well as from the people, process and technology perspective – the main points are further discussed.

The main benefit of IT integration is the ability of information sharing and reuse. The construction industry is an information intensive industry and information is vital in business processes. The ability to share and reuse information overcomes various problems in accuracy, reliability, cost and much more. The establishments of standard data exchange made the ability to share and reuse information with IT is very important and attractive in the construction industry (Aziz and Salleh, 2013; Lou and Alshawi, 2009).

Information exchange standards are the core to IT integration. The ability to share, reuse, publish, download, upload and other exchange processes starts from the exchange standards. Today, the only barrier for information exchange is the incompatibility of systems, either hardware or software or both. For example, drawings in Autodesk Revit could not directly be opened or edited in other software, unless a common platform or technology is used. This prevents systems from 'talking' to each other and exchange of information and data is virtually impossible. The non-existence in standards for data and information exchange is caused by the different developing standards in the fragmented industry. This is also hampered by low commercial interest for system development. Therefore, the common and standard data exchange protocol is vital. The development of the current Industry Foundation Classes version 4 (buildingSMART, 2016) and the UK PAS1192 series (BSI, 2013) is now leading towards an industry-wide standard for data interchange.

The size of the project also influences the growth of integration. With the increase in project size, there are more resources, thus, larger investment could be invested in more sophisticated computing systems and approaches. This could be justified with improved productivity and performance over the life of the project. Larger projects also usually mean that larger and more advanced organisations involved are able to gather and assemble greater depth and system experience among personnel involved. Construction project processes must be scrutinised and made as a standard as a guideline and for other organisations to follow. In this manner, organisations must work together in partnerships and create a 'win-win' culture of. In this environment, it is possible to achieve improved collaboration and the mutual resolution, including IT solutions between organisations. There will be also greater investment in common systems, equitable and transparent sharing of benefits and cost reductions, better sharing of information and also getting sub-contractors into the IT collaboration circle.

Another driver to IT integration is the project information management strategy. Organisations working together could specify and identify the computing and technical specification for data interchange early in the project life. The project information strategy seeks to generate an agreement on how a single project will produce, exchange and manage its information so other participants in the project could benefit from improved sharing and reuse of information.

The use of standard computing products and components could easily create integration in IT. Standard products and components result from the establishment and adoption of common industry standards and do not require reinvention of protocols, saving cost and time. This also incorporates suppliers' experience into the product, improving buildability and whole life costs and integrating the supply chain.

Initiatives have been taken by the UK government with the publication of influential reports in the UK construction industry – starting with the The Egan Report (Egan, 1998) and The Latham Report (Latham, 1994). The Egan Report on Rethinking Construction stresses the need for the industry to change – the culture and structure of the industry as the drivers to improvements in efficiency, quality and safety. The Latham Report highlights the importance of the team approach and invests in high-quality training in order to strive towards cost reductions. More recently, the UK Government Industry Strategy (HM Government, 2012) and the National BIM Report (NBS, 2015) highlighted the importance of IT, specifically through building information modelling, as the way forward for the construction industry. The pressures from globalisation and international competition are increasing. The construction industry in the UK today must compete and battle with other construction industries in the world. World economic factors, such as the emergence of the EC single market and GATT agreements could further erode the UK construction industry in Europe. The emergence of communication technology makes the world a closer place. Dramatic improvements in IT cost, infrastructure and performance are leading the changes in organisational strategy, structure, process, distribution channels and work.

8.2.2.2 Barriers to integration

The construction industry does not change as fast as the computing industry – this result often in inertia in all construction related areas. As for that, today's construction and engineering educational institution do not always equip graduates with the IT skills required operating modern IT system available in the market. Moreover, the pace of IT development makes it very difficult for practicing managers and engineers to keep up with a wide range of tools, methodology and approaches. Also, many senior managerial staff often lacks formal education in computing and information sciences, making it difficult for them to follow trends and cope with day-to-day strategic decision-making. On the other hand, information explosion, globalisation, international competition and new procurement techniques demand solutions that can only be achieved by proper information management. Harnessing IT to take a competitive advantage over competitors is a crucial goal and vital task for many organisations today (Bharadwaj *et al.*, 2013; Sakas *et al.*, 2014).

Barriers to effective IT integration could start even the project have as not begin as there are uncertainties from project inception. The project inception stage is where uncertainties and confusions are at its peak within the project team. Some clients believe that many construction professionals do not fully appreciate the use of IT in construction. The nature of the UK construction industry of being fragmented and traditional does not help in this cause. Virtually, there is a lack of awareness of IT integration within the project team. Projects by their very nature are unique events with unique combination of players and procedures. The consequence of this situation is that it is difficult to find a consistent method of measurable indicators of success on projects. As such, it is impossible to quantity the benefits of IT systems therefore the awareness remains low. In the absence of clear measurable benefits, disciplines such as information systems integration, which may involve many project members, are perceived difficult 'politically' and attract few supporters. This therefore brings the situation where IT project integration often occurs in isolated packets pioneered by enthusiast with a sympathetic client or where a team of advanced IT approach, come together on a project and persuade the client to go this route.

Poor cross-disciplinary communication – a typical project consists of many sub-processes that are carried out by different professionals at different locations. Poor communication has often been identified as a bottleneck for performance improvement and it also re-enforces the confrontational and blaming culture, which is very common in the construction industry. The fragmented supply chain in the construction industry is another barrier for integration. The construction industry consists of hundreds of thousands of firms, of which 90% are small and medium scaled organisations with less than ten employees. Most construction projects undertaken by industrial partners often involve an ad hoc team of fifteen to twenty of small and medium scaled organisations located at disperse places. There is also a lack of industry standards for information interchange. Effective information exchange is as much a prerequisite for a successful conclusion to the construction process as it is for any other business functions. Electronic data exchange involves the computer-to-computer transfer of structured data. Due to the lack of standards for information storage and communication, electronic information exchange between computer programs, tasks and enterprises are still uncommon. There is also a problem with the lack of process transparency. Due to the geographical and organisational behaviours, different project partners often repeat many processes during design and construction. The lack of transparency means knowledge not being shared, thus, resulting in waste of knowledge.

Poor knowledge management are apparent in the industry, enterprise and project levels. Construction is a project based process – a false believe is that because each project is unique, knowledge gained in one project cannot be reused in other projects. Better knowledge management and knowledge capture techniques would enable better re-use, storage and distribution of knowledge at project, enterprise and industry levels. Matters are made worst if the project team have no IT experience or low IT literacy.

Other matters that arises are the technical issues in computing technology – hardware and software compatibility. The quick revolution of computing technologies today is making the hardware and software manufacturers compete among each other. This environment also brought the problem of hardware and software incompatibility. Computer operating systems (Windows, Macintosh, Linux etc.) have different versions of hardware and software requirements. Common standards and protocols are required for different types of computer platforms, operating systems and software applications. This could solve problems concerning aging computers, operating system incompatibilities, aging applications and some mystery files or personally made files.

8.2.3 Extranets

Extranets present a standard platform for all parties involved for communication, data and information exchange, data storage and replication, archiving and much more. Most of all, they initiate a drive for IT integration through data and information interchange and reuse.

Extranets were first introduced in the 2000s with the wider availability of Internet access, which is starting to transform them from expensive, difficult-to-use bits of complex software into cheap and user-friendly business tools. Extranets also deliver a complete integration of information service for asset owners and operators, capturing all information associated with the respective projects or programmes, such as progress reports, up-to-date drawings, risk assessments and others. Through this virtual environment, owner and project managers could operate and maintain facilities more efficiently. One of the main advantages of using extranets is that they ensure that all members of the project team have access to the most up-to-date versions of the various project documents. This means that traditional mistakes generated from someone working from an old document or drawing are removed in theory or at the very least reduced. More crucially project collaboration extranets reduce the opportunity for mistakes and disputes, the biggest causes of waste and inefficiency in construction. Coincidentally the system seem to embrace the non-adversarial ways of working, emulating the Latham (1994) and Egan (1998) reports, may go much further in reducing costs and disputes (Robinson and Udeaja, 2015).

Extranets have the opportunity to significantly improve the way the construction industry works, without needing to make real changes in the structure or practice of the industry. The better tools seek to simplify what good project teams are already trying to do and because they are automated, remove the bureaucracy that often comes with modern working practices (Bin Zakaria *et al.*, 2013; Singleton and Cormican, 2013).

The concepts, benefits and prospects of the use of extranets for construction were quickly identified by organisations in the industry. The industry was beginning to enjoy the benefit of computing technology and the Internet – borderless and unrestricted communication and access to information online 24 hours a day, 7 days a week. Information and data could be exchanged and reused easily among members of the project team. There were massive cost savings in paperwork, communications, courier and postal services. The project team could be seen to collaborate more efficiently and professionally with extranets. With strong collaboration and partnership ties among the consultant team and the construction team, thus reducing design, construction and legislative complications.

The importance of construction project management was recognised by the industry in the Latham and Egan reports. Considering a generic project, the status changes from 'an idea' or 'a concept', going though feasibility studies, execution and finally completion. However, today's construction projects are more complicated than ever before. Construction projects are prone to delays, design errors, material complications, unpredictable weather conditions and many more. Problems could also arise within the management of the construction project and external factors such as new legislations and laws. Project management has taken priority in construction projects – realised as the main path to a successful project. Therefore, there is more focus on project management than ever before. The introduction of the extranets is one major step towards a more organised, effective and intelligent project management tool.

Various extranets software vendors provide extranet applications to support the lifecycle requirements of construction projects. Some examples of vendors (or Software as a Service Providers – SaaS) include BIW Technologies, 4Projects and BuildOnline, among others.

Through the years, extranets have changed everyday business process and the human resource of construction organisations. Extranets were known as 'web-enabled project management' tools whereby the software system is used primarily for data exchange and project management. There were issues of electronic data interchanging standards, types and methods of data interchange, people acceptance and construction process changes. There were also questions about the introduction of automation into management practices, the lack of software integration, electronic versus culture and the lack of integration in the supply chain.

The current collaboration practices in the UK construction industry are heavily document oriented which are in transition to be transformed towards integrated model centric collaboration to fulfil the BIM collaboration requirements. The available collaboration systems, such as project extranets, have significantly improved document collaboration in recent years; however their capabilities for model collaboration are limited and do not support the complex requirements of BIM collaboration. At present, a wide range of BIM applications are available to create intelligent building information models that have improved the visualisation, coordination and management of project life-cycle information in the construction industry. Different industry roles are using BIM tools to create discipline specific building information models, where coordination is limited to visualization and clash detection. However, this situation is improving with the emergence of model collaboration systems (MCS), such as model servers, with the ability to exploit and reuse information directly from the models to extend the current intra-disciplinary collaboration towards multi-disciplinary collaboration. A model server is a type of database system built upon a set of server applications that host model data and allows multiple users to perform collaboration operations on model data using a common platform (Jørgensen *et al.*, 2008). A BIM hosting model server is expected to facilitate exchange of information in a multi-model environment supporting the various applications involved in a building project life-cycle including design tools, analysis tools, document management systems, facility management tools, and so on (Singh *et al.*, 2011).

Managerial practices in industrial relationships between companies have been changing continuously within supply chains in all industries for a long time. In recent years, the wide set of new Internet based tools, such as electronic catalogues, electronic auctions, virtual exchanges and collaboration platforms, contributed to those changes by offering many different opportunities for companies to improve their performances. The manufacturing industry relays much on effective supply chain management, materials product control and inventory control; thus, industrial relations between various different organisations are vital. Communication and trust is the core a successful partnership along the supply chain. The Internet and computing technologies once more presents the perfect solution for the manufacturing industry – simple, cheap and effective way of communication, secured and trustworthy communication networks and archiving tools to ensure contractual promises and terms are met with.

The principles and benefits extranets are slowly being adapted by other industries. The rapid progress of information and network technologies is now changing the situation of the worldwide market competition. Globalisation is becoming a trend for both present and future industrial practice. One of the factors to go to success under this competitive environment is the ability to quickly respond to market and provide the high-quality and low-cost product manufacturing without the limitation of working sites. Unfortunately, the traditional manufacturing modes and enterprises cannot satisfy these requirements well. Therefore, these organisations face a fatal revolution, which often deals with changing their running modes, re-configuring their resources and organisational structures and so on. One solution to meet the needs of this revolution is to establish a web-based online manufacturing mode that is market-driven and has a fast responsive capability to market.

Over the last few years, companies to communicate with their customers, to advertise their products and even to offer online transactions, have extensively used the Internet. To compete effectively, however, these companies have to implement solutions that will allow them to interactively communicate information related to product design, development and manufacturing within their own infrastructures. Internet-based software solutions could offer scalability, easier implementation and compatibility across diverse information technology platforms and thus reduce incremental infrastructure investments. Furthermore, this would allow companies to cut product development and manufacturing costs and greatly increase collaboration among partners internally and externally.

8.3.1 Extranets for other industries

The idea of extranets could have well started within the automobile industry, with the Japanese carmaker, Toyota, as the leader in this field. The manufacturing industry soon began to use Toyota's long-standing and proven principles in lean construction and supply chain management strategies. The supply chain management and relations must be well organised and managed as to achieve 'just in time' and quality materials management. To manage lean construction and a long list of supply chain suppliers, a simple yet effective tool must be created for this purpose. Solutions from the information technology and computing industry presented the solution for communication and information sharing and reuse within the supply chain. The IT solutions were also cost effective and reduce the volume in paperwork, thus pushing up profits. This was the start of the idea and concept of for extranets. The principles and concepts of supply chain management and lean construction was the core for extranets. The Japanese carmaker, Toyota is credited with several innovations in automotive manufacturing including 'kaizen' and 'kanban' – supply chain management and inventory control, including just-in-time as well as quality processes were pioneered by Toyota over the decades. Therefore, the extranets in the automobile manufacturing industry focuses on the supply chain management, inventory control and the lean construction process.

8.3.2 Challenges in construction

Extranets is an extension of an organisation's intranet that allows limited, controlled and secured access between an organisation's internal networks and assigned, authenticated users from remote locations. The most common business applications give suppliers, customers, partners and remote employees' access to internal applications and live databases for safe communication, collaboration and commerce. Industry analysts predict that project extranets will overtake electronic data interchange (EDI) in the next 5 years. Extranets have the same advantages as intranets in terms of platform independence, scalability and low cost of set-up and ownership. This is a big advantage, especially for smaller companies that often could not justify the cost of privately owned, value-added networks. While many businesses have embraced Internet technologies such as Websites and e-mail, some small and medium scale enterprises have not seen any return on their investment, other than the status of having their own web presence. The growing use of extranets, especially for electronic commerce, may change this situation, especially in the construction industry.

Good communication is vital to the procurement and consultation process and electronic communication has revolutionised the means of communication available. However, the take up of electronic communication as part of the procurement process in the UK Construction Industry is a step forward towards a more efficient and economic means of communication between members of the construction project team and communication between construction organisations – better collaboration.

In the 1990s, commonly used data transfer solutions combined the Internet and file transfer protocol (FTP) technologies. Today, there are plenty of commercial applications that can be used for multi-partner projects and are based on dynamic Internet technology or Web2.0. Currently, a centrally accessible information system usually consists of a server to which all project members have the option to access and a common user interface. A number of commercial tools have been created for document management, project information sharing, online communication, design workflow, construction workflow, time control and securing information. These commercial applications have been adopted by industry at a remarkable rate and continue to become more and more commonplace. However, the most widely used application features are fairly simple; in addition, today's applications serve well as centralised document management systems with some integrated work flow models, communication tools, and possibly links to other services, such as printing.

Business-to-Business (B2B) technologies pre-date the early days of the Internet – existed for at least as long as the Internet. B2B applications were among the first to take advantage of advances in computer networking. The EDI business standard is an illustration of such an early adoption of the advances in computer networking. The ubiquity and the affordability of the Internet have made it possible for the masses of businesses to automate their B2B interactions. However, several issues related to scale, content exchange, autonomy, heterogeneity and other issues still need to be addressed.

Project-specific Websites and extranets may present the biggest change to how UK construction companies conduct their day-to-day business. These systems promise to reduce paper consumption, lower costs, improve communications and help to assist in meeting project deadlines. Such technology is a welcome development for the UK construction industry, where postal deliveries amount to a small fortune, where mistakes frequently result from use of out-dated drawings and where projects may involve large numbers of participants scattered all over the country or continent.

Many companies have been using computerised systems for years to manage and schedule projects. However, today's project Websites and extranets claim to provide greater opportunity for consistent document review, multi-party collaboration and expanded communications, both on-site and in the office. For example, companies can post drawings and documents on the system so everyone can easily access and share the latest changes and additions. With most programmes (known as interactive collaboration), users can mark up documents online without changing the original drawings, allowing for resolution of design and engineering queries on-site without the need for CAD/BIM software installed in their offices.

Some online applications are more complex; combining the interactive collaboration features with a workflow tracker that posts and records communications and other documents between architects, engineers, contractors and subcontractors. These systems allow for quick responses to requests for information and change orders, streamlining the site processes, preventing disputes and even moving forward the project's deadlines. Furthermore, the Website or extranet becomes a common depository for communications, creating an accurate and comprehensive virtual audit trail for the project.

Corporations are today becoming largely distributed and deeply founded on networking technology allowing employees to share and access information in different locations. Meanwhile, computer-based information systems, such as Autodesk Revit and Microsoft Project, have become the spinal chord of modern enterprises and new appropriate information tools satisfying fast reactive business requirements and offering a strategic corporate advantage are occurring (Haque and Rahman, 2009). Now, virtual environments or extranets bind everything together – a fragmented and geographically spread set of partners collaborating together. For this, there is still a need for new powerful frameworks to support business models, concurrent enterprising, access to large corporate data sources and multimedia information management, within intranets, extranets and even the Internet. The vital information for the future business of companies must be easily accessed and manipulated in a safe and comprehensive way by multiple actor-oriented applications, thereby satisfying the need for improved customer service, on-time delivery, quality management and project coordination. The development for extranets is increasing in complexity along with an intensifying market competition. The software developments for the construction industry based extranets are in the same boat as the industry itself – fragmented development. The software is developed in individual islands and in isolation. These led to the scenario of the lack in interoperability, no standards and systems not able to 'talk' to each other.

The literature and case study evidences on collaboration motives and benefits suggests that implementation of extranets has inherited difficulties; therefore the amount of effort involved in integrating operational, tactical or strategic levels of separate companies are usually large (Bharadwaj *et al.*, 2013; Robinson and Udeaja, 2015; Sakas *et al.*, 2014; Singleton and Cormican, 2013). However, the same literature also reports that the benefits of such collaboration are considered to be significant. Collaboration should lead to a win-win situation for all parties concerned, participating partners as well as the end-customer. There is no doubt that the motivation for each individual enterprise comes not from the fact that they want to collaborate but from the fact that there are economic advantages to be gained through collaboration.

8.3.3 *Implementation challenges*

For a global organisation, forming close relationships with external organisations and finding an efficient method for teams worldwide to work together has never been so important. At a time when to achieve 'more with less' appears to be a core objective for most businesses, the need for effective collaboration and communication processes is greater than ever. Traditional methods of managing projects have demanded a radical rethink in recent years. Corporate projects have become more decentralised in the way they are managed and projects are becoming more complex and sophisticated. The number of people involved in projects has continued to grow and delivery times continued to decrease. Ways of managing projects need to reflect these changes. To reduce risk and ensure successful project delivery, it is of paramount importance that the whole project team has a common view of the current state of the project and that each team members must know what their responsibilities are. One solution for this is the extranet tool.

Implementation of IT systems and extranets tools must be process-led and not technology-led. Process-led will mean that organisations must identify, learn, re-engineer or recognise processes within the organisation to be completed and computerised by IT. Organisations must not implement IT into processes and force processes to change and suite IT applications – this will eventually fail and lead to a complete organisational meltdown. Implementation challenges include the implementation of extranet software process, timing of implementation, content management solutions, security and confidentiality issues. Effective communication is critical to completing built environment projects on time, ensuring that they remain under budget and maintaining high quality (Angeles, 2001; Ruikar *et al.*, 2005; Wang, 2014; Vlosky *et al.*, 2000). Project collaboration software improves design and construction processes through functions that expedite project communications, reviews and approvals; reduce errors and mitigate risk; improve work practices; and accelerate project processes from start to finish.

The extranet software, like any other software, must be implemented in stages and in carefully selected stages in accordance to the organisation's capabilities to adapt change. Different stages, processes and tests of implementation must be decided by system analysts – the prototype approach, linear sequential (waterfall method), the incremental model, the spiral model or the concurrent development model. Test plans, such as module testing, integrated testing and acceptance testing, must be carried out to ensure the persistence and capabilities of the system. Organisations might also have conversion plans, to convert from an older version to a newer version. Conversion plans include procedure conversion, file conversion and system conversion. System conversion plans include the parallel, direct, phased or prototype techniques.

One of the key features of any successful extranet is that it must enable relevant employees to be able to publish information straight on to a Website from 'native file format', such as Word or Excel, in which the information was created. Unfortunately, many solutions do not give this flexibility, resulting in information being held in a logjam, waiting for a web master to publish it into structured format, such as HTML or XML, on to the Website. Time is critical, if information (such as financial results or news) is delayed by days or even hours, the value of it being published can be greatly diminished with profound implications for an organisation's entire web strategy. Another key issue to consider when implementing a content management solution is the speed of implementation. Many extranet solutions require extensive consultancy work, which can take months to properly implement. This will add to implementation costs increase. Content management is set to propel all organisations that depend upon instant and secure flows of information, into the on-line economy. It can provide the framework for real time information, the exchange of ideas; eliminate the unnecessary duplication of effort or resources; provides a powerful way to distribute information both internally and externally on any device. Using the right technology and the right implementation techniques, the reality of this kind of knowledge-enhanced organisation can be achieved today.

Basic network security issues have changed very little over the past decade. Protecting the confidentiality of corporate information, preventing unauthorised access, and defending against external attacks remain primary concerns for the extranet. What has changed, however, are new technologies and business practices that make these old concerns a far more formidable challenge. Deployment of extranets and wireless local area networks (WLANs) are turning networks inside out from a security perspective. Smarter, deadlier Web-enabled worms and viruses are launching attacks from within networks. Disgruntled and dishonest employees are becoming more computer-savvy and capable of perpetrating mischievous and illegal acts. Threats can come from anywhere. To combat these escalated threats, organisations must find better ways to resolve these vulnerability issues and find new ways to address these escalating threats. Security solutions that are sufficiently flexible and scalable to protect against attacks from all sources, whether internal or external and can easily adapt to the security requirements of emerging technologies are the only answer: providing more granular levels of permission to network resources and allow attack protection to be available anywhere inside or outside the network. New support for multiple security zones and multiple physical interfaces deliver added control and segmentation. Physical interfaces that can all independently have firewall and denial-of-service protections activated. Virtual Private Network (VPN) tunnels that can be terminated to any device, enabling extranets and WLANs to be supported more easily. However, these new security measures must not impact the performance of the network. One of the fascinating things about the Internet is that nobody has had time to establish any rules about design or content management. It has all happened too quickly, the result is 'complete confusion'.

8.4 e-Tendering

The tendering process in the construction industry is deemed to be the most critical and important throughout the life of the construction project. The tendering stages in the construction project will shape the contractual and legislative agreements between the client, consultant team, contractor and other members in the construction project. Therefore, the tendering stages are regarded as most important (Brook, 2012; Elhag *et al.*, 2005; Eriksson and Westerberg, 2011; Knudsen, 2003).

Tendering in the UK construction industry is information incentive and involves much paperwork. Tender documents include the invitation to tender, form of tender, architectural drawings, Bills of Quantities, health and safety agreements and others. Tender documents are paper incentive, not portable and expensive, tedious and troublesome to produce). The introduction of electronic tendering solves all the problems. e-Tenders are not paper-intensive but software- (word processing, spread sheets and drawings) intensive and have editing capabilities. e-Tenders are also portable, inexpensive and simple to compile. The main attraction of e-tenders is the capability of sending it via the Internet through a secure and effective manner. Internet and communication technology today presents the perfect platform for e-tendering processes. e-Tenders could be sent to prospective contractors in an effective, quick and accurate manner. The Internet and IT platform are also cheap, economic and cost-effective to set up and maintained throughout the project life and beyond. Secured networks could be prepared to ensure security, confidentiality and integrity of the tenders sent out. The overall process is also transparent and free from 'red tape'. However, there could be some problems and hitches in the compatibility and interpretability of the hardware and software used among members of the construction project. Problems may also arise in the scalability and reliability of the overall system. Issues such as data exchange standards are much in discussion. Examples of e-tendering systems are such as Sarcophagus, Prologic e-Procurement, BravoSolution, In-Tend, and many more.

e-Tendering is not new. The construction industry, being fragmented and traditional, does not accept new concepts and new workflow frameworks instantly. The industry has been introduced to this concept in the early 2000s and it is slowly being accepted in the industry. Although the concept of electronic tendering was introduced in the mid-1990s, the construction industry could not adapt nor use existing working frameworks from other industries, such as the manufacturing industry. It was not until the uptake of the extranets that the e-tendering concept is being accepted. To accelerate the change and acceptance, vendors and developers of extranets are trying to absorb e-tendering as a module into the current and existing extranets.

The e-tendering environment in the UK is still in its preliminary stages – a way of managing the entire contract letting process electronically. All the documentation is distributed to tenderers via a secure web-based system. This avoids the need for collating all the paperwork and sending it by courier or post. In due course the tenderers return bids the same way. During the tender period, updates and queries are exchanged through the same IT system. This means that all the key information is at the fingertips of those involved in the project team. The client or the consultant team uploads the notice and/or invitation to tender onto the system. Notification is sent out and the suppliers download the information and complete their responses. The purchaser can access the tenders only after the deadline has passed. All the information is held in a central database – easily searchable and can be fully audited, with all activities recorded. It is essential that tender documents cannot be read or submitted by an unauthorised party. Users of the system must be properly identified and registered, with controlled access to the service. Security has to be as good as or better than in manual systems. Data is encrypted and users are authenticated by means such as digital signatures, electronic certificates or smartcards. It is also important to ensure that neither party can deny having sent or received tender documents. All parties must be assured that no undetected alterations can make to any tender. The tenderer is able to amend the bid right up to the deadline – while the client cannot get access until the deadline has passed. The client has the facilities to assess the bids and the extranets software generally contains tools that help in comparison of the tenders.

The UK government is one of many key clients spearheading the adoption of e-tendering, through the Office of Government Commerce. A number of initiatives and pilot schemes are under way. These are aimed at developing its use whilst addressing the issues of security, ease of use, compatibility and legality.

8.4.1 e-Tendering in other industries

Tendering is a method of entering into a sales contract. It is a long and complex business process and generates a series of contractually related legal liabilities. Today, the tendering process is moving forward by transferring the tendering process to be done electronically. Industries today are changing tendering methods. However, the change is moving quicker in other industries besides the UK construction industry.

Other industries are moving towards e-tendering processes to save time and cost, ensure efficiency and accuracy, also to ensure security and confidentiality of the tender documents. The largest enforcer for e-tendering will be governments all over the world. E-government is the use of information technology to support government operations, engage citizens and provide government services. Governments are encouraging and persuading organisations to change and use the e-tendering process to ensure better productivity and efficiency in organisations in the private sector. The government is changing internally to engage in e-tendering activities and to help the private sector to embrace this change. In some cases, the government enforces regulations to ensure bidders for tenders to have e-tendering facilities before the bidding process begins. This therefore, creates the air to change, although some not willing, to adopt change or risk losing out. e-Tendering in local governments is largely influenced and implemented to support government operations and provide government services to citizens. The e-tendering process between the local governments and the private sector is vital as governments could accelerate or delay changes in all industries. Therefore, it could be concluded that the government is the driving force in e-tendering implementations across all industries.

The manufacturing sector involves complex process planning and logistics that require a more sophisticated solution than early procurement systems have delivered. Through the Internet, manufacturers, suppliers and distributors are shrinking costs, ramping collaboration and speeding up processes along the entire value chain across distributed manufacturing industries. One of the most compelling reasons for the manufacturing industry to adopt e-tendering solutions is the reduction in cost and process cycle time. Manufacturing encourages direct and indirect purchase of material in huge quantities and this is where enormous savings are found. Supply chain management in many ways controls the manufacturing process. This streamlines the surge of goods from suppliers to manufacturers, distributors and finally customers. e-Tendering could offer solutions and services to reduce inefficient and maverick buying; revamp redundant processes, non-strategic sourcing and all other ailments of imperfect procurement practices. Manufacturing companies often need a wide range of software and hardware and also skilled personnel to manage and support their infrastructure. Outsourcing the tendering process by electronic processes is the most cost-effective means of handling the costly turnover. e-Tendering could also lower manufacturer's costs of purchasing goods and materials through systematic outsourcing, as well as streamlining the order to invoice procedures.

The textile industry is one of the oldest existing industries and one that has progressed and kept pace with technology advancements and changing market conditions. A consistent effort towards extensive market coverage, improving technical capabilities and an attractive and extensive merchandise, has paid rich dividends to this industry. The textile industry has taken a step further by adopting e-tendering solutions to shorten procurement cycle time and to reduce costs. The e-tendering process makes procurement in the textile industry more systematic and convenient. Benefits include shorter sourcing cycle, accurate comparisons of bids, ability to negotiate dynamically and guarantees complete confidentiality. In addition to that, e-tendering facilities are extremely simple to use and requires no additional investment in hardware or software beyond a computer and an Internet connection. e-Tendering could also drastically reduce the time involved in negotiating cycles; advantages include the suppliers' ability to emphasise on the value they can provide.

8.4.2 e-Tendering challenges

The continuing expansion of electronic business provides opportunities for improved business processes, which are more efficient and responsive, reduce the reliance on paper transactions and lead to reduced costs. e-Tendering is one such opportunity.

The electronic exchange of information has been normal practice throughout the construction industry for some time, indeed tender documentation has often been supported by the use of USB drives, CD-ROMs or even, in some cases, e-mails. In the context of this dissertation, electronic tendering is more fundamental, namely, 'the electronic conduct of tender exercises from advertisement through to contract placement, including the exchange of all relevant documentations for tender submissions'. The benefits of e-tendering include the modernisation of working processes and improved efficiency in the way people work. e-Tendering could also lead to improved commercial relationships with suppliers and reduce costs for suppliers. On the whole, e-tendering could improve the ability to manage supply chain more efficiently, to be more cost effective and ensure better productivity.

e-Tendering could be defined as the automation of the steps involved in the tendering process. It deals with the electronic preparation and exchange of tender documents and includes inviting, receiving and evaluating offers from suppliers. e-Tendering is part of a set of 'Best of Breed' procurement tools that enable purchasing professionals to operate more efficiently and effectively. There are fundamentally two distinct types of e-tendering, firstly and in its most basic format a non-dynamic solution that relies on users maintaining information and data in their current format and the e-tendering system simply acting as an email system to move documents. In its more complex and dynamic form, the e-tendering solution is intelligent and can become involved in scoring and tender appraisal. It is acknowledged that there is an internal market for both variants. The potential of e-tendering will only be realised when the points below are acknowledged and appreciated by the administrators and users of the e-procurement and e-tendering systems (Arslan *et al.*, 2006; Diabagate *et al.*, 2015; Du, 2009; Lavelle and Bardon, 2009; Lou and Alshawi, 2009; Vaidya *et al.*, 2006):

- Facilities are focused most on meeting the recipient's or user's individual requirements and not just on the sender's need to communicate. The wealth of information now available electronically results in a poverty of attention unless it is well structured.
- There is high speed, low cost access. Delays and complexities in accessing information result in low use of e-tendering facilities. It is important to recognise that the power of electronic communications to reach most people in most locations is limited at present. The potential capabilities are huge; but a realistic assessment is needed of the facilities required by each end user to make full use of the potential services.
- Data are collected, stored and disseminated the minimum number of times.
- Data need to be consistent and accessible at all key points in each value chain.
- There is a high level of security, and confidence in the safety of applications, data and communications.

The general concept of electronic tendering is the exchange of documentations during the construction project tender stage through electronic methods, for example, through the Internet and other electronic means. e-Tendering is much more than that. Governments all around the world are promoting and pushing 'electronic' methods of transaction among its citizens and the private sector (Ronald and Omwenga, 2015; Turban *et al.*, 2015).

Electronic methods of transactions today reflect the advancement and technological push of a country towards modernisation of working process within the government and with citizens and the private sector. Developing countries like Malaysia are pushing for electronic transactions and IT throughout the government and transactions done with the government. The Malaysian government is placing its hopes for continued growth on a strategy of government-led policies and initiatives aimed at attracting high-end foreign investment and a transition to a knowledge economy. The best-known element of the Malaysian ICT strategy has been the Multimedia Super Corridor (MSC) and Putrajaya, an ultra high-technology business city built outside Kuala Lumpur and now home to more than 540 companies (Hassan and Abu Talib, 2015; Yigitcanlar and Sarimin, 2015).

e-Tendering is currently one of the biggest challenges facing the public sector procurement process in the UK. Although the Government targets stipulate that 50% of government tenders should be electronic by the end of 2001, the Byatt Report (2000) echoed the recommendations of many e-commerce experts with its advice that local government should tread carefully into the e-tendering market. The introduction of an e-tendering system that fails to deliver can be costly not only in operational terms but in the potential loss of tendering competition. Manual tender processes can be long and cumbersome, often taking three months or longer, which is costly for both buyer and supplier organisations. e-Tendering has the ability to replace manual paper-based tender processes with electronically facilitated processes based on best tendering practices to save time and money. Users can reuse (cut and paste) data from the electronic tender documents for easy comparison in spreadsheets. Evaluation tools can provide automation for a more detailed and extensive the comparison process. Suppliers' costs in responding to invitations to tender are also reduced as the tender process cycle is significantly shortened. e-Tendering offers an opportunity for automating most of the tendering process – from help with preparing the tender specification, advertising, tender aggregation, to the evaluation and placing of the contract. In short, the e-tendering presents the opportunity to automate the tendering process from the inception stage to the award of contract to the contractor and also at the post-contract stage if the need arises. Benefits of e-tendering to a business include the following points (Diabagate *et al.*, 2015; Eadie *et al.*, 2010; Khorana *et al.*, 2015; Lavelle and Bardon, 2009; Vaidya *et al.*, 2006):

- Reduced tender cycle-time.
- Fast and accurate pre-qualification and evaluation, which is enabled by automated processes.
- Rejection of suppliers that fail to meet the tender specification.
- Faster response to questions and points of clarification during the tender period.
- Reduction in the labour intensive tasks of receipt, recording and distribution of tender submissions.
- Reduction of the paper trail on tendering exercises, reducing costs to both councils and suppliers.

- Improved audit trail increasing integrity and transparency of the tendering process.
- Improved quality of tender specification and supplier response.
- Provision of quality management information.

e-Tendering is a relatively simple technical solution based around secure e-mail and electronic document management, however, it may differ from one service provider to another. In general terms, it involves uploading tender documents on to a secure Website with secure login, authentication and viewing rules. Tools available in the current market offer varying levels of sophistication. A simple e-tendering solution may be a space on a web server where electronic documents are posted with basic viewing rules. This type of solution is unlikely to provide automated evaluation tools, instead users are able to download tenders to spreadsheet and compare manually but in an electronic format. Such solutions can offer valuable improvements to paper-based tendering. More sophisticated e-tendering systems may include more complex collaboration functionality, allowing numbers of users in different locations to view and edit electronic documents. They may also include e-mail trigger process control, which alerts users for example of a colleague having made changes to extranets to tender or a supplier having posted a tender. The most sophisticated systems may use evaluation functionality to streamline the tender process from start to finish, so that initial invitations to tender documents are very specific and require responses from vendors to be in a particular format. These tools then enable evaluation on strict criteria, which can be completely automated.

8.4.3 e-Tendering in extranets

The tendering process in the UK construction industry is relatively traditional and has not changed since the past decade. As for that, changing the traditional paper-based process to electronic terms is not easily accepted by the industry; the industry is difficult to change. Now, extranets providers are continuously introducing e-tendering solutions and upgrades into existing system. The e-tendering solutions are either provided in the extranets solution or as a separate module.

The extranets industry leaders are now rolling out plans for further expansion of the current system – one expansion is the e-tendering module. Extranets providers are offering a more integrated, secure and confidential Web-based tendering system. The inclusion of the e-tendering module to the current extranets enables a streamline for the once paper-reliant processes of issuing tenders and managing communication with bidders. The whole tendering process will be much faster, more efficient and less expensive. Bidders could use the same system used by the design team to view or download drawings and documents, to submit queries and to receive information about design changes relevant to the tender. During the tender period, tender managers only have to use one system to manage queries and notify changes with the help of a full audit trail. There will be a faster and more efficient method of communication between the project team and the bidders, which could mean shorter tender periods. Submitting tenders will also be easier. The instant the tender is issued, bidders can download and print out all relevant information via the Web. Thus, few anxious last-minute rushes to complete paperwork and organise couriers. Bidders will use the system to submit responses and the final tender more quickly and simpler with complete privacy and security in the knowledge that the tenders can only be seen by the tender manager.

The e-tendering process is relatively similar with the traditional paper-based process, except that it is done electronically via the Internet. A typical e-tendering module in extranets include tender administration, tender issue process, accessing the tender, tender communication and final submission of the tender.

The tender administration stage involves nominating the tender managers, nominating the bidders and assigning access rights to the bidders. The project administrator assigns the rights to the tender manager to publish tenders relating to the project. The project administrator will also assign rights to the potential tender bidders. In some cases, the client may have a shortlist of preferred suppliers who may be invited to tender. In other cases, an invitation to tender will be issued first. Bidders will be then given access rights to the e-tendering system. Crucially, all tender recipients have exactly the same level of restricted access.

The tender issue processes include the creation of tender, selection of potential bidders and issuing the tender. The e-tendering system will create the tender package from various documents complied as the project tender. The tender manager will have final checks and publish the tender. The tender manager will also complete a tender publishing letter, which includes the detail of the title of tender, project description, closing date of submission, specification of tender documents and drawings and other relevant information. The tender manager will select potential bidders or a shortlist of preferred bidders from the client. Once complete, bidders will be issued project tenders via the Internet.

Bidders will access the tendering environment with the access rights given by the tender manager. The tendering environment will include all relevant information about the project, project drawings and documentation to be downloaded, dateline for submission and others.

The major advantage of the e-tendering solution is the tender communications. Bidders can seek clarifications before submitting bids by tender queries. Once the bidder enters a query, the tender manager is notified and a response is given. The queries are done within the e-tendering module, where all communication and interaction is recorded in audit trails. The objective is to record and manage all queries – detailing what information has been requested by whom and when; who, when and by whom the queries were resolved. Moreover, the exchanges are recorded in the system for potential future use and help to deliver better-focused responses, benefiting both the bidder and the client. The tender manager could do tender amendments. The items are published and managed in the same way as issuing the tender.

Finally, the submission of final tender. While preparing the bid, bidders could access all information that has published in the system, allowing them to collaborate in the production of the final tender response. Once the bidder is satisfied with the project tender bid, submission is done via the Internet within the e-tendering module.

8.5 Conclusion

Public and private sector construction reports indicated that the underpinning factor for the construction industry to move forward would be the use of IT. The power on the Internet changed the way we work and introduced myriad technologies for the built environment industry. Despite being the earlier adopters of technology, the industry has been playing catching up. Extranets started as part of a local area network, expanded into a secured area network – and are now only naturally extending their capability into cloud computing to accommodate the demands of the users. The move towards cloud computing will bring another set of challenges in the areas of data security, new cyber legislation and procurement/contract changes. The revolution and evolution of Extranets has now interweaved into our everyday lives. Extranet systems could now be customised to provide a more flexible and secure system for users. The volume of data will need to be transformed from data to information, information to knowledge, knowledge to intelligence for the user. Intelligent extranet systems could also control these applications where content could be automatically filed, indexed and stored.

The state-of-the-art Extranets will be its use beyond the current Internet of Things (IoT) and Industry 4.0 (I4.0). The technology race today is not of being able to create the biggest storage space or fastest processors – it's about how we optimise and use technology to its fullest. The potential of IoT and I4.0 is yet to be fulfilled but legacy of extranets is the underlying technology for the future.

References

Alshawi, M. and Goulding, J. (2008), 'Organisational e-readiness: embracing IT for sustainable competitive advantage', *Construction Innovation: Information, Process, Management*, 8(1), doi:10.1108/ci.2008.33308aaa.001.

Alshawi, M., Goulding, J.S., Khosrowshahi, F., Lou, E.C.W. and Underwood, J. (2008), *Strategic Positioning of IT in Construction - An Industry Leaders' Perspective*, Construct IT for Business, Salford, UK.

Angeles, R. (2001), 'Creating a digital marketplace presence: lessons in extranet implementation', *Internet Research*, 11(2), 167–184.

Arslan, G., Tuncan, M., Birgonul, M.T. and Dikmen, I. (2006), 'E-bidding proposal preparation system for construction projects', *Building and Environment*, 41(10), 1406–1413.

Aziz, N.M. and Salleh, H. (2013), 'Case studies of the human critical success factors in information technology (IT) implementation in Malaysian construction industry', *Journal of Building Performance*, 5(1), 1–9.

Basu, A. and Jarnagin, C. (2008), 'How to Tap IT's Hidden Potential', *Sloan Management Review/Wall Street Journal Business Insights Series, The Wall Street Journal*, 10 March 2008.

Bharadwaj, A., El Sawy, O.A., Pavlou, P.A. and Venkatraman, N. (2013), 'Digital business strategy: toward a next generation of insights', *MIS Quarterly*, 37(2), 471–482.

Bin Zakaria, Z., Mohamed Ali, N., Tarmizi Haron, A., Marshall-Ponting, A.J. and Abd Hamid, Z. (2013), 'Exploring the adoption of Building Information Modelling (BIM) in the Malaysian construction industry: a qualitative approach', *International Journal of Research in Engineering and Technology*, 2(8), 384–395.

Brook, M. (2012), *Estimating and Tendering for Construction Work*, Routledge, Oxon UK.

Bryde, D., Broquetas, M. and Volm, J.M. (2013), 'The project benefits of building information modelling (BIM)', *International Journal of Project Management*, 31(7), 971–980.

BSI. (2013), *Specification for Information Management for the Capital/Delivery Phase of Construction Projects Using Building Information Modelling (PAS 1192)*, British Standards Institute, London.

buildingSMART. (2016), *Industry Foundation Classes Release 4 (IFC4)*, building SMART International, http://www.buildingsmart-tech.org/ifc/IFC4/final/html/ [Accessed 3 January 2016].

Business Wire. (2008), 'Survey cites dissatisfaction with incumbent ERP vendors', *Business Wire*, www.sys-con.com/node/594952 [Accessed 2 January 2013].

Byatt, I. (2000), *A review of local government procurement in England: main report*, Office of the Deputy Prime Minister, The Stationery Office, London.

Cabinet Office (2011), Government Construction Strategy, London: HMSO.

Carty, V. (2010), 'New information communication technologies and grassroots mobilization', *Information, Communication and Society*, 13(2), 155–173.

Dehlin, S. and Olofsson, T. (2008), 'An evaluation model for ICT investments in construction projects', *Journal of Information Technology in Construction, special issue, Case studies of BIM in use*, 13, 343–361.

Diabagate, A., Azmani, A. and El Harzli, M. (2015), 'E-tendering: modeling of a multi-agents system integrating the concepts of ontology and big data', *Global Journal of Computer Science*, 5(2), 80–89.

Du, T.C. (2009), 'Building an automatic e-tendering system on the Semantic Web', *Decision Support Systems*, 47(1), 13–21.

Eadie, R., Perera, S. and Heaney, G. (2010), 'Identification of e-procurement drivers and barriers for UK construction organisations and ranking of these from perspective of quantity surveyors', *Journal of Information Technology in Construction*, 15, 23–43.

Egan, J. (1998). Rethinking construction: The report of the construction task force, London: HMSO.

Elhag, T.M.S., Boussabaine, A.H. and Ballal, T.M.A. (2005), 'Critical determinants of construction tendering costs', *International Journal of Project Management*, 23(7), 538–545.

Emmitt, S. and Gorse, C.A. (2009), *Construction communication*, John Wiley & Sons, Oxford UK.

Eriksson, P.E. and Westerberg, M. (2011), 'Effects of corporative procurement procedures on construction project performance: a conceptual framework', *International Journal of Project Management*, 29(2), 197–208.

Gann, D.M. and Salter, A.J. (2000), 'Innovation in project-based, service-enhanced firms: the construction of complex products and systems', *Research Policy*, 29, 955–972.

Goulding, J.S. and Lou, E.C.W. (2013), 'E-readiness in construction: an incongruous paradigm of variables', *Journal of Architectural Engineering and Design Management*, 9(4), 265–280.

Haque, M.E. and Rahman, M. (2009), Time-space-activity conflict detection using 4D visualisation in multi-storied construction project, *Visual Informatics: Bridging Research and Practice*, Springer, Berlin and Heidelberg, 266–278.

Hassan, I.E. and Abu Talib, N. (2015), 'State-led cluster development initiatives: a brief anecdote of multimedia super corridor', *Journal of Management Development*, 34(5), 524–535.

HM Government. (2012), *Building Information Modelling, Industrial Strategy: Government and Industry in Partnership*, URN 12/1327, HMSO, London.

IET. (2014), Digital engineering and project controls in the construction industry, *The Institution of Engineering and Technology*, http://www.theiet.org/sectors/built-environment/files/Laing-digital-casestudy.cfm [Assessed 19 October 2014].

Internet World Statistics. (2012), Internet usage statistics - the big picture, *Internet World Statistics*, http://www.internetworldstats.com/stats.htm [Accessed 17 January 2012].

Jørgensen, K.A., Skauge, J., Christiansson, P., Svidt, K., Sørensen, K.B. and Mitchell, J. (2008), *Use of IFC Model Servers-Modelling Collaboration Possibilities in Practice*, Department of Production, Aalborg University, Aalborg, Denmark.

Khorana, S., Furguson-Boucher, K. and Kerr, W.A. (2015), 'Governance issues in the EU's e-procurement framework', *Journal of Common Market Studies*, 53(2), 292–310.

Kiviniemi, A. (2011), The effects of integrated BIM in processes and business models, *Distributed Intelligence in Design*, Wiley-Blackwell, Oxford, UK, 125–135.

Knudsen, D. (2003), 'Aligning corporate strategy, procurement strategy and e-procurement tool', *International Journal of Physical Distribution and Logistics Management*, 33(8), 720–734.

Krigsman, M. (2009), IT project failure. Worldwide cost of IT failure: $6.2 trillion, *ZDNet*, http://blogs.zdnet.com/projectfailures/?p=7627&tag=col1;post-7695 [Assessed 19 March 2014].

Krigsman, M. (2010). Understanding Marin County's $30 million ERP failure. ZDNet, Retrieved September 13, 2010, from http://www.zdnet.com/blog/project failures/understanding-marin-countys-30-million-erp-failure/10678?tag=nl.e539

Latham, M. (1994). Constructing the team. London: HMSO.

Lavelle, D. and Bardon, A. (2009), *E-tendering in construction: Time for a change?*, Northumbria Working Paper Series: Interdisciplinary Studies in the Built and Virtual Environment, Newcastle, UK.

Lee, S.K. and Yu, J.H. (2012), 'Success model of project management information system in construction', *Automation in Construction*, 25, 82–93.

Lou, E.C.W., and Alshawi, M. (2009). Critical success factors for e-tendering implementation in construction: People and process issues. ITcon, 14, 98–109.

Lou, E.C.W., Goulding, J.S., Alshawi, M., Khosrowshahi, F. and Underwood, J. (2012), 'Leveraging IT-based competitive advantage: UK industry perspective', *Journal of Architecture, Planning and Construction Management*, 2(1), 27–62.

NBS. (2015), *National BIM Report 2015*, National Building Specification, London.

Peansupap, V. and Walker, D.H. (2006), 'Information communication technology (ICT) implementation constraints: a construction industry perspective', *Engineering, Construction and Architectural Management*, 13(4), 364–379.

Retik, A. and Langford, D. (2001), *Computer Integrated Planning and Design for Construction*, Thomas Telford Limited, London.

Robinson, H. and Udeaja, C. (2015), Reusing knowledge and leveraging technology to reduce design and construction costs, *Design Economics for the Built Environment: Impact of Sustainability on Project Evaluation*, John Wiley & Sons, Chichester, UK, 227–239.

Rodrik, D. (2014), 'The past, present, and future of economic growth', *Challenge*, 57(3), 5–39.

Ronald, N.K. and Omwenga, J.Q. (2015), 'Factors contributing to adoption of e-procurement in county governments: a case study of County Government of Bomet', *Internal Journal of Academic Research in Business and Social Science*, 5(10), 233–239.

Ruikar, K., Anumba, C.J. and Carrillo, P.M. (2005), 'End-user perspectives on use of project extranets in construction organisations', *Engineering, Construction and Architectural Management*, 12(3), 222–235.

Sakas, D., Vlachos, D. and Nasiopoulos, D. (2014), 'Modelling strategic management for the development of competitive advantage, based on technology', *Journal of Systems and Information Technology*, 16(3), 187–209.

Salah, Y. (2003), IT Success and Evaluation: A General Practitioner Model, PhD Thesis, Research Institute for the Built Environment (BuHu), University of Salford, UK.

Saxon, R.G. (2013), Growth through BIM, *Construction Industry Council*, London.

Singh, V., Gu, N. and Wang, X. (2011), 'A theoretical framework of a BIM-based multi-disciplinary collaboration platform', *Automation in Construction*, 20, 134–144.

Singleton, T. and Cormican, K. (2013), 'The influence of technology on the development of partnership relationships in the Irish construction industry', *International Journal of Computer Integrated Manufacturing*, 26(1–2), 19–28.

Thompson, G. (2013), *The Economic Impact of EU Membership on the UK*, House of Commons Library, London.

Turban, E., King, D., Lee, J.K., Liang, T.P. and Turban, T.P. (2015), Overview of electric commerce, *Electronic Commerce*, 8th Edition, Springer International Publishing, Switzerland, 3–49.

Vaidya, K., Sajeev, A.S.M. and Callender, G. (2006), 'Critical factors that influence e-procurement implementation success in the public sector', *Journal of Public Procurement*, 6(1/2), 70.

Virkar, S. (2015), 'Globalisation, the Internet, and the nation-state: a critical analysis', in Sahlin, J.P. (ed.), *Social Media and the Transformation of Interaction in Society*, IGI Global, Hershey PA USA, 51–66.

Vlosky, P.R., Fontenot, R. and Blalock, L. (2000), 'Extranets: impacts on business practices and relationships', *Journal of Business and Industrial Marketing*, 15(6), 438–457.

Wang, M.T. (2014), 'The design and implementation of enterprise management system based on ERP', *Applied Mechanics and Materials*, 655, 6221–6224.

Wirtz, J., Tuzovic, S. and Ehret, M. (2015), 'Global business services: increasing specialization and integration of the world economy as drivers of economic growth', *Journal of Service Management*, 26(4), 565–587.

Won, J., Lee, G., Dossick, C. and Messner, J. (2013), 'Where to focus for successful adoption of building information modeling within organization', *Journal of Construction Engineering and Management*, 139(11), doi:10.1061/(ASCE) CO.1943-7862.0000731.

Wright, J., Brinkley, I. and Clayton, N. (2010), *Employability and Skills in the UK: Redefining the Debate*, The Work Foundation, London.

Yigitcanlar, T. and Sarimin, M. (2015), 'Multimedia Super Corridor, Malaysia: knowledge-based urban development lessons from an emerging economy', *VINE Journal of Information and Knowledge Management Systems*, 45(1), 126–147.

9 Transforming policy documents into intelligent three-dimensional collaboration tools

Alan Redmond and Mustafa Alshawi

9.1 Introduction

The construction industry has a considerable array of silo data relating to materials and health policy documents. However, the problem statement (gap analysis) is the difficulty of transferring this data into knowledge that can assist decision makers in defining and choosing the most efficient (health conscious) materials at the early design stage and retrofitting. In order, to access such data an automated process for maintaining building-related performance criteria would require an assemble store (data modeling) and disseminate (object hyper linking) building and construction-related environmental, health and technical data. Building Information Modeling (BIM – developing a model virtually before building it physically) has the capability to organise the teamwork around a shared digital mock-up where involved parties can retrieve the information they need, perform their tasks and update the mockup with their work. The conceptual analysis is to evaluate published reports and extract key statements as business rules in order to provide BIM objects with intelligent semantic links, i.e. performance metrics of materials or Indoor Air Quality (IAQ) regulations.

The process of identifying business rules is often iterative and heuristic, where rules begin as general statements of policy. In the US, Corporate Social Responsibility (CSR) statements can contribute to a company's policy on community development and environmental projects but have little or no relevance to the firm's core business model. Chandler and Werther (2014), recognizes the 1987 Brundtland Report (named after its main author, Gro Harlem Brundtland Norwegian prime minister and chair of the UN's World Commission on Environment and Development) for identifying the term 'sustainable development' and the importance of sustainability for firms (CSR). However, the popularized term ('meeting the needs of the present without comprising the ability of future generations to meet their own needs') has become vaguely interpreted from the original response to resource utilization particularly the unsustainable rate of utilization. In practice and from a business viewpoint reports increasingly suggest that there is a growing financial risk to corporations not seen to be conducting business in an appropriate manner. A 2009 article in the Harvard Business Review argued that for many firms, 'sustainability is now the key driver of innovation'.

The solution is to clearly identify sustainability related services that offer innovative technologies that can directly impact a client's progress towards sustainability goals. In recent years techniques such as systems analysis have evolved to provide methods for describing many aspects of a business or government agency. The focus of this chapter is to implement techniques such as business rules associated with semantic knowledge in order to identify appropriate environmental policies on the web. These new classes of intelligent web applications through the assistance of Radio Frequency Identification (RFID – image recognition) will investigate and map existing materials through object hyper linking for viewing in a BIM model. For instance; by implementing strategic techniques associated with policy's for example; Veteran Affairs (VA's) health care system (which includes a network of 150 hospitals, 130 nursing homes, 800 community outpatient clinics serving 6.3 million patients; GAO, 2012) such as 'I CARE' framework can be semantically linked to an appropriate ASHRAE (American Society of Heating, Refrigerating and Air-Conditioning Engineers) document in order to successfully refurbish existing health facilities buildings. **Section 6** (methodology) describes the systems physical architecture demonstrating the functional architecture input and out processing, and user interfaces in order to semanticily link policies through a BIM model.

The economical and financial decision benefits of projecting such documents within an open BIM model by exchanging information via BIM XML and Representational State Transfer (REST) 'Systems-of-Systems' (SoS) is presented via an existing test case ('case scenario-based test that employs combinations of test inputs and conditions to verify an item's ability to accept/reject ranges of inputs, perform value-added processing and to produce only acceptable performance base outcome or results' Wasson, 2006). The Cloud-BIM test-case study (featuring California Community College in section 7.1) is highlighted to illustrate the collaborative benefits of web-based BIM service over traditional standalone model for exchanging data. The final section demonstrates via SEPS how technology is becoming transparent and simple to use, and eradicate information silos while pushing for semantic web applications. The overall findings of this chapter highlights the potential environmental and economical decision benefits of projecting health policy documents within an open BIM model. The contribution of this chapter to eBusiness is based on web services APIs for exchanging partial sets of BIM data in real-time with semantic capabilities. By developing agent-based system architecture ('a network of intelligent agents that share facts with other agents and adapt their behavior in response to shared facts' Clymer, 2009) the future of web services will invoke knowledge in the form of rules that transform policy documents into intelligent three-dimensional collaboration tools for engineering projects. The chapter begins with outlining the importance of policy documents in realtion to economic growth and identifies how such models can act as incentive schemes based on energy performance upgrades.

9.2 Sustainability models – services that offer innovative technologies

The International Finance Corporation (IFC – World Bank Group) outlined the purpose of their 2012 Sustainability Framework based on achieving positive development outcomes through environmental and social sustainability linked to activities such as investments, financial intermediaries and advisory services. A set of performance standards were developed to assist companies, industries and governments to become more competitive, improve corporate governance, or help them become more sustainable. Under Performance Standard 4: community health, safety, and security, it was identified that project activities, equipment, and infrastructure can increase community exposure to risks and impacts. Under the requirement for community heath and safety it stipulated that where hazardous materials are part of an existing project, the client will exercise special care to avoid exposure to the community. The client must apply commercially reasonable efforts to control the safety of deliveries and disposal of hazardous materials and implement measures to avoid or control community exposures to pesticides (IFC, 2012). An energy economic model involves the studying of forces that lead economic agents such as, forms, individuals, governments to supply energy resources in order to convert them into other useful energy forms, i.e. to transport them to users. The type of commodities of interest that lead away from economic efficiency are, gasoline, diesel fuel, natural gas, propane, coal or electricity which are used to provide services such as lighting, space heating, water heating, cooking, motive power and electronic activity. With regards, to valuating the attributes of these relationships normative economics in contrast to positive economics (applying hard science to describe current, past and future economic activity for improving foresight in decision-making) analyses what ought to be in society. For example, deurbanising high energy intensive areas is not the logical solution (Squires, 2013).

Policies have long been recognised as the catalyst for economic growth. Future proofing policy solutions with the potential to deliver social and economic benefits such as health will have a direct impact on basic services and living standards. Atkins and UCL's Development Planning Unit in partnership with the UK's Department for International Development (DFID) developed an integrated diagnostic risk model based on identifying environmental risks, vulnerability, and capacity to respond in order to prioritise opportunities for action. Their integrated approach for responding to risks used a multi-criteria analysis in order to assist decision making and identify the multiple impacts and synergies associated with policy options which are being increasingly used. The concept of multi-criteria is based on quickly identifying and prioritising policy solutions for future proofing 'it can act as a way of identifying (qualitatively) policies which can potentially address current and potential future risks, can respond to urban vulnerabilities and more intermediate development priorities'. Atkins, 2012 identifies an architectural design (system architecture: 'The structure of components, their relationships, and the principles and guidelines governing their design and evolution over time') comprising of five main stages: (i) identifying the solutions relevant to city types, (ii) identifying vulnerabilities addressed and economic development benefits, (iii) identifying the capacity required for implementation, (iv) assessing impact and cost effectiveness and (v) assembling policy portfolios (bringing all of these considerations together, including thinking about how to exploit synergies and management trade-offs between solutions) and the category matrix data system.

According to Kundoo (2015), the pressing need for environmentally sustainable building solutions cannot be overstated. Kundoo recognizes the damage and degradation caused by pursuing growth and development within an Indian context. He is of the opinion that because of the changes in India's demographics from high fertility and low mortality to low fertility and high mortality coupled with improvements in health care facilities, there will be an significant increase in consumption of energy. However; Kundoo views 'certified green buildings' as being one of the main challenges because they are prohibitively priced and out of the reach for the bulk of the population. He suggests that high-tech, high-performance green design solutions should be restricted to a small number of specialized buildings while low-tech green building alternatives need to be developed for most housing and public buildings.

The UK has traditionally taken a hybrid approach to energy efficiency based on a mixture of legislative and regulatory instruments, and market-based financial incentives with key measures such as: building regulations (Part L relating to conservation of fuel and power), code for sustainable homes (measures the sustainability of a new home against categories of sustainable design), CRC Energy Efficiency Scheme (a mandatory scheme aimed at improving energy efficiency and cutting emissions in large public and private sector organizations), and energy supplier obligations (Carbon Emissions Reduction Target – requires all domestic energy suppliers with a base in excess of 50,000 customers to make savings in the amount of CO_2 contributed by the household). In relation to financial incentives, Renewables Obligation (RO) policy instrument places a legal obligation on the electricity supplier to verify that it has supplied a portion of its electricity from renewable energy (Leach *et al.*, 2014). However, recent reports (Element Energy, 2009) cited by Leach *et al.* have suggested that only up to 30% of the UKs population is interested in energy efficiency measures even if they are subsidized.

NIBS (2014) investigated the post occupancy characteristics of small commercial buildings in the 'United States'. They recognize that this type of market is stratified by age, construction characteristics and energy performance. NIBS cite the U.S. Energy Information Administration's (EIA's) 2003 Commercial Building Energy Consumption Survey (CBECS) findings that measure the Energy Use Intensity (EUI) for commercial buildings. Their results show that the highest EUI levels are found in structures larger than 100,000 square feet and that these buildings are more likely to house data-intensive activities that require high levels of electricity consumption, such as hospitals and supermarkets. However, buildings measuring 10,000 square feet or less also recorded relatively high levels of energy intensity. In general all building size classifications, properties constructed from 1960 through to 1989 tend to consume the most energy. The survey noted that from 1990 the majority of structures incorporated energy-saving improvements. Both Kudo and NIBS predict energy consumption at high levels however, these two markets are completely different and require building alternatives. An alternative approach by NIBS is the support for performance energy upgrades investments into the small commercial sector while considering different approaches to designing these policies, because one size will not fit all. NIBS initial strategy is to capitalize on the real-estate market were energy retrofits are estimated to be worth $279 billion across the residential, commercial and institutional property markets. The volume of this market is identified by Deutsche Bank who estimates that investments could leverage $1 trillion in energy savings over 10 years and create over 3.3 million jobs. The following section will define the requirements for developing a more informative system based on health management requirements.

9.3 Health management systems

Rich *et al.* (2013) identifies that a number of leading observers of the U.S. healthcare services have dubbed the system as being unsustainable, especially when comparing outcomes and costs to other countries. They also recognize that the level of resources consumed by US healthcare has grown steadily, rising from 9% to 17.3% GDP in 2012. With regards, to these challenges Rich *et al.* (2013) focus their attention on addressing healthcare leaders who embrace sustainability in order to pursue sustainable performance excellence. In a similar approach to achieving 'a top decile performer in safety or finance' they define such characteristics as developing a sustainability mission and vision statements while engaging in leadership and program innovation as key performance indicators. In contrast to section 2 were 70% of the UK's population was deemed to be uninterested in sustainable measures, Rich *et al.* illustrate the results of The Business of a Better World (500 business leaders) 2011 survey. This survey acknowledges that 84% of the leaders stated that their domestic company will adopt sustainability as part of its core business strategy within the next five years. In conclusion, Rich *et al.* emphasizes that within healthcare organizations (doctors, volunteers and employees) grassroots level of interests and personal goals need to be evaluated in order to connect with the facility's sustainability objectives. In essence; 'stakeholders want to be part of something greater than themselves and sustainability can be the platform'.

The Enhanced Critical Infrastructure Protection (ECIP) Security Survey defined 'dependency' as the reliance of a facility on an outside/external utility or service to carry out operations such as security (e.g. electrical power for closed circuit television (CCTV, scanners, and sensors), provide on-site heat or hot water and the degradation in service of these operations. The ECIP program is based on 222 hospitals since January 2011. Table 9.1 is a combined cross tabulation of two separate tables extracted from NPPD (2014) Sector Resilience Report: Hospitals (Table A – Hospital dependencies and recovery mechanisms and Table B – Critical access hospital dependencies and recovery mechanisms). The noticeably low performer was wastewater treatment with backup or alternate utility source for both dependencies and recovery mechanisms and critical access showing results of 10% and 16%. Some areas of communications also produced poor results with regards to critical access for contingency plan with provider 14% and priority restoration plan with provider 10%. Each of the six utility provider categories indicated low scores for critical access – contingency plan with provider; electric power 24%, natural gas 24%, water 24%, water treatment 24%, communications 14% and information technology (IT) 16%. The most worrying evaluation is the fact that hospitals are dependent upon external products or services pending degradation (address how soon a facility will be affected if the source is lost) from their loss. The following results provided by the DHS and Argonne National Laboratory show within this survey the extent to which this situation is a problem; (i) electric power – degraded 100% after five minutes; (ii) natural gas – degraded 1% to 33% after four hours; (iii) water – degraded 67% to 99% after two hours; (iv) wastewater – degraded 67% to 99% after two hours; (v) communications – degraded 34% to 66% after two hours and; (vi) IT – degraded 67% to 99% after ten minutes. When taking the degradation factor into perspective, the 16% associated with IT's critical access contingency plan illustrates the need for adequate infrastructure planning in order to design a resilient system architecture that assists in maintaining critical infrastructure; cloud computing and BIM (as an innovative technology) are presented in the following section.

9.4 Cloud computing and BIM

The use of Application Performance Interfaces (APIs) for binding heterogeneous applications through a central repository platform such as cloud computing has created a way for different applications to openly interoperate and exchange information. *Cloud computing* refers both to the applications delivered as a service over the Internet, and the hardware and systems software in data centers that provide those services (Ambrust *et al.*, 2009). Onuma (2010) recognises that the Internet is viewed in real-time and that computer devices, coupled with cloud-based tools, should make information more accessible to users. In a BIM context, Onuma acknowledges that traditionally exporting a file from application A and then importing that file into application B was the industry norm for sharing data between

Table 9.1 Cross tabulation of hospital dependencies and recovery mechanisms

Utility provider type	Electric power	Natural gas	Water	Wastewater treatment	Communications	Information technology
Dependencies and Recovery Mechanisms—Dependent upon External Utility Provider (%)	100	78	99	97	95	95
Critical Access—Dependent upon External Utility Provider (%)	100	86	100	100	100	97
Dependencies and Recovery Mechanisms—Backup or Alternate Utility Source (%)	99	58	48	10	72	61
Critical Access—Backup or Alternate Utility Source (%)	100	52	33	16	50	33
Dependencies and Recovery Mechanisms—Contingency Plan with Provider (%)	64	47	51	43	52	59
Critical Access—Contingency Plan with Provider (%)	24	24	24	24	14	16
Dependencies and Recovery Mechanisms—Priority Restoration lan with provider (%)	75	61	61	48	57	62
Critical Access—Priority Restoration Plan with provider (%)	41	36	34	17	10	50

Source: Author's adaption from NPPD, 2014.

applications. However, it resulted in creating multiple copies of data. The identified solution is Service-Oriented Architecture (SOA), where data can remain on application A and be used or modified by application B (or other applications). For Onuma, this approach allows the user to access pertinent data enabling BIM model views to be shared in small data chunks. Kurtalj (2011) defined mashups as a way to integrate into a common system various products and services. For example, using Geographic Mark-Up Language (GML) to define both the geometry and properties of objects that comprise geographic information; for example, adding Google Maps with buildings scattered around the world, it is possible, when clicking on the building, to access a pop-up with a Building Management Systems (BMS) information for that particular building. The mashup concept is highlighted by Kurtalj in relation to Building Automated Systems (BAS) for integrating data points and services from various networks into the same classes/objects infrastructure (as a programming language, such as, a value, variable, function or data structure), usable through the cloud model. Instead of translating the data from one protocol to another, they can all be processed in the same script, as well as returning the data results back to the very network in its native language via so-called 'plug-ins'. BAS and BMS can greatly improve the energy conservation within building operating and maintenance; especially by implementing a centralised heterogeneous network 'cloud'. By exchanging data from interoperable energy consumption analysis applications with so-called 5D costing, via the Internet, in order to calculate the actual costs of running the building, the potential exists for predicting early building performance costs, which would produce alternative economical designs.

Josuttis (2007) recognizes that SOA is a simple and fast way to provide access to data or resources provided by web servers. In analyzing SOA, Josuttis identifies Enterprise Service Bus (ESB – enables asynchronous oriented design for communication and interaction between applications) as the backbone to the architecture and highlights its heterogeneous ability through two approaches; protocol driven, and APIs. The issue with the API-driven approach relates to the tools and/or libraries (set of rules) for mapping the APIs. In order to provide for service calls (communication between two electronic devices over the web) and implementations, the messages to be sent over the ESB have to be defined, which means XML with XSD (XML Schema Definition) must be proven. In contrast, the protocol-driven approach allows the participants to choose their specified tools to send service calls and provide service implementations (connections). Stair *et al.* (2008) emphasizes that the main contribution of APIs is their ability to link application software to the operating system. Oracle (2008) refers to web APIs as Web 2.0 and highlights that its emergence has vastly improved and enriched the user's web experience. Anderson (2007) recognizes that APIs have helped Web 2.0 services expand and have facilitated the creation of mash-ups of data from various sources by providing mechanisms for programmers to make use of the functionality of a set of data without having access to the source code. Web services; which are developed from XML data representation format and Hypertext Transfer Protocol (HTTP) communication protocol, are platform neutral, widely accepted and utilized and come with a wide range of useful technologies and they support SOA (Moller and Schwartzbach, 2006). In 2006 the Open Geospatial Consortium tested new standards to extend to the integration of CAD and GIS technologies on the web. One such Web standard was based on exchanging proprietary BIM data through Web Feature Services (WFS) derived from the GML. CityGML is a champion of the major technical advantage of web services with Industry Foundation Classes for the geospatial environment, as it enables 3D buildings to be displayed and exchanged through web services (OGC, 2007). APIs have allowed developers to connect applications together by extracting rivulets (streamlined) of information from various Web 2.0 sources (read and write data). This process is known as a 'mash-up', where the concept is to open up content production to all users and expose data for re-use (Anderson, 2007). With regards, to transforming policy documents into intelligent 3D collaboration tools, the functional architecture of cloud computing and BIM requires an automated process that can support an integrated service.

9.5 Semantics ingenuity for connecting policy documents

The idea of connecting published policy documents from a website in order to create decision support tool architecture such as, ASHRAE documents relating to refurbishment, requires the provision of semantic knowledge. Policy frameworks such as Atkins (2012) represent knowledge in such a way that it can be understood and processed automatically by a computer. The question is how can we use this obtained information effectively in real applications? Walton (2006) synergy of camera ontology depicts a specification used to classify cameras into different categories. In a similar approach the authors partially illustrates in Figure 9.1 a presentation delivered by Yrjö Engeström (University of Helsinki) at Université de Québec à Montréal ('Concept Formation in the Wild – Avril 2015') demonstrating the conceptualization of three case studies with additional context added by the authors such as linking cognitive activity (process by which knowledge is acquired, including perception, intuition, and reasoning) to collective perspectives and individual perspectives.

The second part of the Concept Map shows the hierarchical breakdown of the basis of ontology (information versus knowledge). The connection is formed between 'interpret concepts' and 'cognitive activity' via perception and learning and reasoning. The boxes contain the concepts (knowledge) with arrows defining the relationships between them. The diagram presents 5 key areas of an ontology; (i) truths, (ii) beliefs, (iii) perspectives, (iv) judgements and; (v) methodologies. For Walton, the root (thing) and subclasses tangible thing (real physical existence) such as a camera and intangible thing (properties) relate to Autofocus. This allows a distinction between the real physical objects (a design conceptual hierarchy highlighting key concepts that are important to photographers). This technique is similar to an

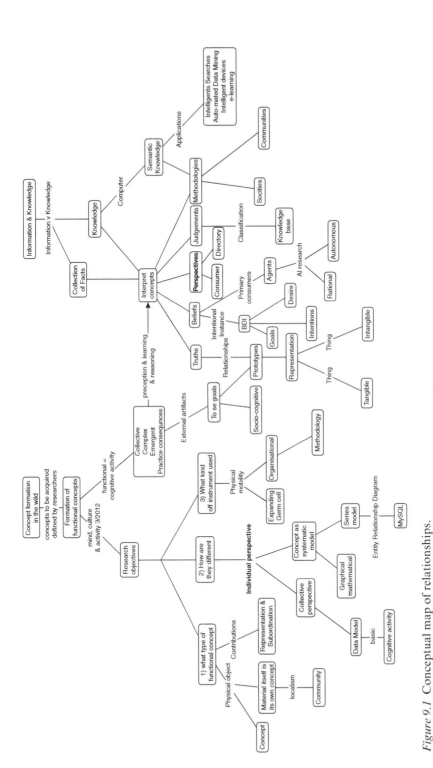

Figure 9.1 Conceptual map of relationships.

Legend: BDI—Belief Desire Intention, AI—Artificial Intelligence.

Source: Authors, 2015.

object-oriented programming language such as Java but in a network of relationships as oppose to a tree like structure. With regards, to the diagram these relationships would be connected to setting goals based on the formation of function concepts derived from, collective, complex, emergent and practice consequence data. The classification process performs an inference such as intelligent devices (mobile phone) having an interchangeable networks which represents knowledge in the form of conceptualization. Walton recognises that representing knowledge in the form of conceptualization is central to the automatic processing of information on the web. Through this method of classifying information representing statements of a logical theory 'ontology' (a specification of a conceptualization by defining a set of objects, and the relationships among them) existing knowledge can result in a more general-purpose representation while increasing the utility of existing knowledge.

International advances of such controlled vocabulary ontology's based on primitive processes are; (i) Community of Interest (COI) Information Exchange Vocabulary (IEV) development – developing universal core schema that enables information sharing in order to promote interoperability and understandability through common semantics and syntax (DoD, 2013) and (ii) CSTB – 'establishment of a set of logical rules that expresses the functional dependencies between renovation components and the existing building (or between renovation components themselves)'. This methodology is based on the semantics of 8 key performance indicators (energy demand, life cycle cost, environmental footprint, comfort assessment, social impacts, health, usage quality and risks) using Bayesian statistical analysis (interpretations of probability) to make energy efficiency decisions (Thorel *et al.*, 2013).

The basis for developing a semantic mechanism whereby immediate ASHRAE documents referring to a particular VA environment can be accessed instantly relates to the Schools Interoperability Framework (SIF – North America). SIF comprises of a data model (common elements and objects in a tabular format) and an infrastructure featuring 'an XML-based messaging framework that allows diverse software applications to interoperate and share and report data related to entities in the K-12 instructional and administrative environment. It can also use Simple Object Access Protocol (distributed application communication)-based transport and corresponding set of Web Service Description Language (WSDL) files to allow web services to fully participate in these interactions'. The technical aspect of SIFs standardized student data in an identical format system to enable comparisons between schools, which can be later analyzed to make informed educational policy decisions, is the same process as their Energy Usage Object design. SIF acknowledges the three main types of systems which can be used to help control energy costs; (i) Building Automation Systems (BAS - regulate HVAC usage and collect usage data derived from room scheduling data and management of sensors), (ii) Utility company services (which supply energy to support HVAC/electrical operations in the school building) and (iii) Energy Monitoring, Data Analytic and Reporting Systems (data supply analysis on energy usage).

However, these three management systems are not linked and SIF through their SIF Zone (services which are defined using the service message types) can enable an Energy Usage Object to be used by one or more systems via loose coupling (SIF, 2012). With regards to BIM and cloud computing for better energy management – Lavelle Energy LCC have started to analyze how to use various energy systems such as heat pumps, HVAC, solar and wind power in order to off-set against peak loads on an education center (Audubon Center 90 miles north of St. Paul, MN). Three independent areas (heat pumps, main building, and dormitory building) were sub-metered and fitted with products such as Wireless Local Energy Meter (WiLEM) connected through Zigbee (a low-cost, low-power, wireless mesh network (consisting of mesh clients (laptops), mesh routers (forward traffic to and from the gateways) standard using a mesh cloud for automatically conveying energy and temperature sensor data. Other services included Cloud servers for data logging (Guru-Plug Linux server running FasBridge vROC communications software) and WordPres (a Content-Management System (CMS) based on PHP and MySQL, which runs on a web hosting service) to be run as a blog that receives previous day's energy information for internal staff reviews (Lavelle, 2014). The graphical diagram of relationships in Figure 9.1 is based on providing the content structure for a Graphical Framework for OWL Ontologies, i.e. providing the intended links for creating a comprehensive machine-readable RDF metadata. However, such an ontology would requires classes (sub-classes based on logical axioms), properties instances (objects created from instance classes) and statements.

9.6 Solution methodology – defining business rules staements

According to Hay and Healy (2000), in 1993 the GUIDE Business Rules Project was organized for identifying rules that define the structure and control the operation of an enterprise. In generic terms 'a business rule is a statement that defines or constraints some aspect of the business. It is intended to assert business structure or to control or influence the behavior of the business. The business rules that concern the project are atomic (that is, they cannot be broken down further)'. The authors approach to interface requirement specification is based on Hay and Healy's style to formulating business rules. The rules are investigated in order to resemble an explicit expression graphically or entity/relationship (or 'object class') diagrams for the purpose of generating a programming code to build a Structured Query Language (SQL) statement.

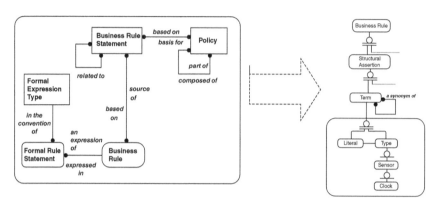

Figure 9.2 (a) The origin of business rules and (b) Kind of term.
Source: Hay and Healy, 2000.

Figure 9.2 (a) shows the implication of a policy representing a general statement of direction for an enterprise. It is composed of more detailed policies, which may be part of one or more general policies. With regards to ASHRAE documents an example of a policy for Indoor Air Quality – Guidelines and Standards might be: 'Ventilation rates are specified for commercial/institutional buildings and low-rise buildings'. A policy forms the basis for one or more business rule statements for example, 'there are 12 schemes labeled as low-Volatile Organic Compounds (VOCs)'. However, each business rule (a statement that defines or constrains some aspect of the business) must be expressed as a formal rule statement.

For example, with regards to determining the ventilation of a low-rise residential dwelling, using ASHRAE 62.2 recommendation for single houses and multifamily structures of three stories or fewer:

```
If low-rise dwelling-2—3 bedrooms–floor-area < 139 then
  Ventilation-air-requirements (L/s.21)
End if
```

Figure 9.2 (b) illustrates (atomic) business rule as a structural assertation; defining concepts of statements for which terms and facts originate from. The use of the word term can be classified as a type (defining abstract categories of documents for example; common indoor contaminates) or as a literal (describing instances of things, such as ASHRAE standard 62.1 and European CR 1752). Figure 9.2 (b) also indicates a particular kind of type 'a sensor' which can be interpreted as a wireless sensor in order to detect and report changing values such as temperature reading or on a more precision basis ventilation rates to speed up the removal of air-borne VOCs (increase outdoor air supply to 30 L/s person 'EUROVEN' from 'ASHARE' 10 L/s). This concept can be attached to Radio Frequency Identifier Tag (RFID – 'a microchip attached to an antenna that is packaged in a way that it can be applied to an object'). In principle a tag can be a barcode print or embedded in plastic but essentially a reader communicates with a transponder which holds digital information in a microchip (Bacheldor, 2013). The introduction of using object tags allows data to be semantically linked to a BIM model (please see section 9.6.2). This technique could be used to determine building materials that may potential exceed the limit developed by the State of California's Chronic Reference Exposure Level (CREL – any chemical at 96-hr not exceed 50% concentration) such as, laminate, linoleum, wooden varnished, and cork/pine flooring, plastic water-based and matt emulsion paints (Missia *et al.*, 2010) containing VOCs such as Acetaldehyde and Benzidine (Charles *et al.*, 2005).

9.6.1 SQL statement

Business Rules and sensors can be connected to ASHRAE documents by a derivation such as a mathematical fact that produces a derived fact according to a specified mathematical algorithm such as, the minimum outdoor airflow rate for whole-house ventilation based on the floor area of the conditioned space and number of rooms:

$$Q_{fan} = 0.05 A_{floor} + 3.5\left(N_{br} + 1\right) \tag{1}$$

or inference where a derived fact is produced using logical induction or deduction (general principles) for example; CREL document is inferred from 'the seven agencies' associated with organic compounds that cover IAQ – Guidelines and Standards (Charles *et al.*, 2005). LogicGem (2006) demonstrates how to build a representation of Business Rules to create a software code.

The document defines decision tables as a map representing the relationships of combinations of conditions such as, is the existing room tagged? (referencing common indoor contaminants possibly containing carbon dioxide, carbon monoxide, ozone, lead and particulates), does the room exceed the recommended limits? To combinations of actions such as, go to ASHRAE/ANSI Standard-62.1–2004. The logic of various Business Rules would be itemized in several different columns all containing different combinations of conditions and actions. In order, to build SQL statements the decision table can be used to generate a code to structure query language statements. The process involves selecting a specific field list from a database containing guidelines and standards for retrieving records based on a series of selection criteria. LogicGem software promotes the decision table as the process of defining logic to build the SQL Where clause part of the SELECT statement. The business requirement is to extract data from an IAQ database in a relational database. The two main fields in the selection

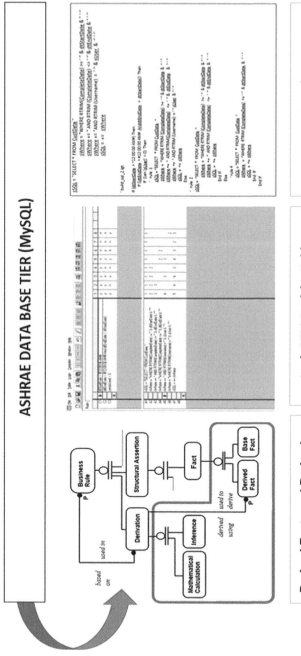

Figure 9.3 ASHRAE data base tier – architectural design.
Source: Redmond *et al.*, 2014.

criteria would be: CompleteDate (relating to document date), and Username (meaning ASHRAE standard). The three program variables of dtStartDate, dtEndDate, and sUser (referring to different year categories of ASHRAE standards) are compared with field values. Figure 9.3 outlines the process in a graphical form starting from designing Business Rules before progressing to LogicGem and completing with a SQL statement and complied code.

9.6.2 *Linking business rules to health and ASHRAE documents*

Bluyseen (2014) investigated existing materials which are known to cause indoor air pollution. Table 9.2 lists some of the main substances that can be located in our indoor environments and the source and type of pollution for example; 'asbestos fibers are a particular type of particle, and the use of asbestos in buildings has been an important route of worker and population exposure. Asbestos is present in many buildings and presents a risk of cancer if fibers are inhaled'.

Wilmering and Mott (2011) acknowledge that in general builders of systems try to design components with well-defined and reliable behaviors and since simple systems have defined components their behavior is often easy to understand. However; engineered systems are typically composed of subsystems and they themselves are elements of a larger system dependent on relationships with numerous other systems. Wilmering and Mott cite Buchanan and Shortliffe (1984) reference to large rule-based systems 'that exhibit intelligent behavior but are composed of numerous atomic rules that taken alone are elementary in nature'.

As part of a proposed function analysis Figure 9.4 shows an example of a process for implementing a system based on hierarchy of components interacting with external services. The base layer comprises of an external mobile device sending data to a web-based BIM service platform.

The data retrieved is obtained from RSS tag or sensor devices that are mapped to a specific library catalogue containing the material identified. This material is labelled and local supplier is sourced. The second row shows the process graphically while also acknowledging that the location of this existing material may need to be eradicated and replaced. The replaced material objects would have to be inserted into the model in a similar approach to embedding point cloud files. The final tier illustrates the external services that can be used to assemble the system together such as; (i) starting the process with a 3D scan to develop a BIM model in Revit; (ii) connecting the data retrieved from the RSS tags or RFID and inserting them into the BIM; (iii) transferring the model in to a BIM web-based service; and (iv) the BIM web-based service automatically links the information objects with Health and ASHRAE documents based on the defined business rules. This process can assist the stakeholder to identify if the existing material is hazardous and suggest an alternative source. To provide such a web services the proposed system architecture will need to easily exchange data openly on an integrated service.

Table 9.2 Main groups of substance and their source known to cause indoor air pollution (author's adaption from Bluyssen, 2010)

Substance	Source	Indoor air pollution
Endocrine-disrupting chemicals	Phthalates; pesticides	(used in vinyl, plastics, building materials); (gardening)
Radon	Radioactive gases	Enters the building from the ground and ingress depends upon factors such as local geology
Inorganic gases	Carbon dioxide (CO_2), carbon monoxide (CO), nitrogen oxides (NO_x), sulphur dioxide (SO_2)	Particles from biological origin, cooking
Volatile organic compounds (VOCs)	Aromatic or halogenated solvents, vinyl chloride (paints), borax	Consumer products including electrical goods such as computers and printers and cleaning products
Very volatile organic compound (VVOC)	Formaldehyde	Adhesives, office furniture, panel systems, a range of building and consumer products
Semi-volatile organic compounds (SVOCs)	Pentachlorophenol, polyaromatic hydrocarbons (PAHs), and phthalates	Polymeric materials such as vinyl flooring and paints
Microbial volatile compounds (MVOCs)	Metabolism	Compounds formed in the metabolism of fungi and bacteria
Ozone	Photochemical reaction	Reaction of ambient air with surface and airborne pollutants to produce new organic compounds and particles
Ultrafine and nanoparticles	Particles sized between 1 and 100 nanometers	Nanomaterials and combustion, such as burning a candle or smoking a cigarette
Asbestos fibres	Crocidolite (blue asbestos), Amosite (brown asbestos) and Chrysotile	Present in many buildings (roofs, ceilings, walls and floors, thermal insulation products) and presents a risk of cancer if fibres are inhaled

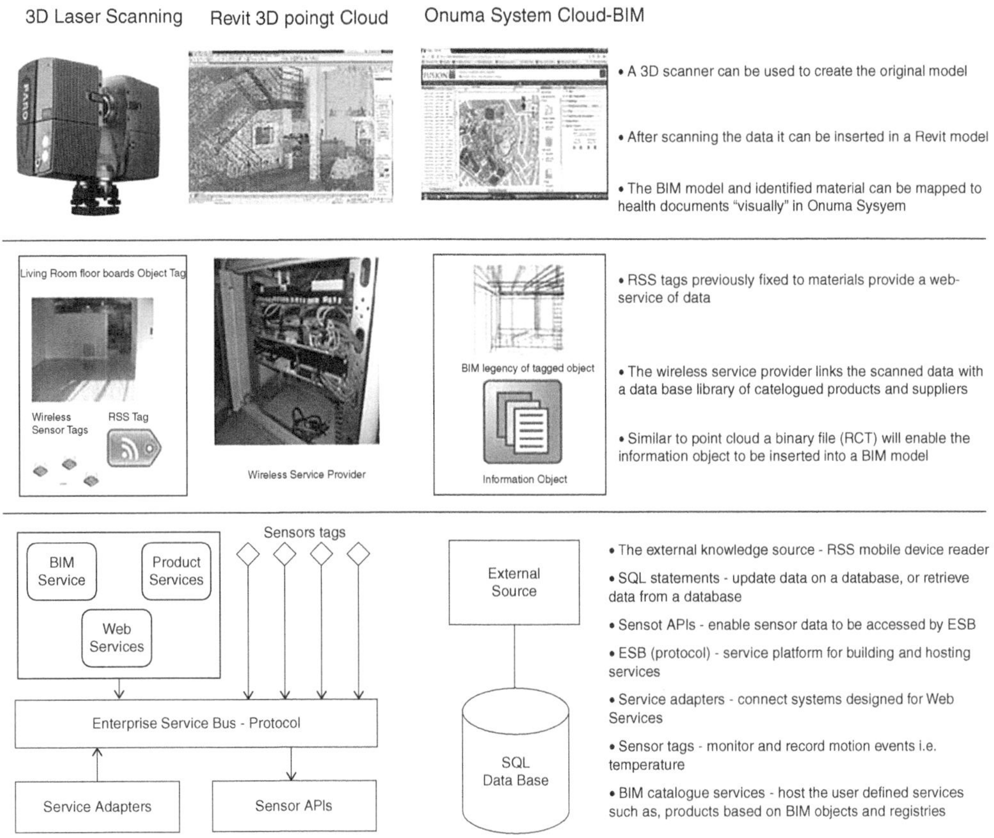

3D Laser Scanning Revit 3D poingt Cloud Onuma System Cloud-BIM

- A 3D scanner can be used to create the original model

- After scanning the data it can be inserted in a Revit model

- The BIM model and identified material can be mapped to health documents "visually" in Onuma Sysyem

Living Room floor boards Object Tag

Wireless Sensor Tags   RSS Tag

Wireless Service Provider

BIM legency of tagged object

Information Object

- RSS tags previously fixed to materials provide a web-service of data

- The wireless service provider links the scanned data with a data base library of catelogued products and suppliers

- Similar to point cloud a binary file (RCT) will enable the information object to be inserted into a BIM model

Sensors tags

BIM Service	Product Services
Web Services	

Enterprise Service Bus - Protocol

Service Adapters   Sensor APIs

External Source

SQL Data Base

- The external knowledge source - RSS mobile device reader

- SQL statements - update data on a database, or retrieve data from a database

- Sensot APIs - enable sensor data to be accessed by ESB

- ESB (protocol) - service platform for building and hosting services

- Service adapters - connect systems designed for Web Services

- Sensor tags - monitor and record motion events i.e. temperature

- BIM catalogue services - host the user defined services such as, products based on BIM objects and registries

Figure 9.4 Example of a BIM system and process for linking material objects with health and ASHRAE documents.

Source: Authors, 2015.

9.7 Open BIM model (operational system) test case

The core issue of cloud computing, being a big data system (data that exceeds the processing capacity of conventional database systems), coupled with BIM, provides challenges to a system that can enable various disciplines to work collaboratively. With regards to developing an Open BIM model connected with a decision base model, requirements analysis and partial synthesis (items and system elements for exchanging information), this type of system architecture has been designed and tested (Redmond and Smith, 2012; Redmond, 2013).

Figure 9.5 shows the design architecture developed under the guidance of Onuma System (Cloud BIM) and FUSION (facility management system) based in California. The process involves six stages: 1) client queries registry via Web Feature Services (WFS – instant sharing capabilities across the Internet) to locate services, 2) registry refers client to subsets of XML documents (various simplified markup languages developed through eXtensible Stylesheet Language Transformation (XSLT) using template rules to connect Onuma System with FUSION), 3) client access subset XML documents (via interface device), 4) Service Oriented Architecture (referencing cloud computing) provides infrastructure to interact with web services such as FUSIONs facility management data, 5) client sends Representational State Transfer (REST - similar to SOAP but less secure) message request and 6) web service returns REST-message response, i.e. Onuma sends a request and FUSION replies while all operations are being conducted in Onuma System. The FUSION+GIS+ONUMA Systems architecture (a potential integration of three independent services) is based on exchanging IFC and SML information seamlessly between FUSION's 'facility management system chosen from California,' GIS 'geographical location,' and Onuma System's 'Cloud BIM'. The actual tested model California Community College indicated significant benefits such as, 'the interoperable capabilities of sharing BIM files on a cloud computing network predicating 80% of the life cycle cost within three days at the feasibility phase') (Redmond, 2013).

9.7.1 Sequence of a lifecycle process – the test-case study of facility building 4200

The Facility Building 4200 at MiraCosta College Oceanside Campus (California Community Colleges CCC), is one of 6,000 buildings in the CCC facility condition assessment program. The Facility Condition Index (FCI) is a deficiency tool (for facility management) used to measure the percentage gauge of a building's condition when determining whether or not the building is worth maintaining. The results of the last assessment, conducted in November 2010, indicated graphically on a facility GIS condition map that Facility Building 4200 had a FCI of greater than 10% (suggesting that it should not be maintained). The FCI ratio is the ratio of the cost of

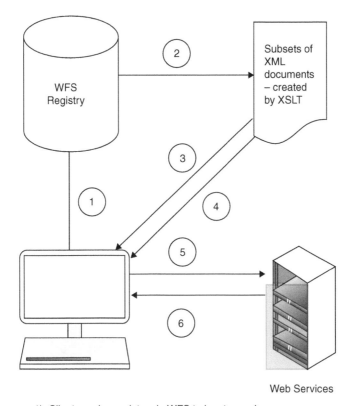

1) Client queries registry via WFS to locate services
2) Registry refers client to subsets of XML document
3) Client Access subset XML documents
4) SOA provides Infrastructure to Interact with Web services
5) Client sends REST-message request
6) Web Service returns REST-message response

Figure 9.5 Exchanging information via BIM XML and REST.
Source: Redmond *et al.*, 2014.

addressing all of the facility's deficiencies versus that facility's replacement value (MiraCosta Comprehensive Master Plan, 2011).

The overall objective of this test-case study was to show the key benefits of Cloud BIM through its capability of exchanging partial sets of BIM data between applications, such as; 'improved communication and collaboration among project participants, enhanced project decision making', more accurate planning and scheduling, greater process standardization, cross-discipline co-ordination / virtual issue resolution and an understanding of the construction environment through visualization from the beginning. Figure 9.6 shows the life cycle model of the participant observation case

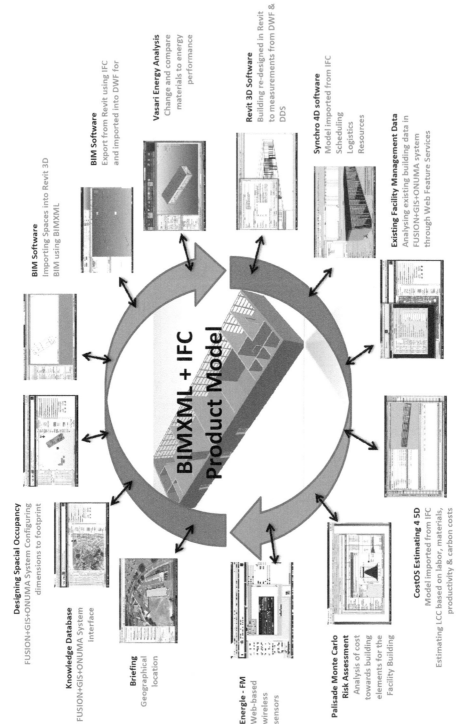

Designing Spacial Occupancy
FUSION+GIS+ONUMA System Configuring dimensions to footprint

BIM Software
Importing Spaces into Revit 3D BIM using BIMXML

BIM Software
Export from Revit using IFC and imported into DWF for

Vasari Energy Analysis
Change and compare materials to energy performance

Knowledge Database
FUSION+GIS+ONUMA System Interface

Revit 3D Software
Building re-designed in Revit to measurements from DWF & DDS

Briefing
Geographical location

Synchro 4D software
Model imported from IFC
Scheduling
Logistics
Resources

Existing Facility Management Data
Analysing existing building data in FUSION+GIS+ONUMA system through Web Feature Services

Energle - FM
Web-based wireless sensors

Palisade Monte Carlo Risk Assessment
Analysis of cost towards building elements for the Facility Building

CostOS Estimating 4 5D
Model imported from IFC
Estimating LCC based on labor, materials, productivity & carbon costs

BIMXML + IFC Product Model

Figure 9.6 BIM XML + IFC product model.
Source: Redmond, 2013; Redmond *et al.*, 2013.

study undertaken at BIMStorm 2011 by the author. The model is a replicate figure of the BIM relationship (NBIMS, 2007) but redesigned based on a BIM XML and IFC exchanging format mode. The original BIM XML schema for Cloud BIM is a subset of XML, which is itself a subset of the Standard General Mark-up Language. BIM XML was developed into several different subclass XMLs (via XML Style Sheet Language – XSL) in order to connect FUSION+GIS+ONUMA System through web Feature Services (connecting individual web services together).

The sequences of events in chronological order were as follows:

- Briefing – there were no available drawings of Building 4200 but there was a reconnaissance sketch outlining the sizes and layout of the rooms.
- Knowledge database – the FUSION database was used as a repository for all facility data about MiraCosta, the opening interface of Onuma System provided access to both FUSION and GIS asynchronously.
- Design spatial occupancy – FUSION information was converted from a basic square with all its associated data (with reference to level of model detail 'basic design' – building footprint, building use, floors and occupancy); and designed to meet the existing layout plan (level of model detail 'medium' – rooms, partitions, open space and access).
- BIM software – this was the first stage of the test that did not use WFS (contrast with using SOAP, the WFS used in the case study are based on REST – a collection of network architecture principles that focus on simple access to resources) the building file was exported using plug-ins and
- Imported into Revit and Vasari (Energy analysis software using Revit code) at this stage the design was formatted taking into consideration the results of the energy analysis such as using 36% fenestration ratio to the buildings floor area.
- IFCs were used to import data into such applications as Autodesk and Synchro in order, to create programs (4D), costing's (5D) and simulations.
- Palisade risk assessment was the only application that represented proprietary software exchange of data.
- *Energle – FM*: This is a web-based wireless sensor connected to the Onuma System interface via web service APIs; the application monitors as-is conditions, such as energy usage based on temperature, humidity, CO_2, and Lux Level.

9.7.2 *The results*

Table 9.3 outlines the performance metrics of the existing test-case study and compares the traditional standalone model for exchanging data with that of Cloud BIM. The capable benefits of transferring nD information

Table 9.3 Performance metrics of FUSION+GIS+ONUMA case study

Performance metrics	Traditional standalone model for exchanging data	Cloud BIM featuring subset of XML for exchanging data
Improved communication and collaboration among project participants	The traditional techniques of using stand-alone or proprietary data exchange mechanisms have been consistent. Projects to date have achieved their required performance indicators. However, the industry recognizes that the supply chain is fragmented, as entry barriers for SMEs exist due to the lack of technical ability (not having the required software).	As the author was the project's only participant, this benefit is hypothetically measured. Certainly the aspect of using WFS to instantly share and use facility data without leaving the BIM server model allows for open and instant collaboration via Internet.
Enhanced project decision making	Ethernets already exist but these collaborative hubs require the building design to be undertaken through standalone applications before being uploaded to the hub.	The ability to analyze information at the earliest stage through the Internet enabled assumptions with a higher amount of detail to be reviewed earlier, such as, deciding on the spatial and design content based on data derived from FUSION.
More accurate planning and scheduling	The actual main software for developing the work and schedule program with a simulation movie was based on a stand-alone application that used the open exchange schema of IFCs to transfer information.	The ability to have instant data allowed the author to have more time when considering certain scenarios. The BPEL work flows had already advanced the planning and scheduling of the project and the level of model detail required for exchanging data.
Greater process standardization	The majority of the stand-alone applications used in the case study was to IFC 4×2 standardization, which does indeed streamline the exchanging of documents (noted not as fast as Cloud BIM).	Both BIM and cloud computing characteristics are based on standards for data interchange – options for saving data or for importing data that is standardized.

Performance metrics	Traditional standalone model for exchanging data	Cloud BIM featuring subset of XML for exchanging data
Cross-discipline co-ordination / virtual issue resolution	Co-ordination between cross disciplines through the traditional technique happens at a slower pace than Cloud BIM. However, some disciplines prefer to only share information when they are ready, as opposed to the instant alternative.	With access to the project in virtual real-time through the Internet, the potential exists for stakeholders and design disciplines to co-ordinate on an open platform at any stage of the project.
Understanding of the construction environment through visualization	3D and 4D BIM is used regularly in the Built Environment but not in collaboration at the feasibility stage and certainly not asynchronously.	The FUSION+GIS+ONUMA Systems use of three services, facility management, geospatial data and BIM collaborating simultaneously enabled the feasibility and design briefing stage to be visually analyzed.

Source: Redmond, 2013.

through subclasses of BIM XML for web services and plug-ins exchanges enabled the case study to be tested successfully.

The results demonstrated the interoperable capabilities of web service APIs for exchanging partial sets of BIM data with real–time constraints at the feasibility design stage. The case study findings highlighted the enhanced decision making capability of using Cloud BIM for the spatial and design content of a building that can factor all of the design team and stakeholders' inputs from the very beginning. Having a service that contributes to sharing information asynchronously with access to rich data, will assist the design team in making educated assumptions on more cost options than previously undertaken within the industry. The ability to produce a model for subset distributions, with a common interface on an open platform containing diverse application files for data exchange, proved to be positive through a Cloud BIM service. However, stand-alone applications were still required to complete the full life cycle costing model.

9.7.3 Shared data services

In 2013–2014 the U.S. DoD Military Health System (MHS) now called Defense Health System, (DHA) funded NIBS to evaluate a prototype of the The Federal Reserve Integrated Facility Management (FED iFM) initiative based on encouraging the process of sharing facility standards and tools (Hagan and Onuma, 2014). As part of an interview session during the Building Innovation 2014 NIBS conference, Onuma acknowledges that there are

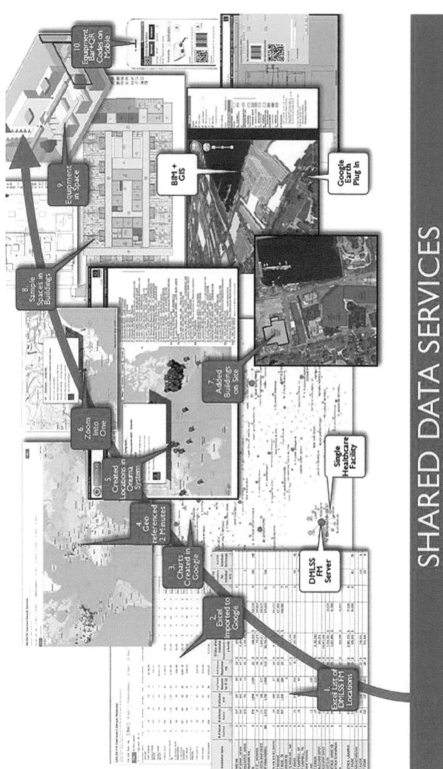

Figure 9.7 iFM and facility life-cycle data.

Source: with permission from Hagan and Onuma, 2014.

four main disruptive technologies directly related to healthcare facilities and facilities management; (i) the mobile Internet, (ii) the automation of knowledge, (iii) the 'Internet of Things', and (iv) cloud technology. He is also of the opinion that because 'technology is becoming transparent and simple to use, information silos are becoming redundant and agile developments such as apps are a minimum viable product that can be quickly deployed'. His theory is based on starting simple and then iterating, but he also refers to the 'White House 21st Century Digital Strategy', which is roadmap with interoperable data at its core.

Figure 9.7 shows the capabilities of utilizing Cloud BIM web service from an excel list of location, geographical referencing in Google, and simple spaces in BIM model that contain populated data on this space, i.e. a particular room would indicate its air changes, materials, kWh etc., to the most recent prototype 'Space and Equipment Planning System (SEPS) were clinic equipment objects are imported from a library into a BIM model.

Previous case studies with Onuma System have shown that the main requirements of an actual policy document can be uploaded in to a BIM model such as CCC, were FUSION contained all of the data in a two dimensional environment before a synchronically adapting it to a web-based multi-interactive BIM service. The authors of this chapter believe that within the context of future shared data services, health management systems are essential to every city and in relation to energy performance upgrades existing facilities should be investigated to determine their economic, environmental and sustainable benefits to the community. The capability of tagging materials and logging the data into a web-based BIM service platform would significantly speed-up the retrofitting decision making process. It would also enable BIM e-Collaboration commerce (Business Process Model – BPM) for areas such as, sales, marketing, production, procurement, recycling, distributing and accounting rules to be semantically linked to various categories of materials based on costs and their location.

9.8 Conclusion and future direction of cloud-BIM

The overall objective of this chapter was to identify a semantic technique based on reviewing policy's that can be linked to health and environmental documents through the use of web-based BIM. Existing research undertaken by the DoD and CSTB was briefly examined to identify how logic rules are being implemented successfully on similar refurbishment concepts. The Solution Methodology section highlighted in detail how the origin of business rules associated with ASHRAE documents can be transformed into SQL statements via decision tables, in order to produce programming codes. The SQL code used for extracting the correct ASHRAE data would be intilially generated from a RFID tag sensing the resistance changes of an environment that can be measured and processed into temperatures, in order to highlight possible hazardous materials. The final section identified the novel concepts of BIM XML and REST for exchanging information on an open BIM model, thus creating 3D virtual representation models connected to policy documents. The iFM and Facility Life-Cycle Data case study showed how agile developments linked to digital strategies are changing the perception of the health services. The chaper acknowledges the need to develop service ontologies using web service interfaces to automate exchanges based on a semantic composition approach such as energy knowledge. There are requirements to not only link GIS with BIM building models but to push the boundaries to include and link material objects with library catalogues. This will enable a more defined supply chain and customer relationship management process to better invoke policies such as CRC Energy Efficiency, with all attributes semantically linked as multi-agents on a BIM web-based platform.

The authors vision is to expand on how multiagents (groups of agents represented as web services) topology can cooperate on specific tasks based on existing web service technology that enables different reasoning techniques to be used. BACnet (Building Automation Control Network – a standard network protocol for building automation systems developed by ASHRAE) provide control systems for HVAC and more commonly Demand Response (DR) for reducing peak loads by network monitoring systems (Butler, 2007). Future research will investigate web-based BIM services to form multi-agent system utilizing existing ontologies (such as BAS) and building upon existing web service interfaces such as Onuma System and Open Building Information Exchange (oBIX) uniting web services using XMLs and URIs to develop a centralized architecture for transforming data into knowledge output.

Challenges – If the existing construction material is not readily detected via RFIDs etc., then further investigations will be required such as the use of a Micro bolometer. It is based on the notion that a bolometer can be used as a detector in a thermal camera for infrared radiation with wave lengths between 7.5–14µm (µm = micrometers = 1/1,000,000 of a meter) striking the dector material and changing its electrical resistance. These resistance changes can be measured and processed into temperatures in order to create an image (Wang *et al.*, 2005; Wikipedia, 2013).

The limitation of infrastructure in relation to increase energy consumption was mentioned in this chapter with a detail review of the ECIP program survey. IT was identified as an area that needs adequate infrastructure planning, however this chapter does investigate issues associated with cloud computing infrastructure. In previous published papers by the authors cloud computing is examined in detail Redmond *et al.* (2012), where issues such as security, contracts, performances, integration, capabilities are analyzed from a construction industry perspective. The major problems associated with BIM and cloud computing is both knowledge and security. Open BIM will be subjected to attacks but open collaboration will mean several companies/individuals working together at the same time can solve any such aggression. However; this is an area that requires further investigation.

9.9 Acknowledgements

The authors would like to say thank you to Prof. Srinath Perera (Western Sydney University), Ms. Megan Smith (Taylor and Francis Group) and Prof. Daniel Forgues (École de Technologie Supérieure).

NOMENCLATURE
Q_{fan} = fan flow rate, cfm
A_{floor} = floor area, ft^2
N_{br} = number of bedrooms; not to be less than one

References

Anderson, P. (2007). 'What is Web 2.0? Ideas, Technologies and Implications for Education,' JISC Technology & Standards Watch, http://www.ictliteracy.info/rf.pdf/Web2.0_research.pdf. Accessed on 25 July 2009.

Armburst, M. *et al.* (2009). 'Above the Clouds: A Berkeley View of Cloud Computing, Electrical Engineering and Computer Sciences'. University of California at Berkeley, http://www2.eecs.berkeley.edu/Pubs/TechRpts/2009/EECS-2009-28.pdf. Accessed on 2 December 2010.

ATKINS. (2012). Future Proofing Cities, Risks and Opportunities for Inclusive Urban Growth in Developing Countries, Department for International Development and University of London: Development Planning Unit, Atkins, Epsom, UK.

Bacheldor, B. (2013). Tageos Apparel RFID Label Certified by the University of Arkansas, RFID News Roundup, RFID Journal Europe 2013, http://www.rfidjournal.com/articles/view?11116. Accessed on 28 June 2014.

Bluyssen, P.M. (2014). *The Healthy Indoor Environment, How to Assess Occupants' Wellbeing in Buildings*, Routledge, New York, p. 283.

Buchanan, G. and Shortliffe, E.H. (1984). Rule-Based Expert Systems: The MYCIN Experiments of the Stanford Heuristic Programming, Addison-Wesley, Reading, M.A. Sourced from Wilmering, T.J. and Mott, C.D. 2011. Health Management Systems Engineering and Integration, *System Health Management with Aerospace Applications,* p. 96.

Business Transformation Agency. (2009). Vocabulary-Driven Enterprise Architecture Development, Guidelines for DoDAF AV-2: Design and Development of the Integrated Dictionary, pp. 11–12, http://dodcio.defense.gov/Portals/0/Documents/DODAF/Primitives_AV-2_Guidelines.pdf. Accessed on 22 March 2013.

Butler, J. (2007). Automated Demand Response and BACnet: Article by Jim Butler, Cimetrics, from AutomatedBuildings, Buildings that have BACnet-based building control systems are particularly well positioned to participate in DR programs, Cimetrics Inc. 2011, http://www.automatedbuildings.com/news/sep07/articles/cimetrics/070831101606cimetrics.htm (accessed 12 November 2013).

Chandler, D. and Werther, W.B., Jr. (2014). *Strategic Corporate Social Responsibility, Stakeholders, Globalization, and Sustainable Value Creation,* 3rd Ed., Sage Publications, Inc., Thousand Oaks, CA.

Charles, K., Magee, R.J., Won, D. and Lusztyk, E. (2005). Indoor Air Quality Guidelines and Standards, National research Council Canada, Final Report 5.1—CMEIAQ-II: Consortium for Material Emission and IAQ Modelling II, http://nparc.cisti-icist.nrc-cnrc.gc.ca/eng/view/fulltext/?id=c597c638-536c-4ed9-b99c-20eb102a3bc0.

Clymer, J.R. (2009). *Simulation-Based Engineering of Complex Systems,* 2nd Ed., John Wiley & Sons, Inc., Hoboken, NJ, pp. 9–11.

Element Energy. (2009). Uptake of Energy Efficiency in Buildings. Report for the Committee on Climate Change, Final Report, 11 August 2009, Sourced from Leach, M., Deshmukh, S. and Ogunkunle, D. 2014. Pathways to Decarbonising Urban Systems, *Urban Retrofitting For Sustainability, Mapping the Transition to 2050,* p. 200.

Government Accountability Office (GAO). (2012). VA and DOD Department-Level Actions Needed to Assess Collaboration Performance, Address Barriers, and Identify Opportunities, GAO-12-992 VA and DOD Collaboration, United States Accountability Office, Washington, DC.

Hagan, S.R. and Onuma, K. (2014). FED iFM and iFM, Integrated Facility Management for Federal Agencies and the Broader Facility Owner Community, *The Future of BIM Vision-Integration-Savings, Journal of The National Institute of Building Sciences,* 2 (6): pp. 26–29.

Hay, D. and Healy, K.A. (2000). Defining Business Rules—What Are They Really? The Business Rules Group formerly, known as the GUIDE Business Rules Project, Final Report revision 1.3, http://www.businessrulesgroup.org/first_paper/BRG-whatisBR_3ed.pdf.

International Finance Corporation (IFC). (2012). IFC Sustainability Framework, Policy and Performance Standards on Environmental and Social Sustainability, Access to Information Policy, International Finance Corporation, Washington, DC.

Josuttis, N.M. (2007). *SOA in Practice, The Art of Distributed System Design,* O'Reilly Media, Inc., Sebastopol, CA.

Kundoo, A. (2015). Affordable housing, rethinking affordability in economic and environmental terms in india, *Inclusive Urbanization, Rethinking Policy, Practice and Research in the Age of Climate Change,* Edited by Shrestha, K., Ojha, H., McManus, P., Rubbo, A. and Dhote, K., Routledge, New York, pp. 108–119.

Kurtalj, N. (2011). 'Mashup, Data World Data Never Sleeps,' AutomatedBuildings.com, http://www.automatedbuildings.com/news/jan11/articles/brightcore/101230115707brightcore.html. Accessed January 2011.

Lavelle, M.R. (2014). *Cloud Computing for Better Energy Management, Pending Publication, Technical notes,* Lavelle Energy LLC, Indianapolis, IN.

Leach, M., Deshmukh, S. and Ogunkunle, D. (2014). Pathways to Decarbonizing Urban Systems, *Urban Retrofitting For Sustainability, Mapping the Transition to 2050,* Edited by Dixon, T., Eames, M., Hunt, M. and Lannon, S., Routledge, New York, NY, pp. 191–207.

LogicGem. (2006). LogicGem 3.0 QuickStart, Catalyst Development, Copyright © 1996, 2006 Catalyst Development Corporation.

MiraCosta Comprehensive Master Plan. (2011). Comprehensive Master Plan MiraCosta Community College District, http://www.miracosta.edu/governance/budgetandplanning/masterplan.html.

Missia, D.A., Demetriou, E., Michael, N., Tolis, E.I. and Bartzis, J.G. (2010). Indoor exposure from building materials: A field study. *Atmospheric Environment* 44 (35): pp. 4388–4395. doi:10.1016/j.atmosenv.2010.07.049.

Moller, A. and Schwartzbach, M.I. (2006). *An Introduction to XML and Web Technologies,* Addison-Wesley, Boston, MA.

National Building Information Modeling Standard (NBIMS) (2007). Transforming the Building Supply Chain through Open and Interoperable Information

Exchanges; Version 1—Part 1: Overview, Principles, and Methodologies, 2007 National Institute of Building Sciences, http://www.1stpricing.com/pdf/NBIMSv1_ConsolidatedBody_Mar07.pdf.

National Institute Building Sciences (NIBS). (2014). Financing Small Commercial Building Energy Performance Upgrades: Challenges and Opportunities, Council on Finance, Insurance and Real Estate, http://c.ymcdn.com/sites/www.nibs.org/resource/resmgr/CC/CFIRE_CommBldgFinance-Final.pdf. Accessed on 30 April 2015.

National Protection and Programs Directorate (NPPD). (2014). Sector Resilience Report: Hospitals, Office of Cyber and Infrastructure Analysis (OCIA) Including the Homeland Infrastructure Threat and Risk Analysis (HITRAC) and the National Infrastructure Simulation and Analysis Center (NISAC), http://www.dhs.gov/office-cyber-infrastructure-analysis.

Onuma, K. (2010). Location, location, location, BIM, BIM, BIM. *Journal of Building Information modelling*, Fall 2010, Cloud, pp. 21–22.

Open Geospatial Consortium (OGC) (2007). OGC Web Services Architecture for CAD GIS and BIM, Reference number of this OpenGIS Project Document: OGC 07-023r2, Version: 0.9, OGC Discussion Paper, Editor: Paul Cote, https://portal.opengeospatial.org/files/?artifact_id=21622. Accessed on 22 August 2011.

Oracle. (2008). Business Process Management, Service-Oriented Architecture, and Web 2.0: Business Transformation or Train Wreck? Oracle Corporation, World Headquarters, Redwood Shores, CA.

Redmond, A. (2013). Designing a Framework for Exchanging Partial Sets of BIM Information on a Cloud-Based Service, PhD Thesis Submitted to Dublin Institute of Technology, School of Real Estate and Management for Degree of Doctor of Philosophy.

Redmond, A., Alshawi, M. and Underwood, J. (2014). Designing Business Rules to Identify BIM Impact on Driving Policies for the Built Environment, The 1st International Conference on Industrial, System and Manufacturing Engineering (ISME'14), Amman, Jordan, November 11–13.

Redmond, A., Hore, A., Alshawi, M. and West, R. (2012). Exploring how information exchanges can be enhanced through Cloud BIM, Automation in Construction Volume 24, July 2012, pp. 175–183.

Redmond, A. and Smith, B. (2012). The use of semantic methods capable of supporting an Urban Sustainability Multi-Attribute Decision Model. The Higher Education Academy, STEM Annual Conference, April 12–13, Imperial College London.

Redmond, A. and Smith, B. (2013). *Designing a Cloud BIM Business Process Model Case Study, AACE International Transactions, BIM-1265*, AACE International, Morgantown, WV, ISBN 978–1–885517–80–7.

Redmond, A., West, R. and Hore, A. (2013). Designing a framework for exchanging partial sets of BIM information on a cloud based service. *International Journal of 3-D Information Modeling* 2 (4): 12–24. IGI Publishing, Hershey, PA.

Rich, C., Singleton, J.K. and Wadhwa, S. (2013). *Sustainability for Healthcare Management, A Leadership Imperative*, Routledge, New York, pp. 17–36.

Schools Interoperability Framework (SIF). (2012). SIF Data Model Implementation Specification (US) 2.6, http://specification.sifassociation.org/Implementation/US/2.6/html/.

Squires, G. (2013). *Urban and Environmental Economics, An Introduction*, Routledge, Abingdon, UK.

Stair, R., Reynolds, G. and Chesney, T. (2008). *Principles of Business Information Systems*, Cengage Learning EMEA.

Thorel, M., Andrieux, F. and Buhe, C. (2013). Knowledge Management Supporting Decision Making in Holistic Building Renovation Design, CIB W078 2013, International Conference on Information Technology for Construction, Tsinghua University, Beijing, China, October 9–11.

Walton, C. (2006). *Agency and the Semantic Web*, OUP, Oxford.

Wang, H., Yi, X., Lai, J. and Li, Y. (2005). Fabricating microbolometer array on unplanar readout integrated circuit. *International Journal of Infrared and Millimeter Waves* 26 (5): pp. 751–762.

Wasson, C.S. (2006). *System Analysis, Design, and Development: Concepts, Principles, and Practices*, John Wiley & Sons, Inc., Hoboken, NJ, p. 735.

Wikipedia. (2013). Microbolometer, http://en.wikipedia.org/wiki/Microbolomete. Accessed on 21 November 2013.

Wilmering, T.J. and Mott, C.D. (2011). Health Management Systems Engineering and Integration, *System Health Management with Aerospace Applications*, Edited by Johnson, S.B., Gormley, T.J., Kessler. S.S., Mott, C.D., Patterson-Hine, A., Reichard, K.M. and Scandura, P.A. Jr, John Wiley & Sons Ltd, West Sussex, UK, pp. 95–113.

10 Impact of collaboration tools and shaping the future of data exchange – A model for BIM communication waste

George Charalambous, Peter Demian, Steven Yeomans and Tony Thorpe

10.1 Introduction

Information communications technologies (ICTs) play a vital role within the information intensive construction sector. As with all ICT-supported industries, advances in cloud computing are gradually leading to the eradication of traditional software architectures as they give way to the paradigm of Service Orientation. Simultaneously, the fast-paced development and propagation of building information modelling (BIM), has meant the term BIM often serves as an umbrella-term that encompasses service-orientation. One vital aspect of such developments is the growing requirement for BIM compliant collaboration tools that can interoperate effectively with various software systems, support required standards and codes of practice, and provide for requirements of construction project information production and management e.g. tools based on model-based workflows.

The pre-requisites of collaboration can be broadly divided into two categories, which are in practice strongly interdependent: (1) coordination of information and responsibilities, and (2) communication. Research described within this chapter recognises recent and continuing efforts to provide effective coordination of information and responsibilities, and aims to support the *communication* aspect of collaboration. Such tools and environments depend on innovative communication tools and novel architectures enabled by advances in cloud computing.

This chapter presents the outcomes of a research project to develop a context-specific conceptual model-ontology, which can support the discourse of requirements engineering and provide a robust and universally applicable framework for evaluating the communication capabilities of BIM collaboration tools. To develop the model, BIM collaboration tools were examined from five phases-perspectives: (1) users and their requirements, (2) vendors, (3) schemata for data interoperability, (4) collaboration tool use patterns and (5) improvement. Each phase-perspective helped to define more specific requirements for the model as well as elements of the model itself. The end result is a 'Model for waste in BIM process interactions', 'WIMBIM'. To evaluate the validity and utility of this model, interviews with BIM experts were conducted. The results indicated that the model is valid and useful as it helps address an aspect that is not usually examined, offering the potential to complement the existing model for BIM maturity. Furthermore, the model provides a useful lens for further academic research into BIM collaboration tools.

Building information modelling promises significant efficiency gains through improved information flow and elimination of the various kinds of waste within the construction process. Cloud collaboration tools such as online collaboration platforms have been used to deliver some of this efficiency improvement. Serving as a central repository for project information, these tools facilitate improved communication and resource sharing between geographically distributed teams. Recent advances have witnessed the incorporation of integrated or bolt-on BIM modules for example, online model viewers offering the functionality to interrogate and communicate around a shared BIM.

This research focuses on the *communication* aspect of BIM collaboration, specifically in developing a conceptual model to support Requirements Engineering for BIM collaboration tools. The overall aim of the research project was to develop a conceptual model to support Requirements Engineering for BIM collaboration tools. As a means to achieve this, the following objectives were pursued:

- To identify and address the key aspects in the process of Requirements Engineering for BIM collaboration tools.
- To identify the challenges faced in the Requirements Engineering for BIM collaboration tools and which specific areas could benefit from improved and more explicit conceptual models.
- To identifying the key elements in this process (which concepts are universal and persist through time).
- To identify the relationships between these concepts and relate them to concepts found in current standards, literature and used in the discourse of Requirements Engineering for BIM collaboration tools.

Both qualitative and qualitative data were utilised. Additionally, the research had a strong ethnographic element since the researcher was based in a collaboration software company, observing the process of Requirements Engineering for a BIM collaboration tool.

10.2 Review of relevant literature

10.2.1 The AEC-FM process and collaboration

On a project level, the purpose of the construction industry is to produce a built artefact as well as the service involved in operating and maintaining it and the information that supports these processes. Koskela (1997) propose conceptualising the design process in the construction industry simultaneously in three different ways; (1) Conversion, (2) Flow and (3) Value Generation. They argue that, for the purpose of waste reduction, the Flow view and Value Generation view can offer more suitable representations over the traditionally more established Conversion view. The ability of concurrent contribution from multiple agents in a construction project is often limited because of the interrelatedness of their inputs (Froese, 2010). As a result, significant bottlenecks in information flow occur. Current practice lacks the early contribution of all disciplines to design decisions, resulting in design rework, constructability issues and construction rework and suboptimal design decisions. BIM technology can address such issues by opening channels of communication and 'instigating' early contribution from agents of different disciplines (Succar, 2009) resulting to a better informed design from the early phases. A BIM model automatically changes communication patterns as it acts as a central building information repository. The traditionally chaotic state of information exchanges would transition to a more ordered state. As noted by Isikdag and Underwood (2010) '… effective collaboration can only be achieved through effective coordination and communication'.

10.2.2 Communication in AEC-FM

The multi-disciplinary nature of construction projects, the transient nature of project teams and the persistent lack of adequate standardisation make project communication particularly challenging. BIM offers the opportunity for new communication paradigms. There are various ways to classify communication within construction ICT systems. These classification approaches can serve as appropriate analysis tools for different purposes. These include classifications according to the content, purpose, project phase, or communication tool/medium. The multiplicity of communication media poses a significant challenge to control and standardisation in project communication.

The last two decades have seen considerable research towards adapting the principles of lean manufacturing for application in construction (Koskela, 1997 and Ballard and Howell, 1998). More recently, the relationship between BIM and Lean has been explored (Sacks *et al.*, 2010 and Dave *et al.*, 2013). The basis of these approaches is to understand the construction process as flow, create systems that favour flow, eliminate waste in time and material and maximise value to client. Despite communication being an essential enabler for 'lean construction', it has not traditionally been the focus of lean approaches as their key objective is to eliminate waste in the form of time and material. However, there has been work on construction communication which lays the ground for equivalent, metrics-based and waste elimination-focused approaches. Communication can be observed, tracked, evaluated (Becerik and Pollalis, 2006) and quantified more distinctly and effectively than collaboration can be. Tribelsky and Sacks (2006) have developed and implemented performance indices for information flow within construction projects.

Communication theory enables researchers to study communication through a more rigorous and universal set of concepts. It views communication explicitly as an act with a purpose and allows, to some degree, the evaluation of the efficiency of a given communication act. The fundamental elements of communication theory, as it has been defined by Shannon and Weaver (1949) are the source (or information source), the sender (or transmitter), the channel, the message, the receiver, the destination and any noise. Dainty *et al.* (2007) adapted these concepts to the context of construction projects, accounting for the relevant traits such as project specificity, transience, unknown organisations, conflicting objectives, referenced information and the chaotic nature of information sources.

10.2.3 *Coordination and coordination tools*

Coordination can be generally understood as 'the orderly arrangement of group effort, to provide unity of action in the pursuit of a common purpose' (Mooney, 1947). Similarly to communication, coordination is a very broad concept whose manifestation could be tracked universally across studies on project management. Isikdag and Underwood (2010) designate BIM coordination issues as versioning, data ownership, model breakdown, information consistency, workflow management and conflict management. Within online collaboration coordination relates to scheduling, user action, user responsibility, model versioning and spatial co-ordination of models (clash detection).

Three main categories of coordination can be identified in the context of BIM: (1) coordination of information, (2) coordination of access to and rights to modify information, and (3) coordination of collaborator effort. These are highly interdependent e.g. well-coordinated project information facilitates coordination of information access and coordination of effort. This research classifies these tools collectively as 'coordination tools'. They exist in various forms from standards to templates including BS 1192 (effort and information coordination), Model Production and Delivery Tables (effort and information coordination), Information Delivery Manual (coordination of collaborator effort by coordinating communication and interaction), Model View Definitions (information coordination), Access Rights tables (information access coordination), BIM Governance Models (Rezgui *et al.*, 2013) (information, access and effort coordination).

10.2.4 *Online collaboration platforms, service-orientation, cloud computing and BIM*

Online Collaboration Platforms (OCPs) are the combination of web-based technologies that create a shared interface, to link multiple interested parties, to share, exchange and store project information in digital form, and to work collaboratively, on the basis of subscription fee, license plus main-tenance, negotiated fixed cost or exclusive business partnership agreement (Liu *et al.*, 2011). OCPS are also closely linked with the concepts of Cloud Computing, Software-as-a-Service (SaaS) and Service-Oriented Architec-ture (SOA). SaaS is summarised broadly as a more user-centric, flexible and modular way of offering software to users. Some implications of the SaaS paradigm to BIM and this study include:

- A need for standards to harmonise the emergence of a range of hetero-geneous applications.
- A characteristic flexibility and modularity which offers the potential for improved services based and added to existing, 'basic' solutions.
- On a more abstract level these technological and business paradigms move the focus on providing a service and improving efficiency rather than providing a software product, hence eliminating some services and processes which are non-value adding.

Isikdag and Underwood (2010) envision that 'cloud computing will ena-ble the next generation of (full state) BIMs' (or BIM 2.0) where the 'digital building model will evolve through the lifecycle of the building'. In this inte-grated environment the Internet will act as the medium through which the BIM model will be continuously updated and open for new information. Beach *et al.* (2011) argue that online collaboration platforms could address the universal BIM adoption issues of 'data sharing, access, and processing requirements'. A study by Liu *et al.* (2011) on the 'marketed functionalities' of OCPs in the UK revealed that communication features are markedly the least satisfied category (the other three categories being System Administra-tion, Document Management and Workflow Management).

Apart from the more traditional paradigm of storing 3D CAD and BIM files on the document management systems, increasingly more OCPs offer online IFC model servers with the ability to view, merge, interrogate IFC models and set-up workflows around them as well as automatically generate COBie spreadsheets. The level of uptake of these OCP BIM modules has not been satisfactory. This is owed primarily to the low reliability of model content as conversions from proprietary BIM software to the IFC standard tend to be associated with considerable data losses.

A number of studies have called upon the need for project collaboration to depart from the document-based paradigm and place the structured model as the focal unit of communication. In fact, model-based working and model-based communication are often seen as indicators for BIM maturity (or its equivalent concept). Aouad *et al.* (2005) have critically described project information as 'unstructured and document based'. Yeomans *et al.* (2006) revealed that the 'single build model' was the least adopted out of eight col-laborative working techniques. In their ICT Vision mapping, Rezgui and Zarli (2006) suggest that document-centric information exchange should be replaced by model-based ICT. Succar (2009) describes progression in BIM maturity by replacing document-based workflows; Isikdag and Underwood (2010) claim that 'the traditional nature of the industry is extremely document-centric'.

The model-based paradigm has a significant effect on the efficiency of communication and coordination. OCPs and collaboration tools in general are the main catalysts for such efficiency improvements as they largely define the way in which users interact with information and interact with each other in reference to that information.

10.2.5 Requirements Engineering

Requirements Engineering is a systematic approach which 'helps determine what to develop, how to develop it, and when it should be implemented' (Aouad and Arayici, 2010). It can defined generally as 'the subset of systems engineering concerned with discovering, developing, tracing, analysing, qualifying, communicating and managing requirements that define the system at successive levels of abstraction' (Hull *et al.* 2005). The basic principles include (Hull *et al.*, 2005; Arayici *et al.* 2006): making decisions traceable, accounting for the whole system in question and not just the technological part, defining appropriate representations/models of systems and sub-systems, stakeholders and requirements, and involving stakeholders throughout the process

In the domain of Requirements Engineering for BIM collaboration tools, two effects emerge as a result of the natural traits of AECFM (project specificity and project-led nature, inadequate standardisation, discipline fragmentation, life-cycle phase fragmentation) and the emergence of cloud-based solutions. These are:

- Cross-project variation in both high-level software configuration (what combination of software to use) and low-level software configuration (which part of each software to use). The vague distinction between the roles of software calls for an approach supporting flexibility (from the perspective of project set-up) and prioritisation (from the perspective of software development).
- Requirements Engineering for cloud-based solutions tends to be a combination of moving existing functionality to the cloud as well as devising novel, 'fit-for-cloud' functionality.

10.3 UK BIM developments from the perspective of a cloud collaboration tool vendor

The decision of the UK Government to 'introduce a progressive programme of mandated use of fully collaborative BIM for Government projects by 2016' (Cabinet Office, 2011) leveraged on the existing drive to utilise the potential of BIM technology to address prolonged industry problems.

There have also been a number of development sin BIM- standards, specifications, guidelines and protocols in the UK, notably:

- PAS 1192-2:2013 (BSI, 2013),
- PAS 1192-3:2014 (BSI, 2014),
- COBie UK-2012 (Nisbet, 2012),
- COBie data drops (Cabinet Office and BSI, 2013),
- BS 1192-4 (BSI, 2014),
- The BIM Overlay to the RIBA Plan of Work (Sinclair, 2012),
- CIC/BIM BIM Protocol (CIC, 2013a) (incorporating coordination constructs/tools 'Level of Detail' and the 'Model Production and Delivery Table'),
- CIC/BIM Best Practice Guide for Professional Indemnity Insurance when using Building Information Models (CIC, 2013b),
- The Employer's Information Requirements (BIM Task Group, 2013), and
- The Government's Soft Landing Policy (Cabinet Office, 2012).

The challenging task for the government to control, maintain, record and act upon a healthy level of communication with industry has been achieved by the formation of the BIM Task Group (BIM Task Group, 2014) which serves as an official BIM hub and includes 'housed' initiatives such as the 'BIM4' groups. The BIM adoption movement can be described as an open two-way discussion between the Government/BIM Task Group and practitioners. This has meant that it has had a strong experimental aspect. The 'early adopter project' on the Ministry of Justice, Cookham Wood facility (MoJ, 2013) produced promising results as well as some lessons for the use of COBie. The, more extensive and hence more challenging, 'Open BIM / COBie trial' on the Gatwick Airport (BRE, 2014) revealed that despite the positive approach demonstrated by leading contracting, design and software companies, some technical issues regarding IFC and COBie were hindering adequate information flow. Users have been reluctant to use IFC-based online BIM tools offered by OCPs. Nevertheless, there has been considerable effort to utilise OCPs as the Common Data Environment as defined in BS 1192:2007 (BSI, 2007). There is however, uncertainty as to the exact role OCPs should have within the BIM process.

10.4 The need for a better conceptual framework – gaps in shared 'BIM constructs'

It is evident from the review of literature and the review of developments in BIM adoption in the UK that most effort has been exercised in creating coordination tools such as the BS1192 and BIM Governance Models. It is proposed that the communication aspect of BIM has not been given the equivalent attention. In Cerovsek's (2011) 'multi-standpoint framework for technological development' the need to recognise this is highlighted. Cerovsek's approach in devising a BIM framework is based on the recognition that BIM is a characteristically multi-aspect domain. Cerovsek identifies two important issues within BIM:

- The need for BIM research and practice to recognise that BIM will always be an evolving field. The implication from this is that BIM frameworks need to be robust enough to accommodate this evolvability i.e. they should not be limited by the capabilities of specific technological paradigms.
- The need to understand that BIM is fundamentally about communication and the resulting need for BIM frameworks to incorporate communication theory.

The review in this study finds that current BIM frameworks do not satisfy the above. A central aim of this study is to provide material to address them. Specifically for OCPs, there is a need for creating and maintaining a shared understanding amongst individuals from different disciplines who normally work in different working environments while an information system is conceptualised. Therefore, shared conceptual models which offer appropriate representations of the system and its intended attributes have an important role. Requirements Engineering offers useful principles for aligning technological paradigms to address such problems. Context-specific 'languages' and performance metrics are required to support the discourse of Requirements Engineering for BIM collaboration tools. These are usually supported by conceptual frameworks-models. There has been significant mobilisation in BIM in the UK, in many ways characterised by uncertainty which has had an observable impact on the OCP software development world. In the context of collaboration tools, the role of OCPs in the BIM process, has not been made explicit neither by the Government BIM roadmap neither by its use in practice.

10.5 Requirements Engineering for BIM collaboration tools through five perspectives

Requirements Engineering for BIM collaboration tools was examined through five perspectives-phases, which helped identify the specific requirements and basic elements of the model.

10.5.1 Perspective 1: understanding AEC-FM practitioners: perspectives, current use and requirements from BIM and OCPs

AEC-FM practitioners were surveyed to capture their views on BIM and OCPs, the way they currently use these technologies, their requirements and the relevant barriers. In parallel, the analysis was used to raise issues regarding the effectiveness of current terminology and the assumed models in communicating about requirements engineering for BIM and OCPs. The survey questionnaire survey consisted of nineteen questions on BIM adoption, interoperability and the drivers for BIM and lifecycle phase-based questions as well as the perceived benefits on the use of web-based collaboration. The main findings included:

- A lack of confidence in IFC data exchanges is a critical barrier to web-based BIM collaboration.
- Graphical User Interfaces (GUIs) for BIM collaboration tools require significant improvement.
- A gap and therefore an opportunity for OCPs to support the early stages of project preparation and conceptual design. This means that collaboration tools should enable GUIs at the Preparation and Design phase where the seamless flow of *intent* is critical.
- Evidence that the researcher, company and respondents had a vague understanding of what they are after, but are not able to explicitly express these requirements.
- Regarding the requirement for flow of intent, the terminology was not fully able to support the communication of this concept in the requirements engineering process.

10.5.2 Perspective 2: understanding the software vendor

The most significant findings from the literature review and the questionnaire survey were used to generate a set of two semi-structured interviews with a senior implementation consultant with fifteen years' experience in construction IT and the company's professional services manager. This aimed to capture the company's perspective and relate it to the perspective of the practitioners and the concepts and proposals. The main findings were:

- The difference in the perspectives of vendors and users and the associated uncertainty in overall Requirements Engineering area require a more robust model for long term development.
- There is significant variation in uses of, and requirements from the collaboration tool studied.
- The importance of politics and high-level agreements within the construction software domain was appreciated.
- The importance of user interface and easiness of use (which was also expressed from users in Perspective 1) was also appreciated.
- The lack of a robust way to communicate about requirements from BIM-enabled OCP within the Requirements Engineering process was observed (as in Perspective 1).

10.5.3 Perspective 3: a closer examination of a web-based BIM tool

There was a general concern that data conversions from native software to IFC omitted information to varying degrees such that organisations actively sought to bypass the use of IFC files. For this reason, a closer examination of an IFC-based online tool was conducted. The examination served as a data fidelity study as well as a study on the efficacy of the examined tool as a communication and coordination tool. The main findings from Perspective 3 are:

- It has been verified through examples that IFC is not producing adequate data transfers and that this stands as a critical barrier for utilising web-based collaboration tools.
- The building information conversion and exchange process is not simple enough to be conducted by non-specialist user effectively. There is a need for specialised knowledge and/or strict conversion protocols.
- There is a need for rigid protocols to guide export and coordination process.
- Apart from data fidelity, which is key, the importance of User Experience, User Interface, easiness of exports and model management are also important barriers i.e. it's not only BIM model data that is lost: Time in BIM workflows is lost, communication of intent is lost. Furthermore, there are also less easily identifiable wastes through information overload.
- Such tools should focus on the intuitiveness and efficiency of BIM model-based communication in order to provide a more useful solution.
- With respect to assumed conceptual models and terminologies: There are various ways in which users can access and exchange data, however, there is no standardised way of referring to them. Furthermore, there is no standardised way of referring to the data and type of data required for them to be effective.

The above process elicited five principles for the requirements from BIM model-based communication:

Principle 1: The model should be placed at the centre of communication. In other words it should act as the focal point of project communication.
Principle 2: The model should be as integrated with associated documents and processes as possible.
Principle 3: OCPs should provide informal communication channels and foster user familiarity. Communication tools should enable the flow of intent and the association of events in face to face communication.
Principle 4: Communication and coordination for effective collaboration cannot be performed distinctly.
Principle 5: Information exchange at the human-to-human communication level should benefit from further standardisation.

10.5.4 Perspective 4: use of software and patterns in digital communication: analysis of communication data and meta-data from BIM collaboration tool workspaces

Communication data and meta-data from projects utilising the examined collaboration platform were collected and analysed in order to explore any relevant patterns in project communication and relate them to concepts such as communication efficiency and BIM maturity. In parallel, identify concepts, which should be included in the conceptual model to be developed would support the requirements engineering discourse. The most central concepts examined in the analysis are explained below:

- User: any project stakeholder who is able to participate in digital project collaboration.
- A container (or resource): anything that could hold information that is relevant to the project. This information could be building information, specifications, requirements, meeting minutes, building regulations etc. A container (or resource) could be in the form of document, a 2D drawing or a 3D model.
- A transmission: Any exchange of information from one user to another. This could be the transmission of project information and/or instructions or opinions in reference to project information or other containers.
  - *Purpose of transmission*: every transmission had a purpose. This was often (not always) explicitly identified within project communication.

Three types of analysis was also conducted:

Statistical (metric-based) analysis of communication meta-data
Data from the reports-spreadsheets was used to generate graphs which illustrated relevant patterns through communication meta-data and related to the five principles proposed in Perspective 3.

Social organisation network analysis on the communication meta-data
Analysis 2 utilised the network properties of the data from the spreadsheets-reports extracted from online workspaces (in every transmission there is a sender and a receiver). In Action Distribution, (sender) users assign an Action in reference to a container to specific users (receivers) while in Commenting (sender) users direct their comments in reference to containers to other users (receivers). The data captured was used to produce network graphs using Social Network Analysis software Gephi (Bastian *et al.*, 2009). Users were represented by the nodes in the network and the interaction between them, the transmissions, are represented by the edges (or ties) in the network. Visual and network metric-based analysis of the networks were used to elucidate patterns in project communication that was facilitated by online workspaces.

Interpretative analysis of communication data (the content of the messages exchanged)

The two predominant ways in which users were able to communicate in reference to uploaded resources (containers) were:

- The Commenting Functionality: users would comment in reference to a specific resource (container) but had the ability to associate other resources (containers) already uploaded or attach a new resource. Therefore, the comment receiver could access the associated or attached resource by clicking on a link that would appear in message screen.
- The, more structured and formal, Form Functionality: forms of predefined structure, typically standardised for the purposes of each project, were used for more structured communication. The form was created independently of any resource (container) but, like in Comments, the user had the option to associate and/or attach a resource (container).

10.5.4.1 Findings from perspective 4

Perspective 4 analysed a type of data set in three significantly different ways. The findings are grouped into the following themes.

10.5.4.1.1 BIM TRANSMISSION (OR 'BIM MESSAGE') AND ITS EFFICIENCY

It is evident that within a project there is a huge number of digital transmissions. Additionally there seems to be considerable amount of waste in terms of communication efficiency and effectiveness. This highlights the potential in eliminating some of the waste in them. This is a matter of whether a transmission should take place as well as how can a required transmission be as efficient as possible.

Waste in transmissions manifests itself as:

- Lag in comments: The 'lag' in responses relates to the 'response latency' as presented by Koskela *et al.* (2013) and (Chachere *et al.* 2009).
- Sending information and instructions to too many receivers
- Information overload
- Lack of immediacy in accessing relevant containers
- Lack of immediacy in referencing parts (e.g. objects) in containers

The interpretation of Comment Content and Form message content revealed the different elements/flows in a transmission:

- Building information: to be modified (including attributes like the state of acceptance of an object or model), to be incorporated, to be consulted.
- Project requirements and specifications
- Industry codes and regulations

- Instructions
- Reference to project event (including other communication events and project actors/ software users)
- Intent (the communication layers necessary for turning project requirements into results).

10.5.4.1.2 DOCUMENT-CENTRIC VS. MODEL-CENTRIC COMMUNICATION
PARADIGM

If the projects studied represent typical communication settings (e.g. number of comments on documents) in construction projects then the challenge for BIM collaboration tool vendors is to create an effective model-based environment which would more efficiently satisfy the collaboration requirements described by the principles expressed in Perspective 3 and ideally eliminate any unnecessary, non-value adding steps within communication.

10.5.4.1.3 SHARED CONCEPTS AND TERMINOLOGIES – REQUIREMENTS FROM
AND ELEMENTS FOR CONCEPTUAL MODEL

This Perspective-Phase has showed that there is significant variation in project software configuration across projects. The existence of multiple software and the resultant need for varied project software configurations has steered the attention away from the fundamentals of communication and contributed to some unintended consequences: poor overall user experience, poor information management and poor knowledge management. It is evident that currently communication tools do not satisfy all communication dimensions: Formal and Informal communications channels, Model/ object-based communication, fostering familiarity, supporting immediacy in communication exchange and supporting transparency in collaborative project information management. Additionally, the analysis has showed that human-human model-based interaction will benefit from further formalisation.

10.5.4.1.3.1 Levels of representation/analysis of communication networks Network Analysis illustrated how project communication can be represented by networks. The networks studied are only two out of many different networks that can be conceptualised. Alternatively, they can be understood as 'layers' (e.g. the action distribution layer) of project-level interaction.

10.5.4.1.4 SEMANTIC TECHNOLOGY – EXPRESSING THE ABOVE DIFFERENTLY

It has been acknowledged that an important type of networks that haven't been analysed are user-container-user networks. In addition there are object-object networks formed by the relationships between model objects.

The project can be represented as a 'knowledge graph'. This provides the link to the utility of semantic technology as it has the power to leverage the semantics within these networks in order improve the efficiency of interacting with project information. In other words, using the project data (given that it is adequately structured) as a knowledgebase.

10.5.5 Perspective 5: mechanisms for improving BIM collaboration tools – an approach for identifying and evaluating opportunities offered from semantic technology to BIM-enabled OCPs

The importance of semantic interoperability has been acknowledged, mostly implicitly, in previous perspectives. Perspective 5 builds a requirements-engineering approach that is specific both to BIM-enabled OCPs Semantic Technology. Through a demonstrated attempt to identify and evaluate opportunities offered to OCPs by semantic technology, a context-specific requirements-engineering process is developed and documented.

'Solving a problem simply means representing it so as to make the solution transparent' (Simon, 1981). Following this notion, Perspective 5 attempts to solve the technology implementation problem by providing suitable representations of different aspects of the problem. The steps followed are outlined as:

1   Deducing the pre-requisites for an effective semantic functionality and the stakeholder context:
2   Understanding the nature of opportunities offered by Semantic Technology in AEC-FM: Their commonality was that they were eliminating a kind of waste that was previously 'not observable' amongst practitioners.
3   Identifying a suitable representation of the role of OCPs in BIM process: BIM Use Purposes were selected as a language for scoping the role of OCPs in BIM and combined with OCP-specific heuristics to devise illustrative use-cases.
4   Identifying a number of illustrative, OCP-specific functionalities.
5   Devising a method for evaluating these functionalities: Their value can be represented as semantic waste elimination and quantified by adapting the *NIST (2004)* framework. Their relative importance can be identified by surveying experts (and users in future work).

10.5.5.1 A fitting representation of opportunity: value as waste elimination

Step 2 helped identify that value to the user can be represented as waste elimination and, in this case, elimination of 'waste in meaning' or 'cost of inadequate semantic interoperability'. Elucidation and evaluation of this waste can be achieved by comparing current technology and process to counterfactual scenarios where semantic interoperability is present. The 'Cost Analysis of Inadequate Interoperability in the U.S. Capital Facilities Industry' (NIST, 2004) provides a useful tool for this approach. Specifically 'Table 4.1: Summary of Technical and Economic Metrics' was used as a basis for evaluating the seven illustrative functionalities identified in Step 4.

10.5.5.2 Evaluating opportunities offered to OCPs by semantic technology – semi-structured interviews for gathering expert opinions

BIM collaboration tool implementation consultants were interviewed in order to evaluate a set of proposed functionalities. Amongst the main findings from the interviews was that the concept of waste, and in this case, waste in semantics, despite at first requiring some clarification was effective for explaining and discussing the capabilities and benefits of a new technological paradigm such as semantic technology.

10.5.5.3 The approach and its utility

The approach that was followed in this Perspective is captured in a flow chart form in Figure 10.1. The captured process can help communicate the approach, track decisions and revise the approach. Within the OCP vendor, it helps compare current ways of working to a semantic technology-enabled state and characterise the natural contribution of semantic technology. Additionally it can serve as a mechanism for communicating gaps and aligning pre-requisites within the industry. Ultimately, the approach can form the basis for an automated requirements elicitation system, given the availability of repositories and codification of resources.

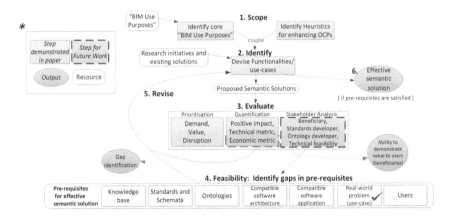

Figure 10.1 Outline of requirements engineering approach followed.

10.5.5.4 Contributions of perspective 5 to the conceptual model

The work in Perspective 5 leads to two main conclusions which contribute to the conceptual model developed in this study:

- Waste provides an appropriate reference concept for an inclusive requirements-engineering process and especially for the purpose of introducing new technological paradigms.
- The capabilities of a container of project information (e.g. semantic richness or the ability to 'understand' or 'explain' the meaning of its content to collaboration systems) significantly impact the efficiency of BIM interactions.

10.5.6 Conclusion – The need for WIMBIM and the emergence of a preliminary WIMBIM

The purpose of the research was to identify the requirements in a conceptual model for use in requirements engineering for BIM collaboration systems (i.e. what it should be used for) as well as to identify its basic elements. The emergent model is called the 'Model for Waste in BIM process Interactions' or 'WIMBIM'. The requirements and elements, as they have arisen from the Perspectives-Stages in are:

1   The language (in the form of shared terms-concepts and metrics) commonly used within practice is not powerful, universal and robust to support the discourse of requirements engineering for BIM collaboration tools effectively. Currently, there is no resource to support this effort.
2   There is an opportunity for OCPs to support BIM communication in early project stages by better facilitating the flow of intent in interactions.
3   Uncertainty in the domain of BIM collaboration tools is a significantly hinders confidence in making decisions for tool development.
4   OCPs should provide informal communication channels and foster user familiarity. This also relates closely to the requirement for the flow of intent as well as to the association/linking of events that have occurred through face to face communication to content in BIM collaboration tools.
5   There is a need to analyse and provide a formal, universal and robust description of a BIM transmission (or 'BIM message' or 'BIM interaction').
6   The different types and levels of representation, including the representations of the various networks in a project have the power to elucidate efficiency (and waste) in different ways.
7   Waste provides an appropriate reference concept for an inclusive requirements-engineering process and especially for the purpose of introducing new technological paradigms.
8   The capabilities of a container (e.g. semantic richness or the ability to 'understand' or 'explain' the meaning of its content to collaboration systems) significantly impact the efficiency of BIM interactions.

10.6 A model for waste in BIM communication

The primary aim of the WIMBIM is the eradication of BIM communication waste through a better understanding of the waste and how it comes about (i.e. its relationship with the WIMBIM elements). By introducing, new, helpful notions of efficiency in requirements engineering for BIM Collaboration Tools, a BIM collaboration tool vendor will more effectively work towards enabling a user to achieve the most from a BIM process interaction. WIMBIM does not aim to impose a way of working in order to eliminate waste but rather aims to make the different kinds of communication waste observable so that BIM collaboration tools can be improved and configured in order to reduce waste.

To achieve this aim, the WIMBIM needs to introduce waste as a more distinct concept within a robust framework (i.e. a framework that is not constrained to specific technology paradigms and specific tools). It follows that it is critical that WIMBIM should effectively provide a common reference, should be concerned with concepts that are universal, and be actionable.

WIMBIM is a set of interrelated concepts which can be used to better describe communication waste within BIM process interactions. The model has the single BIM transmission as the focal unit of analysis and is then built up in a logical way.

10.6.1 BIM flows

'Information' is one of the seven flows in construction that Koskela (2000) identify (Information, Material, Crew, Equipment, External Conditions, Space, and Connecting Previous Works). Ballard *et al.* (2002) defines a 3 Type Model which consists of Directives, Pre-requisites (including design information) and Resources (Figure 10.2).

The flows of Information/Pre-requisites and Directives are broken down to form what WIMBIM calls 'BIM Flows'. These are:

- BIM data, to be:
  - modified
  - incorporated
  - consulted
- Project specific data.
- Non project-specific data. E.g. building regulations
- Context of issue communicated
- Instruction or Response
- Intent

10.6.2 BIM transmissions

A BIM Transmission is the transmission of data relating to one or more BIM flows from a User to another User or from a User to a Data Container. The collective effect of BIM Transmissions is called BIM Interactions.

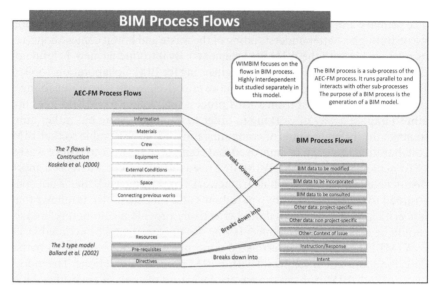

Figure 10.2 AEC-FM process flows and BIM process flows.

10.6.2.1 Main categories of BIM transmissions

As illustrated in Figure 10.3, transmissions can be grouped into:

- User-Data Container-User transmissions.
- User-Data Container transmissions.

Note: User-User transmissions are a false concept since a BIM process is defined as the series of interactions whose collective purpose is to generate a BIM model.

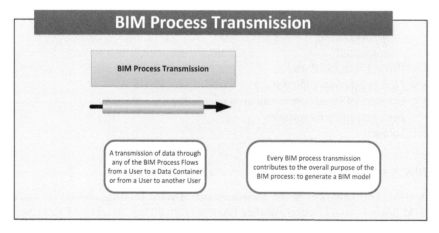

Figure 10.3 The BIM process transmission: the focal unit of analysis of WIMBIM.

10.6.2.2 BIM transmission purpose

Every BIM transmission has a purpose. The purpose defines the required flows and the required subsets of data within these flows.

The transmission purpose always contributes to the ultimate purpose of the BIM process; to generate a BIM model. These purposes relate closely to:

- The 'BIM Use Purposes' (Kreider and Messner, 2013)
- Collaboration information flow concepts such as 'For Information', 'For Acknowledgement', 'For Comment' etc. which are used by collaboration tools such as Asite (2017).
- 'Collect, Create, Correct, Connect' (Coates *et al.*, 2010)
- Modelling, Derivation, Composition (Rezgui *et al.*, 2013)

10.6.2.3 Required and executed transmission

The purpose of a Transmission defines the Required Transmission. In practice, this is typically never the same as the Executed Transmission (Figure 10.4).

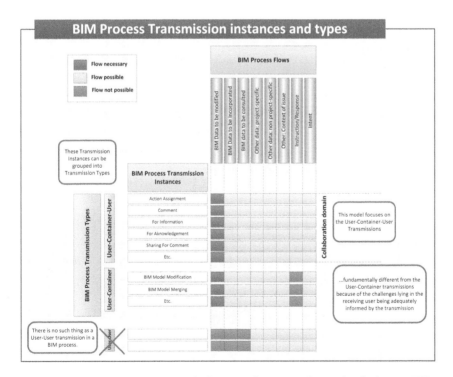

Figure 10.4 BIM process transmissions: main categories and relation to BIM process flows.

10.6.3 *BIM transmission wastes*

BIM Transmission Waste is any discrepancy between the Required Transmission and the Executed Transmission (Figures 10.5 and 10.6).

There are six different types of discrepancies (an example is provided in Figure 10.6):

- Right and unnecessary data in the required flow
- Data in non-required flow
- No data in required flow
- Wrong data in required flow
- Part of right data in required flow
- Data in a non-BIM flow

Note: A required transmission can never be fully understood or executed, i.e. there will always be a level of Communication Waste in practice.

10.6.4 *BIM data containers*

A container (term also used in BS1192) of data that corresponds to BIM Flows. Each Efficiency State (explained later on) offers improved BIM Data Containers. Examples are PDF document, IFC model, IFC object, Revit model, COBie spreadsheet etc.

Container types have attributes such as:

- Structure
- Semantics

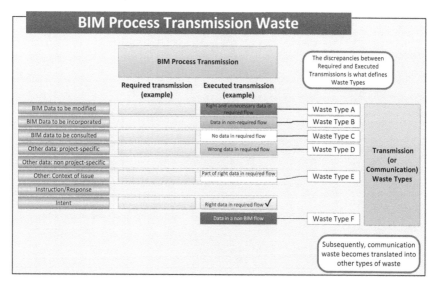

Figure 10.5 BIM process transmission waste: the 6 different types (A–E).

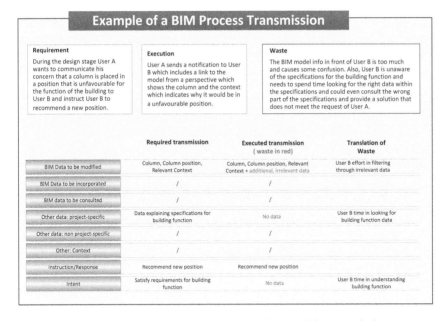

Figure 10.6 An example of a BIM process transmission with transmission waste.

- Granularity
- Interoperability
- Openness/accessibility

These attributes give rise to: 'spectra of fitness' of Containers: spectra which denote the suitability of instances of Containers to specific uses of BIM Collaboration Tools.

10.6.5 BIM transmission media

The media through which the transmissions/interactions take place. Each efficiency state offers improved BIM Transmission Media. These relate closely to:

- BIM software: design, check, coordination, collaboration etc.
- E-mail, telephone.

A BIM transmission can be either a:

- Single medium transmission, or
- Multiple medium transmission

This is examined in detail in the Abdelmohsen (2012) study on Genres of Communication Interface.

10.6.6 BIM coordination tools

The goal of a Coordination Tool is to capture the purpose of any given transmission and allow the transmission of the right data in the right flows. Each Efficiency State offers improved Coordination Tools.

Examples of Coordination Tools in practice are (Figure 10.7 and 10.8):

- Model Production and Delivery Table, MPDT
- Information Delivery Manuals and Model View Definitions
- Semantic Exchange Modules (Venugopal *et al.*, 2012)

10.6.7 Note on dimensions and waste

Figure 10.9 illustrates how Waste can occur in two basic ways.

- Flow type dimension: Required data missing or partially missing from a flow
- Data subset dimension: All the types of Waste can occur as a Data subset dimension fault.

10.6.8 Note on interaction between container, medium and coordination tool

Interaction between Containers, Medium Coordination Tool is important for eliminating Waste. A good container type enables the coordination tool to capture the semantics of the data in order to filter the data for the transmission accordingly.

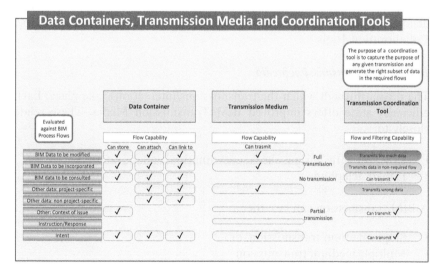

Figure 10.7 BIM data containers, transmission media and coordination tools.

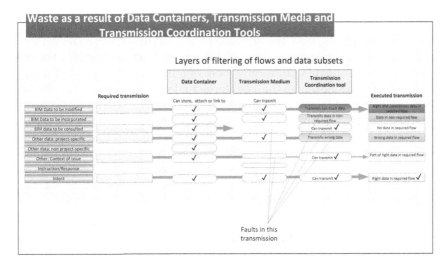

Figure 10.8 BIM transmission waste as a result of BIM data containers, transmission media and coordination tools.

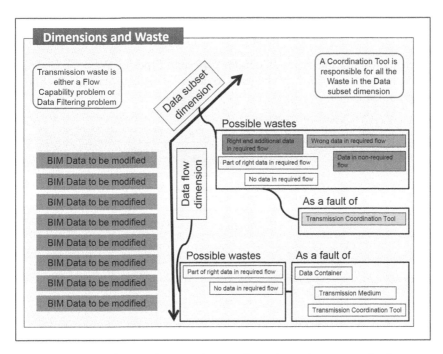

Figure 10.9 Dimensions and waste: the two dimensions of waste are the flow type dimension and the data subset dimension.

10.6.9 BIM efficiency states

BIM Efficiency States are defined by the capabilities and attributes of available BIM data containers, BIM Transmission Media and BIM Coordination Tools.

Each state has a higher maximum efficiency. Each state offers the opportunity to eliminate significantly more of a new type (or types) of waste. States also relate closely to the 'counterfactual scenario' concept defined in NIST (2004) were by a state of improved data interoperability was conceptualised and the relative costs of the then current practices were estimated. As noted above and as Figure 10.10 illustrates, a required transmission can never be fully understood or executed. i.e. there will always be communication waste in practice. The critical level of development of a Container type: The level of development which allows a project to transition to the next state. E.g. IFC being good enough for implementation (Figure 10.11).

10.6.10 BIM transmission/interaction representations

BIM Transmission Representations are methods through which transmissions and interactions can be represented. Examples include:

*   Process Maps (e.g. Critical Path Method)
*   River Model (Bertelsen *et al.*, 2007)
*   True Process Model (Bertelsen *et al.*, 2007)
*   or User-Container-User Interaction Network Graphs
*   or User-User Network Graphs.

Each representation types elucidates different Types of Waste. Properties of these representations can be used to describe differences between States (Figure 10.12).

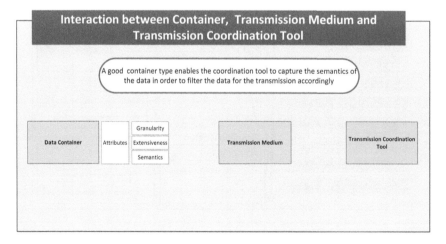

Figure 10.10 Interaction between BIM data container, transmission medium and transmission coordination tool.

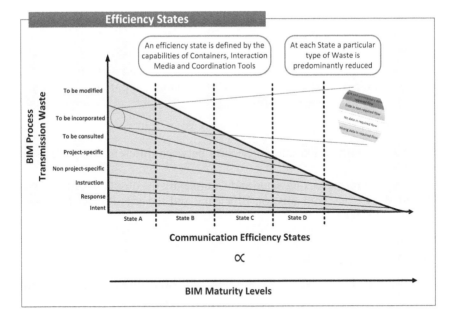

Figure 10.11 BIM efficiency states.

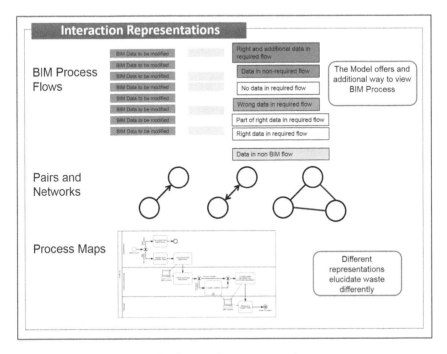

Figure 10.12 BIM transmission/interaction representations.

10.6.11 BIM transmissionlinteraction representation examination lenses: scale and complexity

10.6.11.1 Scale

Progressive levels of magnification at which BIM transmissions/interactions can be examined. At different levels, Waste becomes apparent differently.

10.6.11.2 Complexity

Whether the interaction and conversion between different components of flow is accounted for.

By examining increasing Scale and Complexity the observer's attention is shifted away from Transmission Efficiency and towards Project Effectiveness (Figure 10.13).

10.6.12 Improving the WIMBIM

The WIMBIM was evaluated through semi-structured interviews with four industry BIM experts. It was suggested that the model should be improved by:

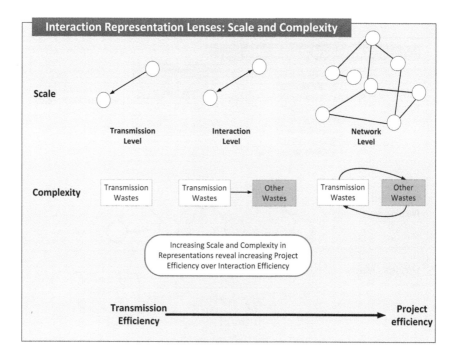

Figure 10.13 BIM transmission interaction representation lenses: scale and complexity.

- Improving robustness by applying it in the different parts of the construction process.
- Providing more examples of transmissions and waste.
- Specifying exactly how communication waste changes through new BIM maturity levels. I.e. address the quantity of communication and the proportion of communication waste across different BIM maturity levels.
- Consolidating the model so that it can 'come down to a level' where it can be used practically.

10.7 Conclusion and recommendations

The presented WIMBIM is based on the following conclusions:

- The language commonly used within research and software development is not able to support the discourse of requirements engineering for BIM collaboration tools effectively. There is no standard or guideline to support this i.e. provide common terms.
- There is a need to analyse and provide a formal, universal and robust description of a BIM transmission (or 'BIM message').
- Discourse within the domain requirements engineering for BIM collaboration tools should be concerned primarily with this question: How can you work towards enabling the User to make the most of a BIM process transmission?
- The different types and levels of representation, including the representations of the various networks in a project have the power to elucidate efficiency (and waste) in different ways.
- Waste provides an appropriate reference concept for an inclusive requirements engineering process and especially for the purpose of introducing new technological paradigms.
- The capabilities of a container of project information (e.g. semantic richness or the ability to 'understand' or 'explain' the meaning of its content to collaboration systems) significantly impact the efficiency of BIM interactions/transmissions.

The WIMBIM can be used to provide a more explicit, software-agnostic and technology paradigm-independent understanding of BIM communication waste. This can be particularly useful during the development of new BIM collaboration tools.

The main principles behind WIMBIM are:

- Its focus on Communication Waste.
- The BIM process transmission being its focal unit of analysis.
- Being built up from the individual transmission (and related to concepts within the domain) in a logical way.

Practical uses of the WIMBIM include:

- Evaluation of BIM collaboration tool configurations.
- Understanding of BIM maturity in terms of BIM communication waste.

The main benefits of WIMBIM as a conceptual model are that it:

- Provides robustness by:
  - 2Being independent of technological paradigms
  - Being independent of what types of software are used within a BIM collaboration tool configuration
- Provides extensibility for the above reasons and by not being bound any strict formalisms.
- Provides a common reference by being involved with concepts that are universal. This is a result of the multi-perspective research conducted in order to generate it.
- Provides exhaustiveness for the above reasons.
- Is a step towards a more scientific understanding of BIM in that it attempts to be partly derived from logical inferences and is therefore constructed in a way that lends itself to falsification.
- Is politically agnostic in that it puts the elimination of BIM communication waste as its target but does not explicitly assign that responsibility to any party within the requirements engineering domain.

10.7.1 Recommendations arising from WIMBIM

The WIMBIM gives rise to a number of proposals. These range from logical implications to recommendations largely based on the reality of change management in the industry. These proposals are useful both as talking points in the model evaluation interviews and for considerations in future research. The main implications and recommendations are outlined below:

In BIM practice, communication waste as it is understood in WIMBIM could be incorporated in the descriptions of BIM maturity levels to complement the existing model e.g. by identifying what the dominant Waste Types are in each BIM maturity level. Additionally, BIM Standards can be complemented based on Communication Theory and BIM communication waste.

The WIMBIM can provide a framework for evaluating BIM collaboration tools, new paradigms and standards i.e. any development should be scrutinised against its potential to reduce communication waste. BIM collaboration tools and collective BIM software configuration can be assessed against BIM flow capabilities. Since a state of zero communication waste cannot be reached, a Transmission best practice guide can be developed. This would help identify critical transmissions, critical *chains* of transmissions and the critical BIM flows for the purposes of critical transmissions and critical transmission chains. This would help prioritise the adequate facilitation and/or 'working around' the waste for Critical Transmission Chains and Critical Flows in Critical Transmissions.

BIM Theory should incorporate Communication Theory: Communication theory is routed in scientific principle and can provide a framework for modelling a phenomenon which is often overlooked by many BIM-related studies: that a communication act starts from a human being as a sender and ends at a human as a receiver (and the executor of an action).

Need for the models used in BIM product development to account for User Interface, User Experience, Human Cognition and Semiotics: A careful application of communication theory which would include the specifics of Human Cognition and Semiotics and how they affect the BIM collaboration tool user's experience is required. These phenomena are often overlooked since because of the structure of the industry and the nature of project software configuration, typically, no party is assigned the responsibility for overall user experience.

10.7.2 Potential observations

This model merely sets a framework for a better understanding of BIM-enabled communication and what principles to follow in order to make it more efficient. Furthermore it was particularly useful in flagging up the need for conceptual models to account for User Interface User Experience, Semiotics was well as Human Cognition. An improved WIMBIM could provide a lens for academic analysis of BIM collaboration tools. The model evaluation process gave rise to a number of important questions in reference to the model:

- What can this transmission-level view of the BIM process offer?
- What waste types do the relevant BIM standards aim to eliminate?
- Are new functionality types eliminating different waste types?
- Are different BIM maturity levels eliminating different waste types?

References

Abdelmohsen, S. (2012). Genres of communication interfaces in BIM-enabled architectural practice. In *6th ASCAAD Conference 2012 CAAD Innovation Practice*, Kingdom University, Manama, Bahrain (p. 81).

Aouad, G. and Arayici, Y. (2010). *Requirements Engineering for Computer Integrated Environments.* Wiley-Blackwell: Oxford, UK.

Aouad, G., Lee, A., and Wu, S. (2005). nD modelling for collaborative working in construction. *Architectural Engineering and Design Management*, *1*(1), 33–44.

Arayici, Y., Ahmed, V., and Aouad, G. (2006). A requirements engineering framework for integrated systems development for the construction industry. *Journal of Information Technology in Construction*, *11* (October 2005), 35–55.

Asite. (2017). 'Asite—Corporate Collaboration, eProcurement, BIM, eSourcing, Contract Management, Building Information Modelling'. Retrieved 2 January 2017 from https://www.asite.com/.

Ballard, G. and Howell, G. (1998). What kind of production is construction? In *Proceedings IGLC 1998*, Guarujá, Brazil, 13–15 August.

Ballard G, Tommelein I, Koskela L and Howell G, 2002. 'Lean Construction Tools and Techniques'. In Best and de Valence (editors), *Design and Construction: Building in Value* (pp. 227–255). Butterworth-Heinemann: Oxford.

Bastian, M., Heymann, S., and Jacomy, M. (2009). Gephi: an open source software for exploring and manipulating networks. *ICWSM, 8*, 361–362.

Beach, T. H., Y. Rezgui, and O. F. Rana. 'Cloudbim: Management of BIM Data in a Cloud Computing Environment'. *Proceedings of the 28th International Conference of CIB W78*, Sophia Antipolis, France, 26. Vol. 28. 2011.

Becerik, B. and Pollalis, S. N. (2006). Computer aided collaboration in managing construction, Harvard Design School, Department of Architecture, Design and Technology Report Series 2006–2.

Bertelsen, S., Henrich, G., Koskela, L. J., and Rooke, J. A. (2007). Construction physics. In *Proceedings of the 15th Annual Conference of the International Group for Lean Construction* (pp. 13–26).

BIM Task Group. (2013). Employers Information Requirements Core Content and Guidance. Retrieved from http://www.bimtaskgroup.org/wp-content/uploads/2013/04/Employers-Information-Requirements-Core-Content-and-Guidance.pdf on 2 January 2017.

BIM Task Group. (2014). http://www.bimtaskgroup.org/. Accessed 7 July 2014.

BRE. (2014). BRE Group: Level 2 BIM on Trial—a buildingSMART UK User Group Event. Retrieved July 24, 2014, from http://www.bre.co.uk/page.jsp?id=3368.

BSI. (2007). BS ISO 1192:2007. Collaborative production of architectural, engineering and construction information—code of practice. Retrieved from http://shop.bsigroup.com/forms/PASs/BS-1192-2007/ on 2 January 2017.

BSI. (2013). PAS 1192-2 : 2013 Specification for information management for the capital/delivery phase of construction projects using building information modelling. Retrieved from http://shop.bsigroup.com/forms/PASs/PAS-1192-2/ on 2 January 2017.

BSI. (2014). Specification for information management for the operational phase of assets using building information modelling. Retrieved from http://shop.bsigroup.com/upload/Construction_downloads/PAS1192-3%20final%20bookmarked.pdf on 2 January 2017.

Cabinet Office. (2011). Government Construction Strategy. Retrieved from https://www.gov.uk/government/uploads/system/uploads/attachment_data/file/61152/Government-Construction-Strategy_0.pdf on 2 January 2017.

Cabinet Office. (2012). Government Soft Landing Policy. Retrieved from http://www.bimtaskgroup.org/wp-content/uploads/2012/09/The-Government-Soft-Landings-Policy.doc on 2 January 2017.

Cabinet Office, BSI. (2013). COBie Data Drops. Retrieved from http://www.bimtaskgroup.org/wp-content/uploads/2012/03/COBie-data-drops-29.03.12.pdf on 2 January 2017.

Cerovsek, T. (2011). A review and outlook for a 'Building Information Model' (BIM): a multi-standpoint framework for technological development. *Advanced Engineering Informatics, 25*(2), 224–244.

Chachere, J., Kunz, J., and Levitt, R. (2009). The role of reduced latency in integrated concurrent engineering. CIFE Working Paper# WP116.

CIC, Construction Industry Council. (2013a). Building Information Model Protocol CIC/BIM Pro. Retrieved from http://cic.org.uk/download.php?f=the-bim-protocol. pdf on 2 January 2017.

CIC, Construction Industry Council. (2013b). *Best Practice Guide for Professional Indemnity Insurance When Using Building Information Models*, 1st ed. London: Construction Industry Council.

Coates, P., Arayici, Y., Koskela, L.J., Kagioglou, M., Usher, C., O' Reilly, K. (2010). 'The limitations of BIM in the architectural process', *First International Conference on Sustainable Urbanization*, Hong Kong, China, 15–17 December 2010.

Dainty, A., Moore, D., and Murray, M. (2007). *Communication in Construction: Theory and Practice*. Routledge: New York.

Dave, B., Koskela, L. J., Kiviniemi, A. O., Tzortzopoulos Fazenda, P., and Owen, R. L. (2013). *Implementing Lean in Construction: Lean Construction and BIM*. CIRIA: London.

Froese, T. M. (2010). The impact of emerging information technology on project management for construction. *Automation in Construction*, *19*(5), 531–538, doi:10.1016/j.autcon.2009.11.004.

Hull, E., Jackson, K., and Dick, J. (2005). *Requirements Engineering* (Vol. 3).Springer: London.

Isikdag, U. and Underwood, J. (2010). Two design patterns for facilitating building information model-based synchronous collaboration. *Automation in Construction*, *19*(5), 544–553, doi:10.1016/j.autcon.2009.11.006.

Koskela, L. (1997). Lean production in construction. In Alarcon, L. (Ed.), *Lean Construction*. Balkema, Rotterdam, pp. 1–10.

Koskela, L. (2000). *An Exploration Towards a Production Theory and Its Application to Construction*. Espoo, Finland: VTT Technical Research Centre of Finland.

Koskela, LJ, Bølviken, T and Rooke, JA (2013), Which are the wastes of construction? in: The 21st Annual Conference of the International Group for Lean Construction., July 31- August 2, 2013, Fortaleza, Brazil, 29 July–2 August.

Kreider, R. G. and Messner, J. I. (2013). *The Uses of BIM: Classifying and Selecting BIM Uses*, Version 0.9, September. University Park, PA: The Pennsylvania State University. http://bim.psu.edu. Accessed 5 May 2013.

Liu, N., Kagioglou, M., and Liu, L. (2011, March). An overview of the marketed functionalities of web-based construction collaboration extranets. In *Information Science and Technology (ICIST), 2011*.

MoJ, Ministry of Justice. (2013). Early Adopters Project-HMYOI Cookham Wood: *New House Block and Education Building, BIM Lessons Learnt, Report Version 3*. Available at: http://www.bimtaskgroup.org/wp-content/uploads/2012/03/Cookham-Wood-Consolidated-Lessons-Learned-version3-with-intro.pdf. Accessed 6 June 2012.

Mooney, J. D. (1947). *The Principles of Organization*. Harper: New York:.

National Institute of Standards and Technology (NIST). (2004). Cost analysis of inadequate interoperability in the US capital facilities industry. National Institute of Standards and Technology (NIST), Advanced Technology Program, Information Technology and Electronics Office: Gaithersburg, MD.

Nisbet, N. (2012). COBie-UK-2012: Required Information for Facility Ownership.

Rezgui, Y., Beach, T., and Rana, O. (2013). A governance approach for BIM management across lifecycle and supply chains using mixed-modes of information delivery. *Journal of Civil Engineering and Management*, *19*(2), 239–258.

Rezgui, Y. and Zarli, A. (2006). Paving the way to the vision of digital construction: a strategic roadmap. *Journal of Construction Engineering and Management*, *132*(7), 767–776.

Sacks, R., Koskela, L., Dave, B. A., and Owen, R. (2010). Interaction of lean and building information modeling in construction. *Journal of Construction Engineering and Management*, *136*(9), 968, doi:10.1061/(ASCE)CO.1943-7862.0000203.

Shannon, C. E., and Weaver, W. (1949). *The mathematical theory of communication*. University of Illinois Press.

Simon, H. A. (1981). *The Sciences of the Artificial*. Massachusetts Institute of Technology: Cambridge, MA:.

Sinclair, D. (2012). *BIM Overlay to the RIBA Outline Plan of Work*. RIBA: London, UK.

Succar, B. (2009). Building information modelling framework: A research and delivery foundation for industry stakeholders. *Automation in Construction*, *18*(3), 357–375.

Tribelsky, E. and Sacks, R. (2006). Measures of information flow for lean design in civil engineering. *Scientific Committee*, 1493. Retrieved from http://centaur.reading.ac.uk/31329/1/CME25-Whole_Procs.pdf#page=1515 on 2 January 2017.

Venugopal, M., Eastman, C. M., Sacks, R., and Teizer, J. (2012). Semantics of model views for information exchanges using the industry foundation class schema. *Advanced Engineering Informatics*, *26*(2), 411–428.

Yeomans, S. G., Bouchlaghem, N. M., and El-Hamalawi, A. (2006). An evaluation of current collaborative prototyping practices within the AEC industry. *Automation in Construction*, *15*(2), 139–149.

11 Towards the establishment of a district information modelling

Sara Moghadam, Patrizia Lombardi, and Jacopo Toniolo

11.1 Introduction

The subject of energy efficient districts is one of the greatest and most challenging of research priorities in the European Union (EU). In order to achieve an effective impact, instead of just concentrating on the improvement in terms of energy efficiency to one particular building, this approach requires challenges to be solved at the neighbourhood level. A new concept, named District Information Modelling (DIM) is introduced which relies upon the integration between Building Information Modelling (BIM) and Geographic Information Systems (GIS) with real-time data, trying to create a prototype of a tool that simplifies the visualization and management of data and the analysis of energy consumption in a Smart City district. This study is part of two ongoing Smart City research studies, an EU VII Framework Program project named District Information Modelling and Management for Energy Reduction (DIMMER) and a national cluster project named Zero Energy Buildings in Smart Urban Districts (EEB). The development of a DIM model is needed in order to optimize energy efficiency, monitoring the whole energy consumption and production process through ICT. The result will be an open platform for real-time data processing and visualization at district level that exploits the information about buildings, the energy distribution grid and user behaviour.

Urban areas are the most important consumers of energy in the European Union. At global level, approximately cities consume 75% of natural resources and 60%–80% of global greenhouse gas emissions are caused by activities in urban areas that is expected to increase (UNEP 2012). Further, more than half of the word population, up to 64%–69%, will settle in cities by 2050, or 5.6–7.1 billion (IPCC 2014). Particularly, the building sector accounts for 30% of global annual greenhouse gas emissions (UNEP 2009) and consumes up to 40% of the total energy consumption in European Union (EU) member states (EPBD 2010/31/EU). Consequently, various studies have shown that the built environment contributes significantly towards global energy consumption and to the production of greenhouse gases, which have an immense impact on climate change.

Energy saving and pollutant emission reduction are main concerns for 21st century sustainable development. As a consequence, policy-makers in Europe have the stressful task of needing to achieve energy security and promote a transition towards decarbonized energy sources without undermining wellbeing and patterns of consumption (Lombardi 2015). Therefore, to avoid a further increase in these values, the EU has determined a new plan for energy efficiency, setting several policy targets, known as the '20-20-20' in order to reduce greenhouse gas emissions by 20% from 1990 levels (Energy Efficiency Plan – COM(2011) 109). Despite the great energy efficiency improvements in buildings, recent energy consumption data analyses show that these targets are unlikely to be reached (Lombardi and Trossero 2013).

Published by the European Commission, Roadmap 2050 aims at moving towards a competitive low-carbon economy, going beyond the 2020 goals and setting out a plan to meet the long-term target of reducing emissions by 80% to 95% compared to 1990 levels (EUROPEAN COMMISSION, COM (2011) 112 final 2012). This approach emphasises the importance of accelerating renovation development in both private and public buildings. It also forecasts how energy providers will be compelled to convince their customers to amend their energy usage.

Traditionally, in the field of building energy performance, experts and researchers' attention has been concentrated upon the single building rather than on wide building stocks. One step towards achieving the desired objectives is to assess energy efficient buildings incorporated into the district, which is one of the most critical concerns of research in this field. In fact, instead of just concentrating on the improvement in terms of energy efficiency to one particular building, this approach requires challenges to be solved at the district level (Koch *et al.* 2012). This fact highlights the significance of targeting building energy consumption as vital to decreasing energy usage at the district level. Therefore, where the purpose is the assessment of the greenhouse gases reduced emissions and global achievable energy savings, it is essential to broaden the focus to the building stock at a territorial scale (Fracastoro and Serraino 2011; Wiel *et al.* 1998).

Previous research has shown that energy consumption profiles at city level have been estimated by first modelling a small group of buildings that is representative of the urban district. For instance, Yamaguchi *et al.* proposed a new district clustering model to define CO_2 reduction scenarios. They modelled more than 500 archetypal buildings by classifying the districts into many types according to urban forms, including Heating, Ventilation and Air Conditioning (HVAC) systems and features of the construction and occupant behaviour, with the aim to better analyse their different energy efficiency and district energy reduction (Yamaguchi *et al.* 2007).

Hence, there are a number of challenges in the design, construction and operation of energy efficiency buildings in the community to be met in order to provide the information model of the district (Sebastian *et al.* 2013). This study presents the initial development of a district information modelling (DIM) for energy consumption reduction. Specifically, the DIM system aims at managing and resolving sensitive problems related to the energy consumption of existing buildings in the neighbourhood (Urban scale), including the processes of management and maintenance of the building envelope and HVAC systems. This may represent a useful tool, especially for existing stocks and historical buildings.

A major contribution of this chapter is a discussion of how to move from the concept of a smart energy building model to an intelligent energy district and the improvements this can bring. Accordingly, the main technologies that can be implemented at district level, starting from buildings, passing through smart energy grids and scaling up to urban level are described. Thus, the new integrated concept of interconnectivity between energy resources into the distributed grids and buildings at a district level is essential.

This approach needs a series of ICT components adjusted to the context of energy management at district level (Lombardi *et al.* 2014). Moreover, the increasing advance of information and communications technology (ICT) in building information modelling (BIM) has enabled the digital organization of building characteristics and parameters. The European Commission has emphasized the significance of ICT for sustainability and energy reduction (Lombardi 2011).

Generally speaking, ICT refers to all the technology used in telecommunications, middleware, network control, audio/visual systems and monitoring functions that allow users to access and operate information. In the last three decades heating, ventilation and air conditioning (HVAC), lighting, fire safety and security systems control have been continuously developed to be implemented in all new buildings and refurbished ones. At the district level, heating and cooling as well as the electricity grid can also be accessed through real-time monitoring by an automated system. Indeed, ICT is one of the crucial issues involved in enhancing the energy efficiency of buildings and districts.

In view of this approach, geographic information system (GIS) and BIM create 3D data models that offer information about buildings and the surrounding environment at a district level. The BIM is a model involving all the information and management of the digital representation of the construction which uses the ICT platform for control and monitoring; GIS is a comprehensive system that can refer to a number of different processes and technologies (i.e. capturing, storing, analysing, managing, and presenting all sorts of spatial or geo-referenced data). The main advantages of using GIS is that the latter can add the environmental information about the surrounding area to the detailed construction information of BIM. Numerous studies in the literature have shown the effectiveness of this combination of GIS-BIM at an urban level (Del Giudice *et al.* 2014; Irizarry *et al.* 2013; Sebastian *et al.* 2013; Sehrawat and Kensek 2014).

Furthermore, some progressive visualization technologies can provide real-time feedback to final users about the influence of their behaviour on energy consumption. Therefore, to raise the final user's awareness it is possible to monitor the whole energy consumption and production process through ICT. It has been proved that end users with efficient control and visualization technology decrease their overall consumption and, specifically, lower the global energy peak requested by the grid (Levin 2015). It is expected, by opening access to database and information on energy consumption and through the feasibility of custom application downloading, that end-users will be able to use their facilities in a smarter way in order to cut their energy consumption. Additionally, energy providers should implement a reformed energy distribution schedule according to the statistical analysis of population profiles in order to increase energy efficiency.

This study is part of two ongoing Smart City research studies, an EU VII Framework Program ICT-Smart city project, named District Information Modelling and Management for Energy Reduction (DIMMER) (http://dimmer.polito.it/) and a national Cluster project, named Zero Energy Buildings in Smart Urban Districts (EEB) (http://home.deib.polimi.it/bolchini/research/eeb.html).

The chapter is organized as follows. Section 2 describes the DIM background study related to various aspects of district models based on the BIM/GIS interface. Section 3 introduces the methodology for the development of the proposed shift from BIM to DIM, where ICT technologies play a major role in order to improve the energy efficiency of any urban community. Section 4 discusses the concept of smart grids as an application of DIM and presents in more detail the DIMMER project methodology which is aiming to provide data in real-time to feed the DIM model. This model can remotely visualize district energy consumption by using 3D models of the buildings and the smart networks of the district. In this way, Smart buildings interact with the grid in order to optimize the energy consumption. Finally, Section 5 provides concluding remarks and future proposals of research.

11.2 Background

In this section, various aspects of district models based on the BIM/GIS interface will be presented, considering that actuators and pervasive sensors integrated with DIM can control the whole energy chain in an efficient way. Moreover, advances in visualization, interaction technologies and 3D modelling and real-time feedback can lead to an increase in energy efficient behaviours (Lombardi *et al.* 2014).

First of all, it is beneficial to know that according to different fields of study, DIM can be interpreted in different ways. For instance, in architectural design processes, DIM means 'Drawing Information Modelling or Design Information Modelling'. This type of modelling is aimed at proposing a group of tools for assisting information modelling and structuring in relation to multi-layered geometric models to support real-time analysis, evaluation, and decision-making (Mostafavi *et al.* 2013). Moreover, in Demolition Engineering DIM means 'Deconstruction Information Modelling' (Ulyatt, http:// www.bimplus.co.uk/management/thinking-beyond-buildings-lifespan/).

Conversely, the term DIM is used here to define a tool, which simplifies the visualization, data managing and analysis of energy consumption in a Smart City District. During the last decade, especially in recent years, the BIM concept and workflows have become a significant research area in order to tackle difficulties in terms of data interoperability and management. The Associated General Contractors Guide (2006) determined BIM '*as a data-rich, object-oriented, intelligent and parametric digital representation of the facility, from which views and data appropriate to various users' needs can be extracted and analyzed to generate information that can be used to make decisions and improve the process of delivering the facility*'. Two definitive features of BIMs were introduced during the study conducted by Isikdag *et al.*, where BIMs emerged as being 'a facilitator of data interoperability' and 'exchange between software applications' (Isikdag *et al.* 2008).

Conceptually, BIM is a strong digital tool with the practical and physical capabilities of a facility management system that plays a fundamental role in the construction sector. Although the BIM model is an efficient tool in order to develop an integrated digital infrastructure because of its effective modelling abilities, visualization, analysis, and data-based simulation, the problem is how to move from the building to the district level.

To this regard, the DIM system proposed tries to extend the BIM methods to a neighbourhood level in order to remotely visualize and collect district level energy usage and integrate it with real-time data (Osello *et al.* 2013). Furthermore, in the last few decades, research has mainly focused on the transition phase from computer aided-design (CAD) to building information modelling (BIM), where BIM could replace CAD not just to provide the information required but also to resolve concerns about combining heterogeneous data sources (Moon *et al.* 2015). On the other hand, in urban planning Geographic Information Systems (GIS) play a leading role (Ramesh *et al.* 2013).

The GIS presented by Huxhold is '*an information system that is designed to work with data referenced by spatial or geographic coordinates*' (Huxhold 1991). Hence, the DIM system proposed combines the benefits of BIM and GIS, merging them into a single model, with the aim of keeping track of the visualization of district energy production and consumption chain status, environmental conditions and user feedback data. Transforming the BIM into its corresponding geo-referenced model allows it to support responses to numerous environmental questions (Bansal 2011).

Generally, the most fulfilling method of investigations and research relies upon the applicability of the BIM/GIS approach to provide district information models. For example, in 2014 the study conducted by Wolisz *et al.* presented city district information modelling (CDIM) as an assessment of smart city approaches. CDIM is a data management notion for city districts that is presented to support the simulation and conception of interdependent energy systems. It can be used for energy management and optimization at a city district level. CDIM information has demonstrated which buildings are connected to an existing district heating (DH) network, and so the energy source could be considered during the construction phase (Wolisz *et al.* 2014).

Moreover, another interesting approach of the district information model (ee-DIM) as presented by Hippolyte *et al.* (2014) is an ee-district information model aiming to support smart coordination between energy loads, energy generation units and energy storage systems at a district level (Hippolyte *et al.* 2014). Stojanovski conceptualized city information modelling (CIM), which is conceived as a system of blocks with dynamic relations or connections. According to this study, CIM is a BIM analogy in urbanism, where it is formed by symbols in 2D and 3D spaces (Stojanovski 2013). Two analyses, namely view and shadow, have been carried out using the semantic information within the geo-referenced BIM model. These analyses demonstrate the value of integrating BIM and spatial data (e.g. shadow analysis plays significant roles for diverse purposes, such as the energy sector.) (Rafiee *et al.* 2014).

Improving the visual monitoring of the construction supply chain by integrating BIM and GIS (Irizarry *et al.* 2013) has led to the development of a BIM-GIS platform, trying to create a graphical database able to visually monitoring and collect data about landscape, buildings and city (Del Giudice *et al.* 2014; Elbeltagi and Dawood 2011). Among the many advantages of GIS and BIM for asset management are the facilitation of data collection, processing and display; the integration of asset mapping with project management and budgeting tools; maintenance, inspections, and expenses (Zhang *et al.* 2009). The Ubiquitous Space Information Model (USIM) developed by Choi *et al.* which the outcome BIM can be shared among indoor GIS applications (Choi *et al.* 2008).

In 2010, Peña-Mora *et al.* have also shown that different IT systems, like GIS and digital building information, require to be integrated in a single comprehensive platform for emergency response operations and managements (Peña-Mora *et al.* 2010). Recently in 2014, Sehrawat *et al.* provided a technique to simulate and collect the energy consumption data of buildings at city level, where it should be considered a notable number of buildings, based on conceptual energy modelling in a BIM, and then, simplified geometry of the buildings and inputs in a GIS database (Sehrawat and Kensek 2014).

The models in BIM (architectural scale) and GIS (urban and geographic scale) are created in two different databases that consider different Levels of Development and different Levels of Detail (LOD) for collecting and sharing information. Therefore, it is required to specify the LODs and the parameters' range for urban planning (Bishr 1998). Actually, the data connection between these two applications is complex and for this cause, many studies have been conducted to resolve interoperability problem between GIS and BIM.

For instance, Isikdag *et al.* employed the proof-of-concept to transfer of the geometric and semantic information of the building basics into geospatial environment. The results of their research confirmed that BIMs offer an adequate level and amount of information (about the building) for the seamless data management automation tasks in the site selection and fire response management (Isikdag *et al.* 2008). Another approach for transforming information from IFC to CityGML framework in the case of water utility network within a GIS context is studied by Hijazi *et al.* (2009). Hagedorn and Döllner (2007) presented a method useful for visualizing and analysing 3D building information models within virtual 3D city models. They integrated the data from CAD, GIS, and BIM tools into a 3D city model by using the data integration capabilities of IFC to CityGML transformation. In the same year, Wu and Hsieh (2007) provided a method to convert the geometric information of an IFC model into geometric objects for the GML model.

The literature review described in this section pointed out a number of studies on the use of BIM/GIS concepts, aiming at creating an information model for the entire city. As one can note, Geographic Information System (GIS) approach is able to integrate individual buildings model linked together (BIM approach) with the energy urban network. However, the DIM technology is used to convey meaning, which represents an integrated BIM-GIS model for visualizing the supply chain process, monitoring, and control the whole energy chain. In the next section, the BIM and GIS methodologies will be further investigate and discuss in relation to the development of a DIM tool.

11.3 The DIM methodology

This section proposes a methodology for DIM to provide a smart management and control based on an ICT infrastructure made of actuation devices and heterogeneous monitoring. An original web service oriented software infrastructure has already resolved those problems by an intelligent ICT-based service monitoring and managing the energy consumption. Conceptually, a DIM system could be developed by the integration of BIMs into the district level, when they are distributed in the network monitoring system (Figure 11.1).

The DIMMER project addresses many technological challenges to build a real-time district model and information management concept for city districts. From the literature review in the previous section, in order to fully capture the environmental and building information parameters that could

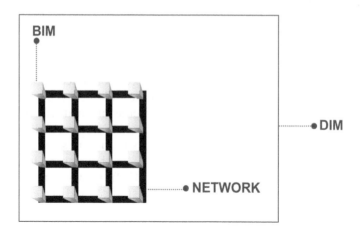

Figure 11.1 Conceptual scheme of DIM, integration of BIMs and network monitoring system.
Source: Author's own elaboration.

be created by BIM/GIS systems in smart districts relied on innovative ICT
technologies, this concept is examined in the DIMMER, which finally aims
at developing a web-service oriented open platform able to visualize real-
time district level data. It lets open access with personal devices and Augmented Reality (A/R) visualization of energy-related information to client
applications for energy and cost-analysis, tariff planning and evaluation,
failure identification and maintenance, energy information sharing.

Accordingly, DIMMER intends to improve the energy efficiency of smart
urban districts by means of user awareness through using Smart devices.
The project will develop the components of the DIM system, enabling
the integration of BIM/GIS. In this concept, the ICT-based systems allow
end-users to have a much deeper perception of the behaviour of the power
systems (Figure 11.2).

Besides the management system in order to represent the information in
real-time, one of the possibilities of DIMMER is to create an advanced 3D
urban modelling of the buildings characteristics (envelope characteristics,
materials, orientation) the energy networks of the district (i.e. heating/cooling/
electricity) and the surrounding environment (Infrastructures, Actual
Weather, etc.) in order to optimize the use of data at district level which
leads to enhance energy efficiency behaviours. Accordingly, one of the main
challenges is to be able to use all this information about buildings, energy
distribution grid (both thermal and electric) and user behaviour.

The occupant behaviour influences significantly on energy consumption
(Torabi Moghadam *et al.* 2015). Therefore, providing more intelligent energy policies based on building thermal performance characteristics and
occupant behaviour in public buildings and enabling an automatic schedule

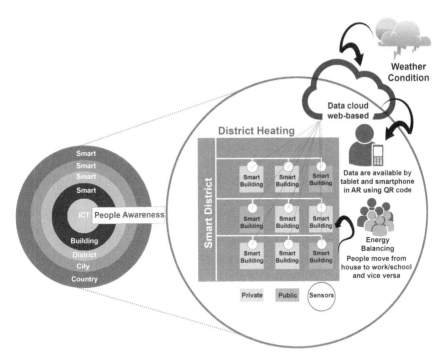

Figure 11.2 Representation of the DIMMER' concept.
Source: Project public document, DIMMER (2015) and EEB (2014), elaborated by authors.

functioning of the smart appliance in the most convenient hours accordingly to energy tariffs in private ones is one of the Examples of possible optimizations. To achieve this target, a software system where information about buildings, their usage as well as user requirements is collected and feedbacks are provided to the stakeholders must be realized. Therefore, in the optimization phase, two levels of policies are considered.

- The first refers to energy optimization systems; which are based on smart grids sensors on the electric grid (e.g. ENEL Distribution system operator or other smart-info providers) or on the district heating/cooling network (from IREN or other DH provider) and building characteristics from BIM/GIS models.
- The second refers to speculative and statistically policies; which are based on user profiling and feedback actions, analysis of building utilization and predicted user behaviours (interface to ANDROID and IPAD).

The DIMMER system will manage energy efficiency by establishing groups of sensors based on power levels and proximity at district level. A group-leading sensor, which can communicate collective data, is chosen for each cluster, while other sensors are only activated, when updated data is needed.

In other words, real-time information is collected through sensors placed on the network and in the buildings and connects to final user feedback in order to analyse the energy-related behaviours. That means the client is allowed to visualize energy related information through his/her personal devices (e.g. personal computers, smart phones, PDAs, ad-hoc displays, etc.) for energy and cost analysis. These are all running while; the whole system should be built by Real-time data collection, advanced middleware technology for data integration, virtual visualization, user/social profiling, simulation, feedback through QR codes, energy efficiency and cost analysis engine and web interface and interaction.

The middleware interfaces with sensor nodes and provides a set of web services to allow the client applications to access sensor data in a hardware independent way. It enables the interpretation of different languages used by sensors to interoperate between them and translate them into a common language. Numerous projects at European level, such as Smart Energy Efficient Middleware for Public Spaces (SEEMPubS) and Inspection of HVAC systems through continuous monitoring and benchmarking (iSERV cmb), has focused on the energy efficiency solution and reduction of carbon dioxide emissions by using ICT tools for monitoring to physical variables of the environment. SEEMPubS used the LinkSmart middleware to act as an interpreter between the heterogeneous languages used by the sensors (Osello *et al.* 2013), while iSERV used the existent BMS of the building, integrated with additional consumption meters (Toniolo *et al.* 2013).

The DIMMER Platform is also formed by the Middleware based on LinkSmart OpenSource, which is linked with Sensor technologies; and Smart City Services, which is connected with the District Information Models. This model can be exploited for many goals including design and refurbishment of buildings, maintenance and monitoring of energy consumption. The idea of interoperability can interact with a common platform that shares different databases information. In this context, data can be extracted from a particular calculation in different applications. A compound ICT-based platform, which integrates BIM and GIS, is an optimal factor to create a digital infrastructure leading to the development of the DIM (Osello *et al.* 2013). ICT systems can efficiently control from the energy production, to consumption and storage (Kelly and Pollitt 2010).

Accordingly, the study conducted by DIMMER research team in 2014 investigated the above method, in order to establish a connection between the GIS and BIM methodologies, with the aim to create a district information model, which could let the final users (e.g. costumers, providers, administrative etc.) to visualize data related to urban planning and power and thermal distribution networks. Following this approach, it will be feasible to do tests related to energy efficiency, considering the district level and not only the single buildings. Finally, the study concluded that the DIM model could be applied to both building and urban scale, starting from one of them to another one and vice versa, as represented by Del Giudice *et al.* (2014).

Truly, different levels of communication relying upon the same database with different operators allow a one to one building data access, optimization, transferability and interconnectivity, augmenting the exchange information improvement based on ICT, the creation of a 'smart' digital archive unique for the city and giving support for the strategic plan of the city.

In tandem with dealing with these issues, the project will bring the following impacts to energy efficiency in smart cities:

- Make households be able to monitor and visualize their energy consumption, receive real-time guidelines about proper behaviours through innovative interaction instruments (e.g. augmented reality, ambient visualization, QR Codes), adapt to tariff changes, control loads by scheduling ON and OFF switching of their plug loads.
- Enhance information exchanged quality on ICT new platform and database, converging IT data flow and classification, building occupational profile and data-focused services.
- Create a 'smart' integrated unique digital network and cloud archive for the city, using existing Smart Grid (intelligent energy distribution system, both thermal and electricity) that comprises the networking and intelligent generation, storage, consumers control and interconnected elements of energy distribution and transmission systems by the means of ICT.
- Maximize the profits of the smart distribution network, ensuring proper levels of interoperability among sensors, devices and feedback information, data availability and security, and finally decision support system for both citizens and city managers.

The conclusion of this section is to describe the creation a parametric digital model to control all information on a district that includes including visualization of energy consumption trends and feedback from end-users, who could benefit from information on energy-saving aspects.

11.4 An application of DIM: smart grids

In this section is shown how the smart grids complexity could be managed with a specific DIM system. Low energy buildings and areas require appropriate combinations of heating, cooling and electrical services. Since one of the emissions' reduction factors is the enhancement utilization of production from renewable energy sources (RESs) fed to the grid (Colak *et al.* 2015), therefore, it is indispensable to promote the expansion of RESs and reduce the dependence on fossil fuels in order to meet the challenge of energy efficiency in the district.

A Smart grid comprises a modernized energy grid that can intelligently integrate the actions of all connected users, consumers and suppliers, in order to improve the reliability, cost-effectiveness, and efficiency of distribution of energy and production. It provides innovative services together with intelligent technologies for the monitoring, control and communication of information. It provides consumers more information and choices that can significantly reduce the environmental impact and increase the reliability, resiliency, and safety of the entire energy system.

A Smart Grid generally refers to an electric system, which is in widespread use in many cities around the world. Undoubtedly, more efficient information exchange leads to managing the electricity network smartly. Actually, smart grids integrate renewable energy sources with innovative digital technologies in order to monitor the entire network and ensure real-time interventions. It is able to distribute the power efficiently and answer to conditions that arise anywhere in the grid, such as transmission, distribution, power generation, consumption, and adopt the corresponding strategies, using modern information technologies (Fang *et al.* 2012).

As an example, it can describe a project developed in the Italian marketplace named Energy@home (http://www.energy-home.it/SitePages/Home. aspx). This project is a non-profit association founded by ENEL, Electrolux, Indesit and Telecom Italia aimed at increasing the energy efficiency of the whole house system. The project consists of an integrated platform that provides the value-added services, based on the exchange of the information connected to energy usage/cost in the home area. The platform includes different devices: '*Electric Meters; Smart Appliances; Smart Plugs and a Home Gateway*'. This innovative kit based on ICT services, i.e. Smart Services, is based on then hypothesis that territories and, in particular, cities achieve a significant level of greater energy efficiency through real-time energy consumption information. The project represents a further step towards the development of the so-called 'smart grid', which, in the framework of a smart grid, allow continuous real-time bi-directional information exchange between energy usage and appliances in houses to enable each customer to 'self-manage' his/her energy behaviour depending on both power supply availability and price (Energy@home 2012).

It is an extremely integrated network, merging sensors, communications technologies, and intelligent control systems with the physical grid. It can handle all the innovative technologies with progressive monitoring, control and communication technologies. In the future, every single home probably will become an interactive part of the grid and so every person will be able to manage his/her energy consumption. Smart grids significantly reduce the environmental impact of the electricity sector, contributing to the achievement of the European Commission's environmental goals and a reduction in CO_2 emissions. In this way, renewable energy sources could be easily integrated into a smart grid that leads to manage in a better way the native intermittency of RES.

According to the DIMMER project, district heating networks is an additional typology of energy to be managed in a smart and innovative way. In particular, the project aims at better controlling and managing existing district heating networks. Since district heating networks are less widespread (and the district cooling ones even less), a smart grid for district heating is an innovative concept (Lund *et al.* 2014). Two main ideas are the basis of this concept: The heat generation could be intermittent and distributed (e.g. RES); the users connected to the grid have smart HVAC control system.

This concept, partially developed in the framework of the DIMMER project, permits reliability and energy efficiency to be improved and the global amount of buildings connected to be increased, without increasing the thermal power installed in the central generation station. Potentially it could also decrease the costs of thermal energy, in accordance with energy availability. Future development will include end user's self-generation of heating power that can deliver the excess energy to the district heating grid. To achieve the latter hypothesis, a third party public grid is necessary, as is the case of private trains on public railroads. This scenario is still far from current situation but has already been anticipated and foreseen by some authors (Grohnheit and Gram Mortensen 2003; Lund *et al.* 2014).

On the other hand, a smart grid could feed customers with persuasive information to improve their awareness of energy behaviour and allow them to optimize their consumption while improving the operation of the overall system. It would let consumers decide if and when they want to consume energy or pump their own energy into the grid, increasing the reliability of the whole network. A major goal of smart grids is, therefore, to increase the 'logical' use of energy, through active participation and the customer's self-awareness regarding their electricity consumption, and so lead users towards more energy saving behavioural profiles. Therefore, the section illustrated that the Integration with electricity system, interoperability with district heating and smart grid sensors, enhanced scalability for large number of sensors.

11.5 Conclusion remarks and future work

BIM technology is definitely spreading among building construction, demonstrating its usefulness to unify all building information into a single framework. Nevertheless, as BIM technology is extremely costly, buildings with a traditional geometry plan and ordinary mechanical systems (HVAC, electrical, etc.) do not normally use it. On the other hand, GIS has a decennial history of success that makes it widespread in Civil engineering and territorial planning. Smart cities require the collection specific building and territorial data (e.g. energy flows, traffic flow, building stock, etc.) within the same framework, thus making claiming for the creation of a DIM possible.

This chapter has focalized on building energy consumption, as a field of research that is significantly important and it will probably provide in next years the commercial DIM solutions. DIM framework presents innumerous benefits, as the updated flow of energy in a certain district and the correct information about building and systems ages. This information could be used to assess in a pretty realistic way District Heating and Cooling network, refurbishment incentive campaigns for inefficient buildings, HVAC system safety control. On the other side, DIM could also be used to evaluate the impact of such measures in a short time period.

Even if DIM seems necessary for city and energy planning, there still exist some obstacles to its complete implementation. The first problem is represented by data collection. Data availability, reliability and privacy are matters that requires specific solutions. Although 'smart' sensor network is already available on the market, 'Internet of Things' is spreading among our houses, we still do not have 'smart contracts' for data privacy and data publishing. Nowadays the problem is addressed in its most restrictive way: the user permission is needed to use even the less significant energy data.

Another issue is represented by communication standards. Smart cities will require open protocol to communicate between different sensors, actuators, meters, etc. Cloud solutions, developed in the last decades are ready to face this issue, that is not the case of some HVAC control and metering system but the market probably will force proprietary protocol company to shift some application to open protocol.

DIM solutions represent a substantial research challenge that include Information technology, energy systems, legal framework and social aspects. However, the development and management of a District Information Model in real world, as explained in this chapter, needs cooperation between private and public enterprises as well as academic sector.

Acknowledgement

The authors of this study wish to acknowledge the contributions of a number of colleagues and institutions involved in the EU VII Framework Program ongoing project, named District Information Modelling and Management for Energy Reduction (DIMMER), including: Politecnico di Torino, CSI-Piemonte, Ove Arup and Partners Int. Ltd, D'Appolonia S.p.A. Clicks and Links Ltd, ST-Polito s.c.a.r.l., Fraunhofer FIT, CNet, IREN Energia S.p.A., The University of Manchester, Association of Greater Manchester Authorities (AGMA) – Oldham Metropolitan Borough Council, Istituto Superiore Mario Boella and Università degli studi di Torino.

References

AGC Contractors' Guide to BIM. 2006. Associated General Contractors Guide. Retrieved from http://www.agcnebuilders.com/documents/BIMGuide.pdf/.

Bansal, V.K., 2011. Application of geographic information systems in construction safety planning. *International Journal of Project Management*, 29, 66–77.

Bishr, Y., 1998. Overcoming the semantic and other barriers to GIS interoperability. *International Journal of Geographical Information Science*, 12, 299–314.

Choi, J.W. *et al.*, 2008. Developing ubiquitous space information model for indoor GIS service in ubicomp environment. In *Fourth International Conference on Networked Computing and Advanced Information Management*, 2, Gyeongju, Korea, IEEE, pp. 381–388.

Colak, I. *et al.*, 2015. A survey on the contributions of power electronics to smart grid systems. *Renewable and Sustainable Energy Reviews*, 47(1), 562–579.

Del Giudice, M., Osello, A. and Patti, E., 2014. BIM and GIS for district modelling. In *eWork and eBusiness in Architecture, Engineering and Construction, Proceedings of the 10th European Conference on Product and Process Modelling, ECPPM*, Vienna, Taylor & Francis Group: London, pp. 851–855.

District Information Modelling and Management for Energy Reduction (DIMMER). Available at: http://dimmer.polito.it/ [Accessed April 15, 2015].

Elbeltagi, E. and Dawood, M., 2011. Integrated visualized time control system for repetitive construction projects. *Automation in Construction Journal*, 20(7), 940–953.

Energy@home. 2012. 'Energy@home Project'. Retrieved December 27, 2015 from http://www.energy-home.it/SitePages/Home.aspx.

EPBD, 2010. Directive 2010/31/EU of the European parliament and of the Council. *Official Journal of the European Union*, L 153, pp. 13–35.

European Commission, 2011. *Energy Efficiency Plan 2011, COM(2011) 109 Final.* Retrieved from http://eur-lex.europa.eu/LexUriServ/LexUriServ.do?uri=COM:20 11:0109:FIN:EN:PDF.

European Commission, 2012. *A Roadmap to Low Carbon Economy in 2050, COM(2011) 112 final.*

Fang, X. *et al.*, 2012. Smart grid—the new and improved power grid: a survey. *IEEE Communications Surveys and Tutorials*, 14(4), 944–980.

Fracastoro, G.V. and Serraino, M., 2011. A methodology for assessing the energy performance of large scale building stocks and possible applications. *Energy and Buildings*, 43(4), 844–852.

Grohnheit, P.E. and Gram Mortensen, B.O., 2003. Competition in the market for space heating. District heating as the infrastructure for competition among fuels and technologies. *Energy Policy*, 31(9), 817–826.

Hagedorn, B. and Döllner, J., 2007. High-level web service for 3D building information visualization and analysis. In *Proceedings of the 15th Annual ACM International Symposium on Advances in Geographic Information Systems—GIS '07*. ACM Press, New York, p. 1.

Hijazi, I. *et al.*, 2009. IFC to CityGML transformation framework for geo-analysis: a water utility network case. In *4th International Workshop on 3D Geo-Information*, Ghent, Belgium, conference proceeding, pp. 123–127.

Hippolyte, J.L. *et al.*, 2014. ICT for a low carbon economy, an ee-district ontology to support the development of the ee-District Information Model of the RESILIENT project. In European Commission DG (ed.), *EEBuilding Data Models Energy*

Efficiency Vocabularies & Ontologies, Proceedings of the 4th Workshop organised by the EEB Data Models Community ICT for Sustainable Places, Nice, France, European Commission, pp. 106–119.

Huxhold, W.E., 1991. *An Introduction to Urban Geographic Information Systems*, Oxford University Press, New York.

IPCC. 2014. Climate Change 2014: Mitigation of Climate Change, The Intergovernmental Panel on Climate Change. Cambridge, United Kingdom and New York, NY, USA. Retrieved from http://www.ipcc.ch/report/ar5/wg3/.

Irizarry, J., Karan, E.P. and Jalaei, F., 2013. Integrating BIM and GIS to improve the visual monitoring of construction supply chain management. *Automation in Construction*, 31, 241–254.

Isikdag, U., Underwood, J. and Aouad, G., 2008. An investigation into the applicability of building information models in geospatial environment in support of site selection and fire response management processes. *Advanced Engineering Informatics*, 22(4), 504–519.

Kelly, S. and Pollitt, M., 2010. An assessment of the present and future opportunities for combined heat and power with district heating (CHP-DH) in the United Kingdom. *Energy Policy*, 38(11), 6936–6945.

Koch, A., Girard, S. and McKoen, K., 2012. Towards a neighbourhood scale for low- or zero-carbon building projects. *Building Research & Information*, 40(4), 527–537.

Levin, A., 2015. Customer incentives and potential energy savings in retail electric markets: a Texas case study. *The Electricity Journal*, 28(3), 51–64.

Lombardi, P., 2011. Managing the green IT agenda. *Intelligent Buildings International*, 3(1), 41–45.

Lombardi, P., 2015. Local experiencies in energy transition. *ENEA, Energia Ambiente e Innovazione, LCS-RNet Transition and Global Challenges towards Low Carbon Societies*, 61, 55–58.

Lombardi, P. and Trossero, E., 2013. Beyond energy efficiency in evaluating sustainable development in planning and the built environment. *International Journal of Sustainable Building Technology and Urban Development*, 4(4), 274–282.

Lombardi, P. *et al.*, 2014. Web and cloud management for building energy reduction: toward a smart district information modelling. In Z. Sun (ed.), *Demand-Driven Web Services: Theory, Technologies, and Applications*. IGI Global: Pennsylvania, USA, pp. 340–355.

Lund, H. *et al.*, 2014. 4th Generation District Heating (4GDH). Integrating smart thermal grids into future sustainable energy systems. *Energy*, 68, pp. 1–11.

Moon, H. *et al.*, 2015. BIM-based construction scheduling method using optimization theory for reducing activity overlaps. *Journal of Computing in Civil Engineering*, 29(3), 130709222650005, pp. 1–16.

Mostafavi, S., Beltran, M.M. and Biloria, N., 2013. Performance driven design and design information exchange. In *31st International Conference on Education and research in Computer Aided Architectural Design in Europe*, Delft, The Netherlands: eCAADe (Education and Research in Computer Aided Architectural Design in Europe), pp. 117–126.

National project Cluster Zero Energy Buildings in Smart Urban Districts (EEB). Available at: http://home.deib.polimi.it/bolchini/research/eeb.html [Accessed April 25, 2015].

Osello, A. *et al.*, 2013. Ugliotti information interoperability and interdisciplinarity: the BIM approach from SEEMPubS project. *Territorio Italia*, 2, 9–22.

Peña-Mora, F. *et al.*, 2010. Mobile ad hoc network-enabled collaboration framework supporting civil engineering emergency response operations. *Journal of Computing in Civil Engineering*, 24(3), 302–312.

Rafiee, A. *et al.*, 2014. From BIM to geo-analysis: view coverage and shadow analysis by BIM/GIS integration. In *12th International Conference on Design and Decision Support Systems in Architecture and Urban Planning, Procedia Environmental Sciences*, 22, Eindhoven—The Netherlands, Procedia Environmental Sciences, ELSEVIER, pp. 397–402.

Ramesh, S. *et al.*, 2013. Urban energy information modelling: an interactive platform to communicate simulation-based high fidelity building energy analysis using Geographical Information Systems (GIS). In *13th Conference of International Building Performance Simulation Association, Proceedings of BS2013: 13th Conference of International Building Performance Simulation Association*, Chambéry, France, Proceedings of Building Simulation, pp. 1136–1143.

Sebastian, R. *et al.*, 2013. Semantic BIM and GIS modelling for energy-efficient buildings integrated in a healthcare district. In *ISPRS 8th 3DGeoInfo Conference & WG II/2Workshop, ISPRS Annals of the Photogrammetry, Remote Sensing and Spatial Information Sciences*, Istanbul, Turkey, ISPR, pp. 27–29.

Sehrawat, P. and Kensek, K., 2014. Urban energy modelling: GIS as an alternative to BIM. In *2014 ASHRAE/IBPSA-USA Building Simulation Conference*, Atlanta, GA, ASHRAE, pp. 235–242.

Stojanovski, T., 2013. City information modelling (CIM) and urbanism: blocks, connections, territories, people and situations. In *SimAUD 2013, Symposium on Simulation for Architecture and Urban Design, Proceeding SimAUD '13 Proceedings of the Symposium on Simulation for Architecture & Urban Design*. Society for Computer Simulation International, San Diego, CA, p. 12.

The Intergovernmental Panel on Climate Change (IPCC), 2014. *Climate Change 2014: Mitigation of Climate Change*.

Toniolo, J., Silvi, C. and Masoero, M., 2013. Energy savings in HVAC systems by continuous monitoring. Results of a long term monitoring campaign on buildings. In *International Conference on Renewable Energies and Power Quality (ICREPQ'13)*, Bilbao, Spain, EA4EPQ, pp. 1–6.

Torabi Moghadam, S. *et al.*, 2015. Simulating window behaviour of passive and active users. *Energy Procedia*, 78, 621–626.

Ulyatt, M., BIM+—thinking beyond a building's lifespan. Available at: http://www.bimplus.co.uk/management/thinking-beyond-buildings-lifespan/ [Accessed April 22, 2015].

UNEP, 2009. *Buildings and Climate Change*, UNEP, Paris, France.

UNEP, 2012. *Global Initiative for Resource Efficient Cities Engine to Sustainability*. UNEP, Paris, France.

Wiel, S. *et al.*, 1998. The role of building energy efficiency in managing atmospheric carbon dioxide. *Environmental Science & Policy*, 1, 27–38.

Wolisz, H. *et al.*, 2014. City district information modelling as a foundation for simulation and evaluation of smart city approaches. In *2014 Building Simulation and Optimization Conference, Proceeding of Building Simulation and Optimization Conference*, UCL, London, UK, University College London, Proceeding.

Wu, I. and Hsieh, S., 2007. Transformation from IFC data model to GML data model: methodology and tool development. *Journal of the Chinese Institute of Engineers*, 30(6), 1085–1090.

Yamaguchi, Y., Shimoda, Y. and Mizuno, M., 2007. Proposal of a modelling approach considering urban form for evaluation of city level energy management. *Energy and Buildings*, 39(5), 580–592.

Zhang, X. *et al.*, 2009. Integrating BIM and GIS for large scale (building) asset management: a critical review. In *The Twelfth International Conference on Civil, Structural and Environmental Engineering Computing*, Funchal, Madeira, Portugal, CC2009, pp. 1–15.

12 Capability maturity modelling of construction e-business processes

Srinath Perera and Anushi Rodrigo

12.1 Introduction

The Construction sector is a major part of the UK economy which contributes some 7% of GDP which is worth about £110 billion per annum, comprising of three main sub sectors: commercial and social, residential and infrastructure (GCS, 2011). However, it has long been seen as an inefficient and underachieving industry (Latham, 1994; Egan, 1998; Fairclough, 2002; GCS, 2011). There has been stream of reports published which identify the need of performance improvement in the construction industry (Latham, 1994; Egan, 1998; 2002; Fairclough, 2002). Many authors recommend that the industry should pay greater attention to processes and organisations should move towards a focus of 'process thinking' in order to achieve desired improvements (Latham, 1994; Egan, 1998; 2002; Atkin *et al.*, 2003; Harris and McCaffer, 2013).

Construction is a unique industry that generally deals with unique output, a specific building within a specific context, conditions and requirements. Each project consists of several phases and number of parties and project teams are involved. Further, the industry is highly fragmented, with over 300,000 businesses (of which 99.7% are SMEs) and over 2 million workers (GCS, 2011). This complexity of construction outputs and processes, and the nature and number of participant involved results in an extensive and complex information flow throughout construction processes (Wilkinson, 2005; Sommerville and Craig, 2006; Anumba and Ruikar, 2008; Chen, 2012). These specific characteristics demand a detailed appraisal and consideration of business activities and advanced methods for information management; where Information and Communication Technology (ICT) applications can be of great assistance (Sun and Howard, 2004).

Electronic business (e-business) is defined as a means of developing new ways of carrying out business activities with the use of Information and Communication Technologies (ICTs) (Li, 2007). It has become a significant source of innovation for modern businesses in every industry including Construction. e-Business is an innovative approach for traditional construction organisations to gain competitive advantages and product, process and performance improvements. It has substantial benefits for every type of organisation. e-Business processes can play important roles in each stage of a construction project and they have been proven to have the potential to improve many industry processes through time and cost savings (Anumba and Ruikar, 2008; Schneider, 2010; Chaffey, 2011). Construction organisations have strategically redesigned their business processes by complementing ICTs in such a way that the processes are more efficient and effective (Love *et al.*, 2004). Keraminiyage *et al.* (2008) confirm a relationship between the IT and some of the existing process improvement initiatives. Improved productivity and efficiency in the construction industry would have significant impact on the national economy.

The implementation of e-business has broadened the boundaries and capabilities of organisations bringing in opportunities to achieve competitive advantages, productivity improvements and efficiency savings. The effective implementation and use of e-business is vital for any organisation to achieve potential benefits and rewards. Appropriate consideration on organisational e-business processes and practices are important to increase and sustain the benefits gained from them (Stockdale *et al.*, 2006). Construction process is defined as a set of activities, methods, practices and transformations that people use to construct and maintain buildings and other civil engineering structures and the associated products (Keraminiyage, 2009). Application of ICT to general construction processes creates the e-business profile of an organisation. Similarly, construction e-business process can be explained as a set of electronically supported activities, methods, practices and transformations that people use to construct and maintain buildings and other civil engineering structures and the associated products.

e-Business capability models and e-readiness models are used as evaluation tools to determine the current status of e-business implementation of an organisation. Capability is defined as power or ability to do something, where in e-business it the ability to carry out e-business activities. e-Readiness models gives an indication of how ready an organisation is to adopt e-business approaches with its current capabilities (Mutula and Van Brakel 2006; Ruikar *et al.* 2006). The limitation associated with e-readiness models is that, even though they provide an indication of how able and ready an organisation is to adapt electronic processes for their day to day business activities; they do not provide possible improvements or systematic procedures for further development.

On the other hand, maturity models are designed to support improvement in processes, products and delivery (Alshawi, 2007). Early maturity model approaches emerged from the area of software engineering (Humphrey *et al.*, 1987). They were developed targeting the need to measure and control processes more thoroughly. The first maturity model in the software development field was the Capability Maturity Model (CMM) of Carnegie Mellon University (Paulk *et al.*, 1993). Maturity models are designed to support improvement in processes, products and delivery (Alshawi, 2007). They aim to help organisations to benchmark themselves and to identify next steps for organisational development. Maturity level indication gives an evidence of the effectiveness and efficiency of an organisation and probable quality of its outcomes. Maturity model concept has been widely adopted by organisations in several disciplines worldwide. Thus, many maturity models have been developed and used by organisations over the past several years.

12.2 The capability maturity model (CMM) concept

12.2.1 History of the CMM

In early 1990s, The first Capability Maturity Model (CMM) was developed in the software development industry by the Software Engineering Institute (SEI) at the Carnegie Mellon University as a framework to inspect capability maturity of software providers (Paulk *et al.*, 1993). Original CMM was based on the 'process maturity grid' developed by the Humphrey *et al.* (1987) and it was evolved by alterations and transformations in the structure of the model. This was initiated by United States' Department of Defence (DOD) as a framework to scrutinize the capability maturity of software suppliers who supply software products to the DOD and US government. Subsequently CMM principles were applied as a process improvement tool in software organisations and it is now one of the most widely adopted process improvement initiatives within numerous industries such as software, manufacturing, drug and cosmetic, etc. Software CMM (SW-CMM) provides a staged based step-by-step framework which permits businesses to assess where they are positioned within the framework and then provides guidelines on what are their process improvement priorities (Hutchinson and Finnemore, 1999).

12.2.2 Fundamental concepts of the CMM

CMM is composed focusing fundamental concepts underlying software process. It defines the software process and explains the basic concepts such as process capability, process performance and process maturity. Table 12.1 below summarises and explains these basic concepts within SW-CMM.

12.2.3 The CMM structure

Advocates of CMM realised that to achieve capability and maturity in organisations, they need to eliminate barriers systematically and structure process improvement initiatives in a methodical way and to achieve long-term rewards from process improvement efforts, it is necessary to design an evolutionary path that increases an organisation's software process maturity in stages (Paulk *et al.*, 1993). CMM provides a road map for continuous process improvement and guides advancement and identifies deficiencies in the organisation.

Table 12.1 Fundamental concepts of SW-CMM

Concept	Explanation
Software process	Set of activities, methods, practices and transformations that people use to develop and maintain software and the associated products.
	As an organisation matures, software process becomes better defined and more consistently implemented throughout the organisation.
Software process capability	The range of expected results that can be achieved by following a software process.
	Provides one means of predicting the most likely outcomes to be expected from the next software project the organisation undertakes.
Software process performance	Represent the actual results achieved by following a software process.
	Within each context which the project is conducted; actual performance of the project may not reflect the full process capability of the organisation.
Software process maturity	The extent to which a software process is explicitly defined, managed, measured, controlled and effective.
	Maturity implies a potential for growth in a company.
	It indicates the richness of organisation software process and consistency with which it is applied in projects throughout the organisation.

Source: Paulk *et al.*, 1993.

CMM demonstrates a staged approach with five maturity levels that have to follow to achieve continuous process improvements. CMM maturity levels are interconnected and each maturity level comprise with Key Process Areas (KPAs), common features and key practices. Each KPA is organised by common features which address implementation or institutionalisation of software process. The common features contain key practices which, when collectively achieved, accomplish the goals of KPAs. The internal structure of CMM is further explained in Figure 12.1.

A maturity level in CMM is described as a precise evolutionary stage toward attaining a mature software process (Paulk *et al.*, 1993). In CMM, maturity levels define an ordinal scale for measuring the maturity of an organisation's process and each level indicates a level of software capability. Achieving each maturity level results an increase in the process capability of the organisation. Types of process capabilities being institutionalised

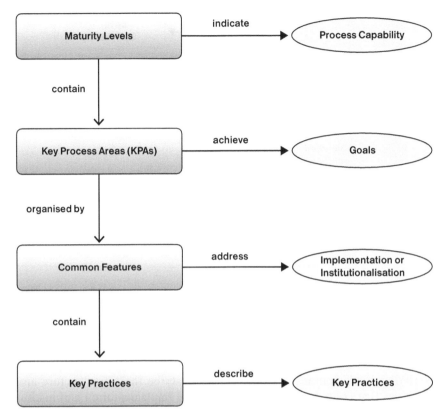

Figure 12.1 Structure of SW-CMM.
Source: Paulk *et al.*, 1993a.

by the organisation at each level of the maturity framework illustrate in Figure 12.2.

Key Process Areas (KPAs) can be identified as key features of an organisation's processes which need to improve at a time to achieve a new maturity level. Each KPA classifies a group of associated activities to reach a set of goals essential for enhancing process capability. They compile key practices which are essential features that much exist in order to accomplish relevant process area (Bate *et al.*, 1994). Following Figure 12.3 illustrates the KPAs which are associated with each maturity level in the SW-CMM.

The goals of CMM structure summarise the key practices of a key process area. These can be used to determine whether an organization or project has effectively implemented the key process area. The goals indicate the scope, boundaries and intention of each key process area (Paulk *et al.*, 1993).

Common features are elements that indicate whether the implementation and institutionalisation of a key process area is effective, repeatable, and

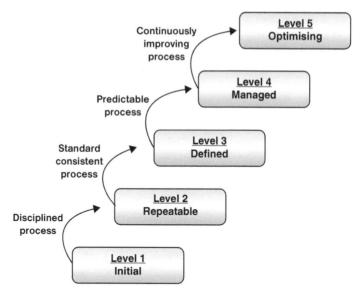

Figure 12.2 Five maturity levels of SW-CMM.
Source: Paulk *et al.*, 1993.

lasting. There are five common features named Commitment to Perform, Ability to Perform, Activities Performed, Measurement and Analysis, and Verifying Implementation. Key practices describe the infrastructure and activities that contribute most to the effective implementation and institutionalisation of the key process area. They describe what is to be done to accomplish the goals of KPAs.

12.2.4 Limitations of CMM

Despite the fact that CMM offers more benefits to process improvements in organisations, it is noticeable that it does not perform perfect all the time. Beside the rewards, CMM was criticised for its weaknesses by some detractors (Weinberg, 1993; Bach, 1994). Jones (1994) presented his own model which developed independently from CMM for by indicating that CMM ignores factors such as individual contribution of engineers towards software productivity. Similarly Bach (1994) pointed out that CMM considers the processes but ignores people. Further he claimed that CMM does not have formal theoretical basis and has only vague empirical support. Besides, there are some possible constraints such as need of specialised personnel, might leading to lack of management commitment, might not feasible for small organisations as CMM is basically designed for large organisations and wrong motive as organisations might have to work on process areas

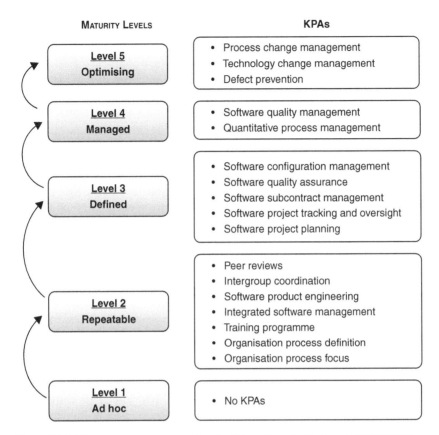

MATURITY LEVELS

Level 5
Optimising

Level 4
Managed

Level 3
Defined

Level 2
Repeatable

Level 1
Ad hoc

KPAs

- Process change management
- Technology change management
- Defect prevention

- Software quality management
- Quantitative process management

- Software configuration management
- Software quality assurance
- Software subcontract management
- Software project tracking and oversight
- Software project planning

- Peer reviews
- Intergroup coordination
- Software product engineering
- Integrated software management
- Training programme
- Organisation process definition
- Organisation process focus

- No KPAs

Figure 12.3 Key process areas associated with each maturity level of SW-CMM.
Source: Adapted from Paulk *et al.*, 1993.

having less important for organisation, but necessary for a maturity level (Shaikh *et al.*, 2009).

Organisation willingness to change and commitment of staff and top management can be identified as essential attempts to overcome the challenges which organisations might face in adopting CMM. Appropriate training and guidance should be provided if required regarding the awareness about e-business and process improvement benefits. Stelzer and Mellis (1999) also identifies staff involvement and setting relevant and realistic objectives as success factors for organisational change in process improvement. In addition accessing appropriate resources, tools and technologies is vital in order to develop and use the CMM concept productively. However, CMM is acknowledged and widely applied for many process improvement initiatives in many other disciplines.

12.2.5 *Capability maturity model integration (CMMI)*

In addition to CMM, Software Engineering Institute of Carnegie Mellon University developed the Capability Maturity Model Integration (CMMI) integrating system engineering, software engineering and product and process development (SEI, 2002). The purpose of CMMI was to provide assistance for improving organisational processes and ability to manage the development, acquisition, and maintenance of products or services.

There are multiple CMMI models available such as systems engineering, software engineering, integrated product and process development and supplier sourcing (SEI, 2002). These models are generated from the CMMI framework and organisations have to decide which CMMI model suits best for their process improvement requirements. Further CMMI has both staged and continuous representations and organisation have to select a representation to use the model. The main components of both the staged and continuous representations of CMMI models are process areas, specific goals, specific practices, generic goals and generic practices. Following Figure 12.4 presents the relationship between these components.

The continuous representation of CMMI uses capability levels to measure process improvement, while the staged representation uses maturity levels to measure process improvement. There are five maturity levels of staged representation of CMMI as Initial, Managed, Defined, Quantitatively Managed and Optimising. These maturity levels indicate an organisations overall maturity. Capability levels which belong to continuous representation of CMMI indicates organisation's process improvement achievement for each process area. There are six capability levels such as Incomplete, Performed, Managed, Defined, Quantitatively Managed and Optimising.

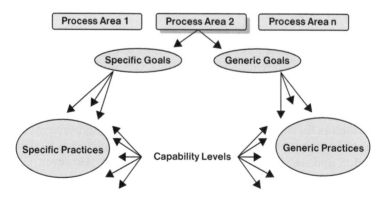

Figure 12.4 Main components of CMMI.

12.3 Process maturity models in construction

12.3.1 Structured process improvement for construction enterprises (SPICE)

Structured Process Improvement for Construction Enterprises (SPICE) is an attempt to use the process capability maturity concept within the construction industry. The capability maturity model development was initiated in Carnegie Mellon University by developing first Capability Maturity Model (CMM) in the software development industry (Paulk *et al.*, 1993). CMMs are designed to support improvement in processes, products and delivery. The basic viewpoint behind process capability maturity is that a when an organisation becomes mature, the processes also mature and become more predictable and reliable (Stewart and Spencer, 2006). Mature organisations carry out processes systematically while immature organisations does not have systematic processes and their success relies on individual efforts and unstructured approaches (Becta, 2005). Capability maturity models help organisations to benchmark themselves and to identify next steps for organisational development. Maturity level indication gives an evidence of the effectiveness and efficiency of an organisation and probable quality of its outcomes.

The SPICE project carried out developed by a research expert team of Salford University, following the fundamentals of the original Software Capability Maturity Model (Sarshar *et al.*, 1999; 2000). Initially the process capability maturity characteristics of lower maturity levels of construction organisations were explored based on the CMM maturity levels one to three (Sarshar *et al.*, 2000; Jeong *et al.*, 2006). Later the process capability maturity characteristics of lower maturity levels of construction organisations were established (Keraminiyage, 2009).

SPICE has borrowed basic concepts from CMM and established a construction specific CMM. SPICE model comprises of maturity levels and key process areas similar to CMM. The five maturity levels of SPICE are illustrated in Figure 12.5.

Each maturity level of SPICE is comprise of key process areas. The ability to perform a key process area is measured against key enablers named Commitment, Ability, Evaluation, Verification and Activities. An organisation should successfully perform all key process areas belong to a particular maturity level in order to achieve the level of maturity. The maturity levels are sequential and organisations cannot skip maturity levels while progressing (Sarshar *et al.*, 1998; 2000).

12.3.2 e-Readiness models

e-Readiness is defined as "the ability of an organisation, department or workgroup to successfully adopt, use and benefit from information and communication technologies" (Ruikar *et al.*, 2006: p. 99). In other terms,

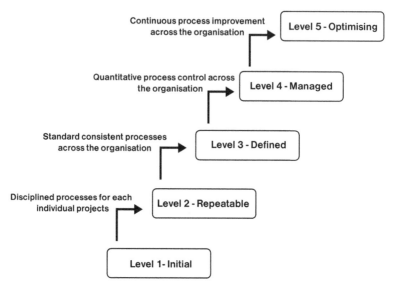

Figure 12.5 SPICE maturity levels.
Source: Sarshar *et al.*, 2000.

it is how able and ready an organisation is to adapt electronic processes for their day to day business activities. Implementation of e-business in an organisation involves modifications in its current practices, systems, processes and workflows. Therefore, organisations need to take measures on their organisational e-readiness to successfully adopt and use e-business tools. Evaluating and measuring organisational e-readiness is important to companies to ensure a productive and beneficial implementation of e-business tools within their businesses.

There are several attempts of measuring organisations' e-readiness and an increasing number of readiness assessment tools have been developed over last few years. These e-readiness models and tools have been developed to assess how ready an organisation, an industry or an economy is to adopt ICTs and e-business (Mutula and Van Brakel, 2006). Key construction specific e-readiness models are examined in detail in the following section.

12.3.2.1 Verify End-User e-Readiness using a Diagnostic Tool – VERDICT

Ruikar *et al.* (2006) presents Verify End-user e-Readiness using a Diagnostic Tool (VERDICT) as an e-readiness assessment model which can be used to measure the readiness of construction organisations for e-commerce. VERDICT is internet based prototype application that assesses the overall e-readiness of end user companies and profiles by considering management, process, people and technology aspects. VERDICT model is developed

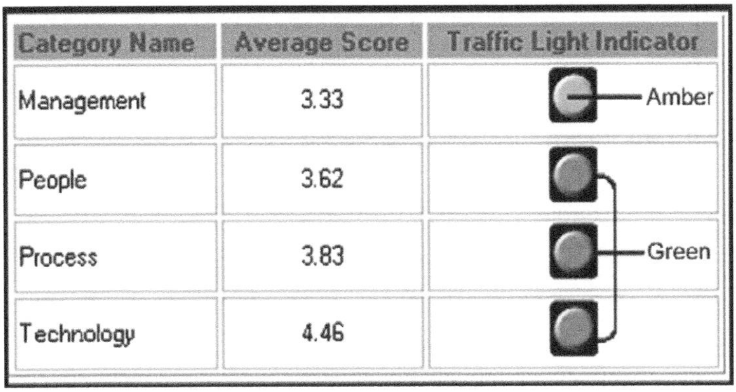

Category Name	Average Score	Traffic Light Indicator
Management	3.33	Amber
People	3.62	
Process	3.83	Green
Technology	4.46	

Figure 12.6 Typical table summarising average scores in each category with traffic light indicators of VERDICT.

Source: Ruikar *et al.*, 2006.

based on the principle that for any company to be e-ready, its management, people, process and technology have to be e-ready (Ruikar *et al.*, 2006). Organisations cannot use VERDICT to measure their overall e-readiness; but can use it to periodically review their progress in achieving e-readiness by using the four categories as key performance indicators. Figure 12.6 presents a typical table summarising average scores in each category with traffic light indicators of VERDICT.

The scores of each category are averaged and presented with traffic light indicators of red, green and amber lights to visually indicate their e-readiness in each category. Red indicates that several aspects within a particular category need urgent attention to achieve e-readiness. Amber indicates that certain aspects within a particular category need attention to achieve e-readiness. Green indicates that the end-user organisation has adequate capability and maturity in these aspects and therefore is e-ready. In addition to this traffic light indicators, VERDICT gives an indication of critical issues that companies need to address to achieve e-readiness and points out organisations' strengths and weaknesses regarding its readiness for using e-commerce tools. Further, being Internet-based, the system is platform-independent and provides all the benefits of using the Internet such as flexibility, accessibility, portability, device independence and eco-nomical. However, VERDICT does not define measures for organisations to address the issues of highlighting areas and the necessary procedures for them to go through to make improvement in such areas.

12.3.2.2 IS/IT readiness model

Alshawi (2007) developed 'IS/IT Readiness Model' to assist organisations in the construction industry to successfully implement information technology and information systems. The Model describes the readiness of an organisation

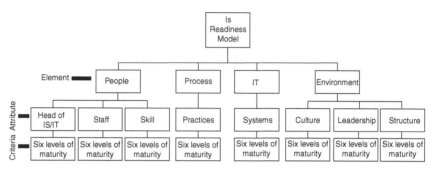

Figure 12.7 The structure of the IS/IT readiness model.
Source: Alshawi, 2007.

for an IS/IT (Information System/Information Technology) project in terms of four categories; people, processes, technology and environment. IS/IT Readiness Model illustrates these four elements in detail using a criteria describing the attributes related with each of them. Attributes of the people category are head of IS/IT function, staff and skills. Process category considers general practices and technology category considers systems in terms of hardware, software, communication and networks as attributes. Environment category consists of culture, leadership and structure attributes. Under each category these attributes are described as they might occur in each of the six maturity levels. However, how organisations can improve their organisational readiness from the lowest to the highest level has not been addressed in the model. The structure of the IS/IT Readiness Model is presented in Figure 12.7.

12.4 e-Business capability maturity models

Maturity is defined as the degree to which organisational processes and activities are executed following principles of good practice (Alshawi, 2007). Maturity concept encourages repeatable and predictable outcomes within a process or practice in an organisation. The fundamental concept behind maturity is that mature organisations carry out processes systematically while immature organisations does not have systematic processes and their success relies on individual efforts and unstructured approaches (Becta, 2005). As a result, mature organisations consistently accomplish schedules and goals and achieve right outcomes in an efficient manner (Harmon, 2004).

Early maturity model approaches were emerged from the area of software engineering (Humphrey *et al.*, 1987). They were developed targeting the need to measure and control processes more thoroughly. The first maturity model in the software development field was the Capability Maturity Model (CMM) of Carnegie Mellon University (Paulk *et al.*, 1993). Maturity models are designed to support improvement in processes, products and delivery

(Alshawi, 2007). They aim to help organisations to benchmark themselves and to identify next steps for organisational development. Maturity level indication gives an evidence of the effectiveness and efficiency of an organisation and probable quality of its outcomes. Maturity model concept has been widely adopted by organisations in several disciplines worldwide. Thus, many maturity models have been developed and used by organisations over the past several years.

12.4.1 General e-Business Capability Maturity Model (EB-CMM)

The e-Business Capability Maturity Model (EB-CMM) was developed by Chen *et al.* (2006) focusing cosmetic and drug industry. It is essentially aimed at general procurement of goods and services. They applied the concepts of original CMM and CMMI for electronic business practices in the cosmetic and drug industry. EB-CMM has both capability and maturity representations and they represent the different focus on process and organization.

EB-CMM is comprised with six maturity levels and seven capability levels. Maturity levels are named Initial, Internal, Integrated, Defined, Quantitatively Managed and Optimising. Capability levels of the model are Incomplete, Performed, Internal, Integrated, Defined, Quantitatively Managed and Optimising. In the model capability representation describes how an organisation does regarding a particular process area and maturity representation describes the entire organization performance on certain e-business related process areas. The model provides what-to-do implementation suggestions for organizations to electronically perform certain core activities for certain process category. Maturity levels and their process areas of EB-CMM are presented in below Figure 12.8.

This model can be identified as a generic e-business capability maturity model for organisations in the cosmetic and drug industry. However the applicability of this model to the construction discipline is debatable as construction is a unique industry and as it has distinctive characteristics from other industries. Therefore in order to consider EB-CMM in construction context, the internal components and process areas have to be carefully investigated and the applicability of EB-CMM process areas for construction organisations have to be confirmed.

12.4.2 National building information model standards' capability maturity model

The Unites States National Building Information Model (NBIM) Standard proposed a CMM for use in measuring the degree to which a building information model implements a mature BIM Standard (NIST, 2007). The idea was to allow BIM users to plot their current position and to set goals for their future operations (NIBS, 2007). NBIMS' CMM is available for use as a standardised tool to assist users with BIM evaluation and development. There are

ML	Level Name	PA	Process Area and the Description
2	Internal	OP	**Order processing:** Managing and tracking orders in order to know the status.
		DM	**Document management:** Electronically managing ad-hoc business documents using IT tools.
		NM	**Network management:** Establishing internal/external network infrastructure.
		ISS	**Information and systems security:** Ensuring secure use of internal information. information system and network environment.
		TS	**Technical solution:** Providing candidate IT solution evaluation.
		SD	**System development:** Describing and managing IT system implementation.
		RSKM	**Risk management:** Identifying and managing potential problems before an IT solution is implemented.
3	Integrated	OLAP	**OLAP, on-line analytical processing.** Performing online multi-dimensional data warehousing analysis.
		INE	**Integrated network between enterprises:** Establishing and integrating business networks between enterprises.
		ICRM	**Integrated customer relationship management:** Establishing and integrating front-end and back-end sales and customer information for offering better, more flexible customized serviced or products.
		EDI	**Electronic data interchange:** Establishing and integrating standards and platforms for information exchange automation.
		EAI	**Enterprise application integration:** Integrating internal or external information systems
		EKMS	**Electronic knowledge management and sharing:** Establishing online information indexing and knowledge querying services
		IISS	**Integrated information and systems security:** Establishing security mechanisms between systems or between organizations.
		IOPF	**Integrated order processing and forecasting:** Integrating sales information among players for better business forecast.
4	Defined	WFM	**Workflow management:** Defining managing, and automating business processes.
		OCE	**Organizational culture and environment for EB:** Developing and maintaining EB implementation culture and enviroment.
		EPF	**Electronic process focus:** Understanding current situation of EP process, planning EB process improvement.
		EPD	**Electronic process definition:** Establishing and maintaining EB process assets for managing existing and composing new EB processes.
		DAR	**Decision Analysis and Resolution.** Establishing evaluation criteria for stakcholders to make valid decision.
5	Quantitatively Managed	QEBP	**Quantitatively EB Performance.** Establishing and maintainning key performance indices for measuring ad-hoc EB processes.
		QSQM	**Quantitatively system quality management:** Establishing and maintaining key performance indices for measuring the performance of ad-hoc information system.
6	Optimizing	CAR	**Causal analysis and resolution:** Analyzing the root causes of EB problems, taking corrective actions and preventing from recurrence.
		OID	**Organizational innovation and deployment:** Choosing innovative information technologies for continually optimizing electronic business performance.

Figure 12.8 Maturity levels and their process areas of EB-CMM.

two versions of NBIMS' CMM. Firs version is a tabular version identifying eleven areas of interests measured against ten levels of increasing maturity. The areas of interest include; data richness, life-cycle views, change management, roles or disciplines, business processes, timeliness/response, delivery method, graphical information, spatial capability, information accuracy, and interoperability/IFC support (NIBS, 2007) (Figure 12.9).

Second version is an Interactive Capability Maturity Model (I-CMM). It is a multi-tab Microsoft Excel workbook based on initial tabular model (NIBS, 2007). Following Figure 12.10 presents the I-CMM. The administration portion of the I-CMM provides categories for scores levels within the model. These categories range from 'Minimum BIM' to 'Platinum'. The scoring levels within the I-CMM reflect the maturity level of an individual BIM as measured against a set of weighted criteria agreed to be desirable in a BIM.

Maturity Level	A Data Richness	B Life-cycle Pices	C Roles Or Discipliaes	D Change Management	E Basiaess process	F Tiacliaessl Response	G Delivery Method	H Graphical Information	I Spacial Gappability	J Information Acceracy	K Interoperability I IFC Support
1	Basic Core Data	No Complete Project Phase	No Single Role Fully Supported	No CM Capability	Separate Processes Not Integrated	Most Response Info manually re-collected - Slow	Single Point Access No IA	Primarily Text - No Technical Graphics	Not Spatially Located	No Ground Truth	No Interoperability
2	Expanded Data Set	Planning & Design	Only One Role Supported	Aware of CM	Few Bus Processes Collect Info	Most Response Info manually re-	Single Point Acess w/ Limited IA	2D Non-Intelligent As Designed	Basic Spatial Location	Initial Ground Truth	Forced Interoperability
3	Enhanced Data Set	Add Construction/ Supply	Two Roles Partially Supported	Aware of CM and Root Cause Analysis	Some Bus Process Collect Info	Data Calls Not In BIM But Most Other Data Is	Network Access w/ Basic IA	NCS 2D Non-Intelligent As Designed	Spatially Located	Limited Ground Truth Int Spaces	Limited Interoperability
4	Data Plus Some Information	Includes Construction/ Supply	Two Roles Fully Supported	Aware CM, RCA and Feedback	Most Bus Processes Collect Info	Limited Response Info Available In BIM	Network Access w/ Full IA	NCS 2D Intelligent As Designed	Located w/ Limited Info Sharing	Full Ground Truth - Int Spaces	Limited Info Transfers Between COTS
5	Data Plus Expanded Information	Includes Contr/Supply & Fabrication	Partial Plan, Design&Constr Supported	Implementing CM	All Business Process (BP) Collect Info	Most Response Info Available In BIM	Limited Web Enabled Services	NCS 2D Intelligent As-Builts	Spatially located w/Metadata	Limited Ground Truth Int & Ext	Most Info Transfers Between COTS
6	Data w/Limited Authoritative Information	Add Limited Operations & Warranty	Plan, Design & Construction Supported	Initial CM process implemented	Few BP Collect & Maintain Info	All Response Info Available In BIM	Full Web Enabled Serices	NCS 2D Intelligent And Current	Spatially located w/Full Info Share	Full Ground Truth - Int And Ext	Full Info Transfers Between COTS
7	Data w/ Mostly Authoritative Information	Includes Operations & Warranty	Partial Ops & Sustainment Supported	CM process in place and early implementation	Some BP Collect & Maintain Info	All Response Info From BIM & Timely	Full Web Enabled Services	3D - Intelligent Graphics	Part of a limited GIS	Limited Comp Areas & Ground	Limited Info Uses IFS's For Interoperability
8	Completely Authoritative Information	Add Financial	Operations & Sustainment Supported	CM and RCA capability implemented	All BP Collect & Maintain Info	Limited Real Time Access From BIM	Web Enabled Services - Secure	3D - Current And Intelligent	Part of a more complete GIS	Full Computed Areas & Ground	Expanded Info Uses IFS's for Interoperability
9	Limited Knowledge Management	Full Facility Life-cycle Collection	All Facility Life-Cycle Roles Supported	Business processes are sustained by CM using RCA and Feedback	Some BP Collect&Maint in Real Time	Full Real Time Access From BIM	Netcentric SOA Based CAC Access	4D - Add Time	Integrated into a complete GIS	Comp GT w/Limited Metrics	Most Info Uses IFC's For Interoperability
10	Full Knowledge Management	Supports External Efforts	Internal and External Roles Supported	Business processes are routinely sustained by CM, RCA and Feedback loops	All BP Collect&Maint In Real Time	Real Time Access w/ Live Feeds	Netcentric SOA Role Based CAC	nD - Time & Cost	Integrated into GIS w/ Full Info Flow	Computed Ground Truth w/Full Metrics	All Info Uses IFC's For Interoperability

Figure 12.9 Tabular BIM capability maturity model.
Source: NIBS, 2007.

The Interactive BIM Capability Maturity Model

Area of Interest	Weighted Importance	Choose your perceived maturity level	Credit
Data Richness	84%	Data Plus Expanded Information	4.2
Life-cycle Views	84%	Add Construction/ Supply	2.5
Change Management	90%	Limited Awareness	2.7
Roles or Disciplines	90%	Partial Plan, Design&Constr Supported	4.5
Business Process	91%	Some Bus Process Collect Info	2.7
Timeliness/ Response	91%	Data Calls Not In BIM But Most Other Data Is	2.7
Delivery Method	92%	Limited Web Enabled Services	4.6
Graphical Information	93%	3D - Intelligent Graphics	6.5
Spatial Capability	94%	Basic Spatial Location	1.9
Information Accuracy	95%	Limited Ground Truth - Int Spaces	2.9
Interoperability/ IFC Support	96%	Most Info Transfers Between COTS	4.8

	Credit Sum	40.0
	Maturity Level	Minimum BIM

National Institute of
BUILDING SCIENCES
Facilities Information Council
National BIM Standard

ADMINISTRATION

	Points Required for Certification Levels		
	Low	High	
	40	49.9	Minimum BIM
	50	59.9	Minimum BIM
	60	69.9	Certified
	70	79.9	Silver
	80	89.9	Gold
	90	100	Platinum

Remaining Points Required For:	Certified	20.0

Figure 12.10 Interactive BIM capability maturity model.

The goals of both tabular and interactive versions of CMM are to help users gauge their current maturity level, as well as plan for future maturity attainment goals through a commonly accepted, standardized approach.

12.4.3 UK BIM maturity model

The UK government has initiated encouraging construction organisations to embrace BIM to increase their potential capabilities in performance. As a result, the report of 'Government Construction Strategy' (2011) announced that the government requires construction organisations to have minimum of Level 2 BIM according to the Bew-Richards BIM maturity ramp. The report stated that "Government will require fully collaborative 3D BIM (with all project and asset information, documentation and data being electronic) as a minimum by 2016. This refers to all centrally procured Government projects as outlined in the GCS including new build and retained estate, vertical and linear" (GCS, 2011).

This BIM maturity model defines maturity regards to the ability of the construction supply chain to operate and exchange information. Further, it also defines the supporting infrastructure required at each level of capability. As presented in Figure 12.11, BIM level 0 indicates that an organisation utilises mainly 2D CAD drafting and no collaboration. Level 1 BIM indicates a mixture of 2D and 3D CAD usage. Level 2 is collaborative working where all parties use 3D CAD models. In this level design information is shared through a common file format and that enables any organisation to be able to combine that data with their own in order to make a united BIM model, and to carry out interrogative checks on it (NBS, 2014). Level 4 BIM

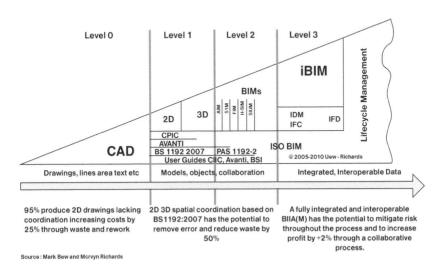

Figure 12.11 BIM maturity ramp.
Source: Bew & Richards, 2008.

represents full collaboration between all disciplines using a single, shared project model which is held in a central repository (NBS, 2014). All project parties can access and modify the same BIM model and it eliminates the possibility of conflicting information in project lifecycle management.

12.4.4 Building information modelling maturity matrix (BIM3)

Succar (2009) presents Building Information Modelling Maturity Matrix (BIM3) which includes a set of maturity levels which indicate the evolutionary improvement of processes, technologies and policies within each BIM Stage. Maturity index of BIM3 was developed by analysing and integrating several models from different industries. It has five maturity levels named Initial/Ad-hoc, Defined, Managed, Integrated and Optimises. Its maturity levels reflect the extent of BIM abilities, deliverables and their requirements as opposed to minimum abilities reflected through capability stages. Succar *et al.* (2012) introduces BIM3 as a knowledge tool which incorporates many BIM framework components for the purpose of performance measurement and improvement. The components of BIM Maturity Matrix are represents in following Figure 12.12.

The BIM Maturity Matrix incorporates maturity index for a set of competencies, capability stages and organisation scales. Competency areas include Technology, Process and People which expand through four granularity levels. Three capability stages are presented as Modelling, Modelling based collaboration and Network based integration. Organisation scales include in the model are micro, meso and macro. Industry practitioners can use the BIM Maturity Matrix to increase their capability across a range of technology, process and policy steps. It provides an assessment tool to

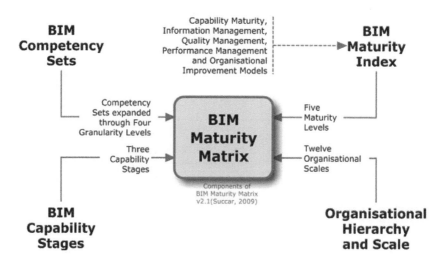

Figure 12.12 Components of the BIM maturity matrix.

measure BIM capability and maturity at selective organisational scales and granularity levels. Organisations can use BIM3 to continuously assess and improve their BIM performance.

12.4.5 e-Procurement capability maturity model (EP-CMM)

Eadie (2009) developed an e-procurement capability maturity model for construction organisations. The concept for this maturity model also borrowed from original SW-CMM (Eadie *et al.*, 2011; 2012). Initially construction related e-procurement drivers and barriers were identified for the development of the model. Then the drivers and barriers were classified using Principal Components Analysis into Key Process Areas to produce a capability maturity model (EP-CMM), which shows how mature or e-ready an organisation is in relation to e-procurement in Construction. The EP-CMM enables an organisation to classify its level of maturity thereby enabling organisations to identify the needs for further development to enhance e-procurement capability. The maturity levels and key process areas of EP-CMM are presented in Table 12.2.

The e-procurement CMM comprises of five maturity levels and 12 key process areas. This model can be used to measure the capability and maturity of e-procurement process in construction organisations. The wide application of the model within construction allows easy adoption by all the disciplines within construction from public sector procurement officials to private sector quantity surveyors (Eadie, 2009). The key purpose of this model is to identify through assessment how an organisation can improve e-readiness in e-procurement. Further this model allows organisations to determine their overall maturity level in relation to e-procurement. This will allow organisations to plan and improve e-procurement practices within their organisations.

Table 12.2 Maturity levels and key process areas of e-procurement CMM

Maturity levels	e-Proc. CMM KPAs
Level 1—Initial	No KPAs
Level 2—Repeatable	Quality Management System
	Cost Management System
	Time Management System
	Operational Analysis
Level 3—Defined	Intergroup Coordination
	Integrated Teaming
	Requirements Development
	Integration Management system
	Organisational Environment
Level 4—Managed	Organisational Change Management System
	Knowledge Management System
Level 5—Optimising	Governance Management System

12.4.6 Construction e-business capability maturity model (CeB-CMM)

Construction e-Business Capability Maturity Model (CeB-CMM) is proposed by Rodrigo *et al.* (2011, 2014) as an e-business process improvement tool for construction organisations. CeB-CMM is developed based on the original CMM concept. However the structure of the CeB-CMM is customised to increase the simplicity and usability of the model. CeB-CMM is comprised of a construction process classification and construction e-business maturity characteristics. The model structure is illustrated in following Figure 12.13.

Process classification of CeB-CMM is comprised of five process categories such as preparation and brief, planning and design, tendering, construction and handover and aftercare. Construction maturity characteristics of maturity levels of CeB-CMM are presented below in Table 12.3.

These e-business maturity characteristics are interpreted for each process category in the CeB-CMM. The model provides e-business maturity level indications for each process category of an organisation by analysing organisational e-business characteristics against e-business maturity characteristics of CeB-CMM. The complete CeB-CMM is presented in Table 12.4.

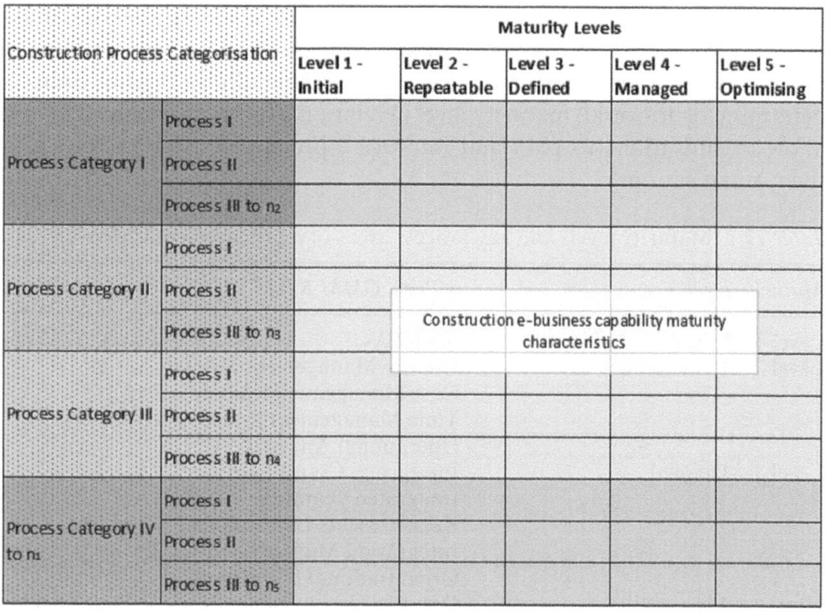

Construction Process Categorisation		Maturity Levels				
		Level 1 - Initial	Level 2 - Repeatable	Level 3 - Defined	Level 4 - Managed	Level 5 - Optimising
Process Category I	Process I					
	Process II					
	Process III to n₂					
Process Category II	Process I					
	Process II		Construction e-business capability maturity characteristics				
	Process III to n₃					
Process Category III	Process I					
	Process II					
	Process III to n₄					
Process Category IV to n₁	Process I					
	Process II					
	Process III to n₅					

Figure 12.13 Structure of the CeB-CMM.

Table 12.3 Construction e-business capability maturity characteristics for CeB-CMM

Maturity level	Characteristics
1—Initial	Most processes operate manually e-Business initiatives within the organisation are unplanned. Organisation does not provide stable environment to support e-business processes.
2—Repeatable	Basic e-business processes are established. Established successful e-business processes are recognisably repeated within organisation practice. Established e-business processes are isolated and are not integrated within the organisation. Provide a stable environment to support e-business processes.
3—Defined	e-Business processes are well established. Standardised e-business processes are used constantly across the organisation. e-Business processes are integrated within the organisation. Provide necessary training and education for staff regarding e-business processes. Updating their existing e-business systems with new advancements.
4—Managed	Administrate and maintain current e-business processes. Establish quantitative and qualitative indicators for measure e-business process performance. Organisation quantitatively and qualitatively measures and monitors e-business process performance. e-Business processes are quantitatively and qualitatively understood and controlled. Organisation e-business processes are compatible and capable to incorporate with other partnering organisations' ICT systems.
5—Optimising	Focused on maintenance and continuous improvement of e-business capabilities. Problem diagnosis and resolution. Identify new innovative technology improvements and deploy suitable approaches.

12.5 Conclusions

e-Business is an innovative approach for traditional construction organisations that enables to gain competitive advantages and product, process and performance improvements. It has substantial benefits for every type of organisations. Successful implementation of e-business is the key to achieve optimum benefits and rewards for an organisation. Therefore careful consideration has to be given while implementing e-business practices into businesses. e-Business strategy plays an important role in ensuring a productive and beneficial e-business implementation in an organisation. Robust e-business strategy will guarantee an effective execution to achieve the desired goals. It is important

Table 12.4 Construction e-business capability maturity model (CeB-CMM)

Construction e-business capability maturity model (CEB-CMM)

Process categories		Maturity levels				
		Level 1—initial	Level 2—repeatable	Level 3—defined	Level 4—managed	Level 5—optimising
Preparation and Brief	Identify business needs; Identify project requirements; Develop project brief; Undertake initial feasibility studies; Initial risk assessment	Most processes of Preparation and Brief category operate manually	Basic e-business processes of Preparation and Brief category are established	e-Business processes of Preparation and Brief category are well established and understood	Administrate and maintain current ebusiness processes of Preparation and Brief category	Focused on continuous improvement of e-business capabilities of Preparation and Brief category
		e-Business initiatives for the processes of Preparation and Brief category within the organisation are unplanned	Established successful e-business processes of Preparation and Brief category are recognisably repeated within organisation practice	Standardised e-business processes of Preparation and Brief category are used constantly across the organisation	Establish quantitative and qualitative measures for e-business process performance of Preparation and Brief Category	Problem diagnosis and resolution
		Organisation does not provide stable environment to support e-business processes of Preparation and Brief category	Established e-business processes of Preparation and Brief category are isolated and are not integrated with other categories' e-business processes	Different e-business processes of Preparation and Brief category are integrated with other categories' e-business processes	Organisation measures and monitors e-business process performance of Preparation and Brief category	Identify new innovative technology improvements and deploy suitable approaches for the processes of Preparation and Brief category

Category	Processes	Level 1	Level 2	Level 3	Level 4	Level 5
(Preparation and Brief, continued)				Organisation provides stable environment to support e-business processes of Preparation and Brief category	Provide necessary training and education for staff regarding ebusiness processes of Preparation and Brief category (if needed) · Update existing e-business systems with new advancements/versions (if available)	e-Business processes of Preparation and Brief category are understood and controlled · Organisation e-business processes of Preparation and Brief category are compatible and capable to incorporate with other partnering organisations' ICT systems
Planning and Design	Determine procurement strategy · Determine contract strategy · Prepare initial project programme · Preliminary cost planning · Preparation of concept design · Preparation of developed design · Preparation of technical design	Most processes of Planning and Design category operate manually · e-Business initiatives for the processes of Planning and Design category within the organisation are unplanned	Basic e-business processes of Planning and Design category are established · Established successful e-business processes of Planning and Design category are recognisably repeated within organisation practice	e-Business processes of Planning and Design category are well established and understood · Standardised e-business processes of Planning and Design category are used constantly across the organisation	Administrate and maintain current ebusiness processes of Planning and Design category · Establish quantitative and qualitative measures for e-business process performance of Planning and Design Category	Focused on continuous improvement of e-business capabilities of Planning and Design category · Problem diagnosis and resolution

(Continued)

Construction e-business capability maturity model (CEB-CMM)

Process categories		Maturity levels			
	Level 1—initial	Level 2—repeatable	Level 3—defined	Level 4—managed	Level 5—optimising
				Organisation measures and monitors e-business process performance of	Identify new innovative technology improvements and deploy suitable approaches for the processes of Planning and Design category
	Organisation does not provide stable environment to support e-business processes of Planning and Design category	Established e-business processes of Planning and Design category are isolated and are not integrated with other categories' e-business processes	Different e-business processes of Planning and Design category are integrated with other categories' ebusiness processes	e-Business processes of Planning and Design category are understood and controlled	
		Organisation provides stable environment to support e-business processes of Planning and Design category	Provide necessary training for staff regarding e-business processes of Planning and Design category (if needed)		
			Update existing e-business systems with new advancements/versions (if available)	Organisation e-business processes of Planning and Design category are compatible and capable to incorporate with other partnering organisations' ICT systems	

Tendering	Most processes of Tendering category operate manually	Basic e-business processes of Tendering category are established	e-Business processes of Tendering category are well established and understood	Administrate and maintain current ebusiness processes of Tendering category	Focused on continuous improvement of e-business capabilities of Tendering category
Preparation of tender documents Invitation to tender Tender submission Tender evaluation	e-Business initiatives for the processes of Tendering category within the organisation are unplanned	Established successful e-business processes of Tendering category are recognisably repeated within organisation practice	Standardised e-business processes of Tendering category are used constantly across the organisation	Establish quantitative and qualitative measures for e-business process performance of Tendering Category	Problem diagnosis and resolution
	Organisation does not provide stable environment to support e-business processes of Tendering category	Established e-business processes of Tendering category are isolated and are not integrated with other categories' e-business processes	Different e-business processes of Tendering category are integrated with other categories' e-business processes	Organisation measures and monitors e-business process performance of Tendering category	Identify new innovative technology improvements and deploy suitable approaches for the processes of Tendering category
		Organisation provides stable environment to support e-business processes of Tendering category	Provide necessary training and education for staff regarding ebusiness processes of Tendering category (if needed)	e-Business processes of Tendering category are understood and controlled	

(Continued)

Construction e-business capability maturity model (CEB-CMM)

Process categories		Maturity levels				
		Level 1— initial	*Level 2— repeatable*	*Level 3— defined*	*Level 4— managed*	*Level 5— optimising*
				Update existing e-business systems with new advancements/ versions (if available)	Organisation e-business processes of Tendering category are compatible and capable to incorporate with other partnering organisations' ICT systems	
Construction	Post-contract project management Post-contract project team coordination Post-contract project planning, tracking and monitoring Post-contract change management Post-contract cost management Quality control Interim payments Site record keeping Claims management Dispute resolution	Most processes of Construction category operate manually	Basic e-business processes of Construction category are established	e-Business processes of Construction category are well established and understood	Administrate and maintain current e-business processes of Construction category	Focused on continuous improvement of e-business capabilities of Construction category
		e-Business initiatives for the processes of Construction category within the organisation are unplanned	Established successful e-business processes of Construction category are recognisably repeated within organisation practice	Standardised e-business processes of Construction category are used constantly across the organisation	Establish quantitative and qualitative measures for e-business process performance of Construction Category	Problem diagnosis and resolution

Category / Processes					
Construction (continued)	Organisation does not provide stable environment to support e-business processes of Construction category	Established e-business processes of Construction category are isolated and are not integrated with other categories' e-business processes	Different e-business processes of Construction category are integrated with other categories' e-business processes	Organisation measures and monitors e-business process performance of Construction category	Identify new innovative technology improvements and deploy suitable approaches for the proceses of Construction category
		Organisation provides stable environment to support e-business processes of Construction category	Provide necessary training and education for staff regarding e-business processes of Construction category (if needed)	e-Business processes of Construction category are understood and controlled	
			Update existing e-business systems with new advancements/versions (if available)	Organisation e-business processes of Construction category are compatible and capable to incorporate with other partnering organisations' ICT systems	
Handover and Aftercare — Project performance review; Update project information; Facilities management	Most processes of Handover and Aftercare category operate manually	Basic e-business processes of Handover and Aftercare category are established	e-Business processes of Handover and Aftercare category are well established and understood	Administrate and maintain current e-business processes of Handover and Aftercare category	Focused on continuous improvement of e-business capabilities of Handover and Aftercare category

(Continued)

Construction e-business capability maturity model (CEB-CMM)

Process categories

	Maturity levels				
	Level 1—initial	Level 2—repeatable	Level 3—defined	Level 4—managed	Level 5—optimising
	e-Business initiatives for the processes of Handover and Aftercare category within the organisation are unplanned	Established successful e-business processes of Handover and Aftercare category are recognisably repeated within organisation practice	Standardised e-business processes of Handover and Aftercare category are used constantly across the organisation	Establish quantitative and qualitative measures for e-business process performance of Handover and Aftercare Category	Problem diagnosis and resolution
	Organisation does not provide stable environment to support e-business processes of Handover and Aftercare category	Established e-business processes of Handover and Aftercare category are isolated and are not integrated with other categories' e-business processes	Different e-business processes of Handover and Aftercare category are integrated with other categories' e-business processes	Organisation measures and monitors e-business process performance of Handover and Aftercare category	Identify new innovative technology improvements and deploy suitable approaches for the processes of Handover and Aftercare category

Organisation provides stable environment to support e-business processes of Handover and Aftercare category	Provide necessary training and education for staff regarding ebusiness processes of Handover and Aftercare category (if needed)	e-Business processes of Handover and Aftercare category are understood and controlled
	Update existing e-business systems with new advancements/ versions (if available)	Organisation e-business processes of Handover and Aftercare category are compatible and capable to incorporate with other partnering organisations' ICT systems

for organisations to scan and position their current situation, and provide a holistic approach to assist them in developing an executable e-business strategy. This demands a thorough understanding of the current status of e-business processes of the organisation to set future objectives and to decide priorities. Evaluation and analysis of current e-business processes is critical and vital.

e-Readiness models and capability maturity models are used to determine the current status of e-business implementation of an organisation. e-Readiness models give an indication of how ready an organisation is to adopt e-business approaches with its current capabilities. The restriction of e-readiness models is that they do not provide potential enhancements or methodical guidelines for improvements. On the other hand, Maturity models bridge this gap by providing an indication of how mature are the e-business processes of an organisation which leads to an understanding of current status of their e-business capabilities together with a systematic approach or procedures for attaining higher maturity levels. Thus, maturity models can be used as the initial or first step in developing an e-business strategy for an organisation. Moreover, e-business capability maturity models are process focused and process improvements leads performance improvements of organisational practices.

There are few construction specific e-business models developed by numerous researchers (Ruikar *et al.*, 2006; Alshawi, 2007; NIBS, 2007; Eadie, 2009; Succar, 2009; Rodrigo *et al.*, 2011; 2014) as described in previous sections of this chapter. Former e-business capability maturity models were either specific to an industry other than construction or focus mainly or only on one e-business approach. There were CMMs developed for specific e-business approaches in construction organisations such as for e-procurement and BIM. Recently, CeB-CMM was developed as a construction specific e-business CMM which focuses on different types of construction e-business processes. It helps construction organisations to identify the potential to improve its e-business process capability which allows them a pathway to further enhance those processes and have productivity and efficiency improvements.

References

Alshawi, M. (2007) *Rethinking IT in Construction and Engineering: Organisational Readiness*. Oxon, UK: Taylor & Francis.

Anumba, C. J. & Ruikar, K. (2008) *e-Business in Construction*. Oxford: Wiley-Blackwell.

Atkin, B., Borgbrant, J. & Josephson, P. (2003) *Construction Process Improvement*. Oxford: Blackwell Science.

Bach, J. (1994) The immaturity of CMM, *American Programmer*, September, pp. 13–18.

Bate, R., Garcia, S., Armitage, J., Cusick, K., Jones, R., Kuhn, D., Minnich, I., Pierson, H., Powell, T., Reichner, A. & Wells, C. (1994) *A System Engineering Capability Maturity Model, Version 1.0*. Pittsburgh, PA: Software Engineering Institute, Carnegie Mellon University.

Becta. (2005) *The Becta Review: Evidence on the Progress of ICT in Education.* Becta ICT Research, British Educational Communications and Technology Agency, UK, http://dera.ioe.ac.uk/1428/ accessed on 23 Dec 2016.

Bew, M. and Richards, M. (2008) *Bew-Richards BIM* Maturity Model, http://www.bimtaskgroup.org/bim-faqs/, accessed on 23 Dec 2016.

Chaffey, D. (2011) *e-Business and e-Commerce Management: Strategy, Implementation and Practice.* 5th ed. Harlow, UK: Pearson Education Limited.

Chen, C., Chen, Y. & Yu, P. (2006) Establishing an e-business CMM with the concepts of capability, maturity and institutionalization, *International Journal of Electronic Business Management*, 4(3), pp. 205–213.

Chen, Y. J. (2012) *Strategic implications of e-business in the construction industry.* Degree of Doctor of Philosophy. Loughborough University.

Eadie, R. (2009) *Methodology for developing a model for the analysis of e-procurement capability maturity of construction organisations.* Degree of Doctor of Philosophy. Ulster University.

Eadie, R., Perera, S. & Heaney, G. (2011) Key process area mapping in the production of an e-capability maturity model for UK construction organisations, *Journal of Financial Management of Property and Construction*, 16(3), pp. 197–210.

Eadie, R., Perera, S. & Heaney, G. (2012) Capturing maturity of ICT applications in construction processes, *Journal of Financial Management of Property and Construction*, 17(2), pp. 176–194.

Egan, J. (1998) *Rethinking Construction: The Report of the Construction Task Force.* London: Department of the Environment, Transport and the Regions.

Egan, J. (2002) *Accelerating Change: Strategic Forum for Construction.* London: Construction Industry Council.

Fairclough, S. J. (2002) *Rethinking Construction Innovation and Research: A Review of Government R&D Policies and Practices.* London: Department of Trade and Industry.

GCS. (2011) Government construction strategy. Cabinet Office Report. [Accessed April 2015]. Available from: https://www.gov.uk/government/uploads/system/uploads/attachment_data/file/61152/Government-Construction-Strategy_0.pdf.

Harmon, P. (2004) Evaluating an organization's business process maturity. [Accessed April 2015]. Available from: http://www.bptrends.com/publicationfiles/03-04%20NL%20Eval%20BP%20Maturity%20-%20Harmon.pdf, accessed on 22 Dec 2016.

Harris, F. & McCaffer, R. (2013) *Modern Construction Management.* 7th ed. Oxford: Wiley-Blackwell.

Humphrey, W. S., Sweet, W. L., Edwards, R. K., LaCroix, G. R., Owens, M. F. & Schulz, M. P. (1987) *A Method for Assessing the Software Engineering Capability of Contractors.* Pittsburgh, PA: Software Engineering Institute, Carnegie Mellon university.

Hutchinson, A. & Finnemore, M. (1999) Standardised process improvement for construction enterprises (SPICE), *Total Quality Management & Business Excellence*, 10(4), pp. 576–583.

Jeong, K. S., Kagioglou, M., Haigh, R., Amaratunga, D. & Siriwardena, M. L. (2006) Embedding good practice sharing within process improvement, *Engineering, Construction and Architectural Management*, 13(1), pp. 62–81.

Jones, C. (1994) *Assessment and Control of Software Risks.* Upper Saddle River, NJ: Prentice-Hall.

Keraminiyage, K. P. (2009) *Achieving high process capability maturity in construction organisatons.* Degree of Doctor of Philosophy. The University of Salford, UK.

Keraminiyage, K., Amaratunga, D. & Haigh, R. (2008) UK construction processes and IT adoptability: learning from other industries. In: Kazi, A. S. (ed.) *ICT in Construction and Facility Management.* Finland: VTT (Technical Research Centre of Finland) and RIL (Association of Finnish Civil Engineers).

Latham, M. (1994) *Constructing the Team: Joint Review of Procurement and Contractual Arrangements in the United Kingdom Construction Industry.* London: HMSO.

Li, F. (2007) *What Is e-Business? How the Internet Transforms Organizations.* Oxford: Blackwell Publishing.

Love, P. E. D., Irani, Z. & Edwards, D. J. (2004) Industry-centric benchmarking of information technology benefits, costs and risks for small-to-medium sized enterprises in construction, *Automation in Construction,* 13, pp. 507–524.

Mutula, S. M. and Van Brakel P. (2006), 'An evaluation of e-readiness assessment tools with respect to information access: Towards an integrated information rich tool', *International Journal of Information Management,* Volume 26, Issue 3, June 2006, Pages 212–223, http://www.sciencedirect.com/science/article/pii/S0268401206000077, accessed on 22 Dec 2016.

NBS. (2014) BIM levels explained. [Accessed June 2015]. Available from: http://www.thenbs.com/topics/bim/articles/bim-levels-explained.asp.

NIBS. (2007) National Institute for Building Sciences (NIBS) Facility Information Council (FIC)—BIM Capability Maturity Model. [Accessed April 2015]. Available from: http://www.facilityinformationcouncil.org/bim/pdfs/BIM_CMM_v1.9.xls.

NIST. (2007) *National Building Information Modeling Standard—Version 1.0—Part 1: Overview, Principles and Methodologies.* Washington, DC: National Institute of Building Sciences.

Paulk, M. C., Curtis, B., Chrissis, M. B. & Weber, C. V. (1993) *Capability Maturity Model for Software, Version 1.1.* Pittsburgh, PA: Software Engineering Institute, Carnegie Mellon University, (CMU/SEI-93-TR-024).

Rodrigo, A., Perera, S., Udeaja, C. & Zhou, L. (2011) Towards a stepwise improvement tool for construction e-business: conceptual approach. *10th International Postgraduate Research Conference (IPGRC),* 14–15 September 2011, University of Salford, UK.

Rodrigo, A., Perera, S., Zhou, L. & Udeaja, C. (2014) Construction process categorisation towards developing an e-business maturity model. *International Conference on Construction in a Changing World,* 4–7 May 2014, Heritance Kandalama, Sri Lanka.

Ruikar, K., Anumba, C. J. & Carrillo, P. M. (2006) VERDICT—an e-readiness assessment application for construction companies, *Automation in Construction,* 58, pp. 98–110.

Sarshar, M., Finnemore, M. & Haigh, R. (1999) SPICE: is the capability maturity model applicable in the construction industry?, *8th International Conference on Durability of Building Materials and Components (CIB W78),* May 30–June 3, Vancouver, Canada.

Sarshar, M., Haigh, R., Finnemore, M., Aouad, G., Barrett, P., Baldry, D. & Sexton, M. (2000) SPICE: a business process diagnostics tool for construction project, *Journal of Engineering, Construction and Architectural Management,* 7(3), 241–250.

Sarshar, M., Hutchinson, A., Aouad, G., Barrett, P. & Golding, J. (1998) Standardised process improvement for construction enterprises (SPICE). *2nd European Conference on Product and Process Modelling*, Watford, Hertfordshire, UK.

Schneider, G. P. (2010) *Electronic Commerce.* 9th ed. Boston, MA: Thomson Course Technology.

SEI. (2002) *Capability Maturity Model Integration (CMMI): Version 1:1.* Pittsburgh, PA: Software Engineering Institute, Carnegie Mellon University.

Shaikh, A., Ahmed, A., Memon, N. & Memon, M. (2009) Strengths and weaknesses of maturity driven process improvement effort. *Complex, Intelligent and Software Intensive Systems, 2009. CISIS '09. International Conference*, 16–19 March 2009, Fukuoka, Japan.

Sommerville, J. & Craig, N. (2006) *Implementing IT in Construction.* Oxon, UK: Taylor & Francis.

Stelzer, D. & Mellis, W. (1999) Success factors of organizational change in software process improvement, *Software Process Improvement and Practice*, 4(4), pp. 227–250.

Stewart, R. A. & Spencer, C. A. (2006) Six-sigma as a strategy for process improvement on construction projects: a case study, *Construction Management and Economics*, 24(4), pp. 339–348.

Stockdale, R., Standing, C. & Love, P. E. D. (2006) Propagation of a parsimonious framework for evaluating information systems in construction, *Automation in Construction*, 15, pp. 729–736.

Succar, B. (2009) Building information modelling maturity matrix. In: Underwood, J. & Isikdag, E. (eds) *Handbook of Research on Building Information Modelling and Construction Informatics: Concepts and Technologies.* Hershey, PA: IGI Publishing.

Succar, B., Sher, W. & Williams, A. (2012) Measuring BIM performance: five metrics, *Architectural Engineering and Design Management*, 8, pp. 120–142.

Sun, M. & Howard, R. (2004) *Understanding IT in Construction.* Oxon, UK: Taylor & Francis.

Weinberg, G. M. (1993) *Quality Software Management (Vol. 2): First-Order Measurement.* New York: Dorset House Publishing.

Wilkinson, P. (2005) *Construction Collaboration Technologies: The Extranet Evolution.* Oxon, UK: Taylor & Francis.

13 e-Business infrastructure and strategic frameworks

Yongjie Chen, Kirti Ruikar and Patricia Carrillo

13.1 Introduction

The implementation of e-business tools in organisations impacts on various facets that concern its people, processes, technologies and leadership. This chapter discusses the development of a novel strategic framework that highlights the aspects to be considered internally (within organisation) and externally (within projects) to enhance collaboration and derive business benefits from the implementation. It focuses on the development and implementation of the framework. The chapter starts from the discussion of the crucial needs of industry organisations when formulating their e-business strategies, and the introduction of the adopted methodology for developing the framework. Then a review of different relevant approaches for strategy formulation and framework development is followed. The main body of the chapter presents the detailed description of the framework and its evaluation. Future implementation consideration is also discussed in the concluding section of the chapter.

13.2 Need for an e-business Strategy

This section discusses the importance of an appropriate e-business strategy for organisations in the construction industry to derive maximum benefit from their investments in e-business technology. It also introduces the adopted methodology for developing the strategic framework. It is followed by the review of different relevant approaches for strategy formulation and framework development.

13.2.1 The strategic consideration of e-business implementation in construction

The concept of strategy was introduced to the e-business domain after the dot-com bubble burst, and organisations realised that technology alone was not adequate for the success of e-business implementation. There was a growing realisation that it must be accompanied with appropriate managerial and organisational practices (Chaffey, 2011; Jelassi *et al.*, 2014; Laudon and Laudon, 2014). Currently there is a tendency for e-business strategy to be incorporated within the functional strategies, for example within a marketing plan or logistics plan, or as part of information systems (IS) strategy (Chaffey, 2011). However, the leaders in e-business have typically defined e-business as an element of their corporate strategy development and the importance of e-business strategy has been recognised by the senior management board (Deise *et al.*, 2000; Norton, 2001; Chaffey, 2011; Jelassi *et al.*, 2014). e-Business strategy defines how organisations connected with external partners as well as how organisations operated within management activities, processes and systems (Zeng and Li, 2008). It enables organisations to promote the alignment of business and IT infrastructure in order to derive the maximum benefit from their investments in technology. According to Chaffey (2011), without a clearly defined e-business strategy, the following problems may result:

- *Missed opportunities*: because of a lack of evaluation of opportunities or insufficient resourcing of e-business initiatives;
- *Inappropriate direction in e-business development:* having no long-term consideration of e-business development and without clearly defined objectives;
- *Limited integration:* at only a technical level potentially resulting in 'silos' (e.g. separate organisational team with distinct responsibilities that do not work in an integrated manner with other teams) of information in different systems; and
- *Resource wastage:* due to duplication of e-business development in different functions and limited sharing of best practice.

Therefore, it is important for all organisations to define an appropriate e-business strategy to guide its e-business implementation and support the overall corporate strategy. Organisations in the construction industry are no exception. However, the strategic consideration of e-business implementation in the construction industry was very limited. Ruikar *et al.* (2008) recommended a strategy development based on the e-readiness measurement and evaluation. Other experts stated that it was important to consider a holistic approach for e-business implementation in the construction industry since the approaches available were mainly 'reactive' and lacked a long-term vision or strategy (Ruikar, 2004; European Commission, 2006; Ruikar *et al.*, 2006; Alshawi *et al.*, 2008). However, the issues of the content of an e-business strategy and how it should be developed have not been discussed. This research is the first step towards bringing the current gap.

13.2.2 Methodology for developing the framework

A multi-methodological research design and a pragmatic mixed-methods approach, involving a combination of both quantitative and qualitative datasets, were adopted to investigate the e-business practices of organisations in the industry. These include an exploratory investigation (an industry survey with 250 industry organisations), and an explanatory investigation (four case studies with specific industry end-user companies). The collected data was analysed and problems were identified; the elements for a holistic approach to manage e-business implementation emerged. Mixed-method designs provide a basis for triangulation but, more often, become the source of different ways of conceptualising the problem, such as different instrument design for data collection (qualitatively and quantitatively), and various methods of data analysis and inference (statistically and qualitatively) (Creswell, 2009; Tashakkori and Teddlie, 2010; Creswell, 2013). Therefore, besides the strengths of complementarity and triangulation, the mixed-method designs have other advantages, such as providing both narrative and numeric data, interpreting a broader and more complete range of questions, increasing generalizability of results, and producing more complete knowledge to inform practice (Traynor, 2004).

As an initial step, an inclusive review of existing literature on the current available approaches (models, frameworks and tools) for e-business strategy formulation and implementation was carried out. The analysis of the industry survey and the multiple-case studies were then reviewed to generalise the shared themes. The emerged main themes revealed the essential elements for organisations in the construction industry to formulate and implement e-business strategy.

The industry evaluation was adopted to gauge the appropriateness and effectiveness of the framework from an industry perspective. Details of the evaluation are described in Section 13.3 of the chapter.

13.2.3 *Review of relevant approaches*

There have been many different views of the contents of an e-business strategy, such as focusing on technology (Anice *et al.*, 2001), business and managerial practices (Norton, 2001; Charlesworth, 2016), human factors (Daghfous and Al-Nahas, 2006), environment (Pai and Yeh, 2008), inter-organisational relationships (Jutla *et al.*, 2001), networks (Sultan and Hussain, 2001), financial considerations (Levenburg and Magal, 2004), and markets (Jarvenpaa and Tiller, 2001). However, as more research work was undertaken in e-business strategy concepts and solutions, researchers started to realise that e-business strategies should not be limited to simply one aspect, but also should include multiple elements (Mohammadian *et al.*, 2010). Furthermore, to clearly define an appropriate e-business strategy, different models or frameworks were developed to support strategy formulation and implementation.

Several approaches for e-business strategy formulation and implementation were reviewed as a part of this study, but none of them were construction-specific. These approaches built on the theory of strategy formulation and management, but each had a different focus. Kalakota and Robinson (2004) defined the Roadmap for e-Business model, which had an emphasis on the continuous review and prioritisation of investment in new applications. However, this model is not suitable for organisations that wish to leverage their resources to improve e-business capabilities and provide an integrated approach for long-term e-business implementation. Jelassi *et al.* (2014) identified the Strategy e-Business Framework, which suggested that organisations should undertake e-business strategy options based on the created value of implementing such options. However, the created value of e-business is difficult to be quantified in monetary terms sometimes when the value is associated with intangible benefits (Al-Mashari, 2002; Mogollon and Raisinghani, 2003; Fink, 2006; Chaffey, 2011). Therefore, it would more persuasive to show the role of e-business strategy in improving the overall performance of the organisations and supporting the corporate goal. Chaffey (2011) established the Generic e-Business Strategy Process Model, which defined the elements of an e-business strategy and its development in a dynamic manner, and could be used as a guide for organisations to determine e-business strategic issues at a high level. However, since it is a generic model for all business sectors, industry-specific elements must be added when planning a holistic approach for organisations in the construction industry.

Other than the above three approaches, four other construction-specific approaches (models, frameworks or tools) relevant to the current study were also reviewed. These approaches included the PIECC Decision-Making Framework (Shelbourn *et al.*, 2006), the e-business readiness assessment tool VERDICT (Ruikar *et al.*, 2006), the IS/IT Organisational Readiness Model (Alshawi, 2007), and the Construction Process Protocol (Cooper *et al.*, 1998). The following gives an overview of each:

- *The PIECC Decision-Making Framework:* The PIECC (Planning and Implementing Effective Collaboration in Construction) Decision-Making Framework was designed to guide organisations in the planning and implementing effective collaborative working (Shelbourn *et al.*, 2006). The framework defines four key aspects and a set of sub-processes for each aspect to work through in order to develop a mutual acceptable collaboration strategy. The framework also defined two activities for reviewing the collaboration strategy (e.g. reflections and feedback of collaboration, and measure the collaboration performance). However, it does not identify the way in which feedback and measures can be shared across the collaboration teams and project teams, so it will be difficult for organisations to learn lessons from previous collaboration practices. Moreover, the framework defines an activity to obtain support externally (i.e. collaboration support from external sources), but the role of the external business partners is not clearly identified.
- *The VERDICT Application:* VERDICT (Verify End-user e-Readiness using a Diagnostic Tool) was designed for organisations in the

construction industry to gauge their e-readiness (Ruikar *et al.*, 2006). VERDICT defines the categories of criteria necessary to assess the ability of an organisation to adopt, use and benefit from e-business. The categories include Management, People, Process and Technology. VERDICT can also help in highlighting areas that must be addressed to achieve e-readiness. However, VERDICT does not define measures for organisations to address the issues of highlighting areas and the necessary procedures for them to go through to make improvement in such areas.

- *The IS/IT Organisational Readiness Model*: this was designed to assist organisations in the construction industry to successfully implement IT/IS (Alshawi, 2007). The Model identifies four categories of criteria to assess the ability of an organisation to successfully implement and evaluate IT/IS: People, Processes, Technology and Environment. The Model also defines the maturity level of each category for units of measurement. However, how organisations can improve their organisational readiness (e.g. improve the maturity level from the lowest to the highest) has not been addressed in the model.

- *The Construction Process Protocol:* this was developed to help construction project participants work together seamlessly through bringing together diverse functions/companies involved in construction projects under the common framework of a structured process (Process Protocol, 2016). The Process Protocol Framework incorporates the concepts of process gate and process review. The process gate concept (soft gate and hard gate) is designed to ensure that the key decision points in the process are respected, and the process review concept (feedback from the results of previous decisions) enables continuous process improvement (Cooper *et al.*, 1998). The framework can be adapted to manage different processes sequentially. However, the framework is not suitable for managing processes that are not sequential and not repeatable (e.g. one-off events or activities).

The framework developed in this chapter combines the aspects of five approaches: Chaffey's Generic e-Business Process Model, the PIECC Decision-Making Framework, the VERDICT application, the IT/IS Organisational Readiness Model and the Process Protocol Framework, and builds on them. The Generic e-Business Process Model was used as a guide to define the phases of the Strategic e-Business Framework, the PIECC Decision-Making Framework helped to identify the main activities of each phase, the VERDICT application and the IT/IS Organisational Readiness Model helped determine the factors responsible for the main activities of each phase in the e-Business Strategic Framework, and the Process Protocol Framework was adopted to identify the layout of the Strategic e-Business Framework. Figure 13.1 illustrates the development of the Strategic e-Business Framework.

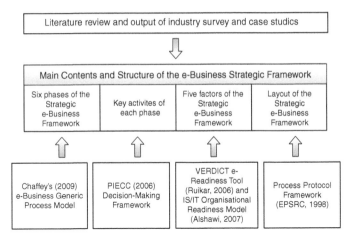

Figure 13.1 Reviewed approaches and their relationships with the strategic framework.

13.3 The strategic e-business framework

This section presents the e-Business Strategic Framework, which includes the structure of the framework and its main components, the detailed description of each component (mainly factors and phases), and the methods, processes and results of the framework evaluation.

13.3.1 The main components of the framework

The main purpose of the Strategic e-Business Framework was to provide a holistic approach for e-business strategy development and implementation in order to achieve the aim of the research. The framework is a comprehensive manual on how to develop e-business strategies for organisations in the construction industry. It provides guidance for organisations in the industry to help them to utilise their available IT resources and maximise the benefits of e-business through strategic practices. The framework is designed specifically for the senior IT management staff (e.g. company Senior IT managers, or corporate IT Directors) to define organisational level e-business strategies and implementation plans. However, the framework also requires the involvement of other staff in the organisation, from the senior management to end-users.

The framework comprises the following six main components (see Figure 13.2).

13.3.1.1 Phases

Phases are the high level elements of the Strategic e-Business Framework, which are the main subjects that organisations in the industry must work through when developing their e-business strategies. The framework

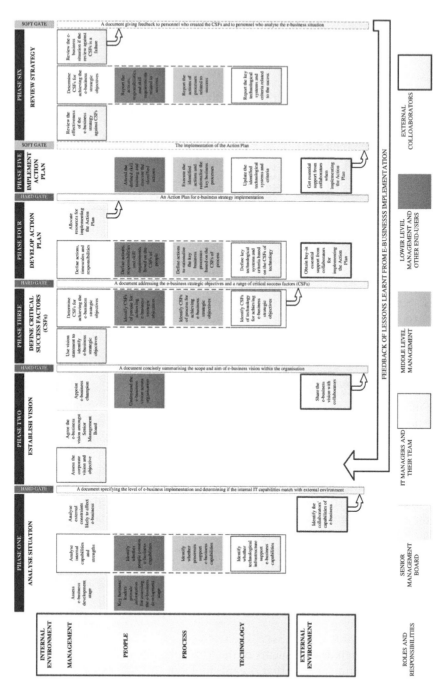

Figure 13.2 The strategic e-business framework for organisations in the construction industry.

consists of six phases: Analyse Situation, Establish Vision, Define Critical Success Factors, Develop Action Plan, Implement Action Plan and Review Strategy. The details of each phase and the included activities will be presented in Section 13.3.3.

13.3.1.2 Activities

Activities are the actions within each phase. Once all the activities have been fully accomplished, the relevant phase has been successfully completed.

13.3.1.3 Factors

Factors refer to the categories used to group the activities within each phase. Factors emphasise the importance of different functions within an organisation to ensure strategic implementation of e-business and its continuous improvement. The framework includes five factors: Internal Environment (Management, People, Process and Technology) and External Environment. The details of each factor will be presented in Section 13.3.2.

13.3.1.4 Sub-activities

Sub-activities refer to the actions within some activities, which are defined when these activities require input from people, process and technology.

13.3.1.5 Phase gates

Phase gates include soft gates and hard gates, which are defined to ensure that the key decision points in the process are respected. The framework consists of four hard gates and two soft gates.

13.3.1.6 Roles and responsibilities

Roles and responsibilities refer to the teams or people identified as the appropriate parties responsible for carrying out the activities. Five groups of people were assigned to the activities and sub-activities: Senior Management Board, IT Manager and their teams, Middle Level Management, Lower Level Management and other End-users, and External Collaborators.

Organisations must go through the framework phase by phase strictly when developing their e-business strategies. After accomplishing all the activities of each phase, they must check whether or not they have worked out the outputs that are displayed in the phase gates. For example, at the end of the Analyse Situation Phase, a document that specifies the level of e-business implementation and determines if the internal IT capabilities match with external environment should be produced when all the activities in the Phase are completed. If the outputs of the Phase are not obtainable, organisations need to review the performance of each activity and determine the reasons that cause the unachievable results.

Moreover, the senior IT management staff (e.g. company Senior IT managers, or corporate IT Directors) are responsible for tracking the accomplishment of each activity before the e-business champion is assigned, which includes ensuring the right people (check the role and responsibilities of each activity) do the right thing (check the performance of each activity) at the right time (check the activity is strictly performed phase by phase). The e-business champion is responsible for the tracking tasks when he/she is assigned.

13.3.2 Factors of the framework

13.3.2.1 Internal environment

A complete e-business strategy requires a concerted effort within an organisation, including the commitment and involvement of senior management, the awareness and skills of employees, the rationalisation of key business processes and the support of technological infrastructure, systems and criteria. Management, people, process and technology are the four factors that were used to categorise the main activities of each phase in the Strategic e-Business Framework. The meaning and function of each factor in the framework are explained in the next a few sub-sections.

13.3.2.1.1 MANAGEMENT

Management can be defined as a set of activities (including planning and decision making, organising, leading and controlling) directed at an organisation's resources (human, financial, physical and information) with the aim of achieving an organisation's goals in an efficient and effective manner (Griffin, 2013). Ruikar *et al* (2006) highlight the role of management in dealing with the strategic change. In the Strategic e-Business Framework, management refers to all the activities related to planning, decision making, organising, leading or directing, controlling, and staffing in order to define, implement and review the e-business strategies. In the framework, the development of e-business strategies heavily relies on such management activities. Management acts as one of the major categories necessary to ensure the accomplishment of each phase.

13.3.2.1.2 PEOPLE

In e-business implementation, the people factor includes the social and cultural aspects related to the people within an organisation (Ruikar *et al.*, 2006). In the Strategic e-Business Framework, the people factor takes into account the awareness, understanding and skill requirements of staff within an organisation when implementing e-business. This factor acts as one of the major categories necessary to ensure the accomplishment of each phase.

13.3.2.1.3 PROCESS

Process means a practice, or a series of actions, done for a specific purpose (Craig, 2004). In e-business implementation, process refers to part of a system that has a clearly defined purpose or objective and clearly defined inputs and outputs (Chaffey, 2009). In the Strategic e-Business Framework, process refers to the key business working rules and procedures used by construction organisations to implement e-business. The function of process in the Strategic e-Business Framework is similar to the people factor, and it is a necessary part of accomplishing each phase.

13.3.2.1.4 TECHNOLOGY

In e-business, technology refers to information and communication technologies (ICT) including both hardware and software availabilities and usage within an organisation (Chaffey, 2009). In the Strategic e-Business Framework, the technology factor refers to e-business tools/applications as well as the technological infrastructure or systems supporting information transaction and sharing. The function of technology in the Strategic e-Business Framework is similar to people and process, being a necessary part of accomplishing each phase.

13.3.2.2 External Environment

The External Environment gives organisations their means of survival but also represents a source of threats (Porter, 1985). In e-business implementation, the External Environment refers to the elements of the external environment that are likely to impact e-business implementation, which includes a consideration of both the micro-environment and the macro-environment (Chaffey, 2009). The micro-environment refers to the immediate competitive environment that a company faces, such as customer demand, competitor activity, marketplace structure and relationships with business partners. The macro-environment refers to the wider environment in which a company operates, which includes economic, social, legal and ethical factors (Johnson and Scholes, 2003). In the Strategic e-Business Framework, external environment also acts as one part that is necessary to complete each phase.

13.3.3 Phases of the framework

13.3.3.1 Analyse situation

Analyse Situation refers to the review of information about an organisation's internal processes and resources and external marketplace factors in order to define and plan e-business strategies (Chaffey, 2009). Organisations need to have clear picture of their available IT (or e-business) resources, the way in which their current internal processes work, and the kind of external environment they compete in. They then must decide when and how to respond to the macro-environment and the competitors (Smith and Taylor, 2004; McDonald, 2008; Chaffey, 2009; Jelassi and Enders, 2009). To define a complete and precise e-business strategy, organisations must design the analysis with a practical purpose in mind (Johnson and Scholes, 2003). This phase provides guidance and techniques to allow organisations to analyse their e-business situation. The outputs of this phase include: (1) a document specifying the desired level of e-business implementation in the future, and (2) a determination of whether or not the company's internal IT capabilities and processes match with the capabilities and processes used by external partners and customers. The process gate at the end of this phase is a hard gate, which means that only when all the activities in this phase have been accomplished, can organisations initiate the activities described in the following phase. Five activities included in this phase: assess e-business development stage, key business leaders provide information for assessing the e-business development stage, analyse internal capabilities and strengths, analyse external constrain likely affecting e-business, and identify the collaborators capabilities of e-business.

13.3.3.2 Establish vision

An e-business vision refers to a concise summary of the scope and broad aims of a company's future e-business activities, including the explanation of how these activities will contribute to the organisation and support its core business activities. Establishing a Vision requires the participation of a broad cross-section of company staff to work out the Vision that is usually captured in the form of a written document. Defining a specific vision can help organisations contextualise e-business in relation to the overall corporate strategies. It also helps place a long-term emphasis on e-business transformation within an organisation (Chaffey, 2009; Jelassi and Enders, 2009). This phase provides guidance to help companies establish their e-business vision. The output of this phase is a document concisely summarising the scope and broad aim of e-business activities within the organisation. The process gate at the end of this phase is a hard gate. Five activities included in this phase: assess corporate vision and objectives, agree the vision amongst Senior Management Board, appoint e-business champion, understand the vision across the organisation, and share the e-business vision with collaborators.

13.3.3.3 Define critical success factors (CSFs)

Critical Success Factors (CSFs) refer to elements that are vital for a strategy's success. Defining CSFs can help organisations to translate their e-business vision into practical actions, investigate the applicability of strategic objectives, and review the effectiveness of the e-business strategy. Previous research has suggested that the CSFs of information technology are primarily technological related elements (Chaffey, 2009). Defining CSFs for e-business implementation should not be limited to considering technologies and systems, but also requires identifying CSFs related to people and process. This phase of the framework provides guidance to help a company define CSFs related to e-business implementation. The outputs of this phase include the e-business strategic objectives and a range of critical success factors. The process gate at the end of this phase is a hard gate. Two activities included in this phase: use vision statement to identify e-business strategic objectives, and determining CSFs for achieving the e-business strategic objectives.

13.3.3.4 Develop action plan

Develop Action Plan refers to identifying the required actions time horizons and resources for implementing the e-business strategic solutions within an organisation. After analysing the situation, establishing the vision and determining the strategic objectives and CSFs, organisations need to plan their actions for e-business implementation. In strategy development, an Action Plan refers to 'a sequence of steps that must be taken, or activities that must be performed well, for a strategy to succeed' (Business Dictionary Online, 2016). A workable Action Plan should provide confidence that the strategic objectives are achievable within the constraints of time and cost (Billingham, 2008). This phase provides a guide for organisations to prepare for carrying out all the tactics that will be used to achieve the strategic objectives. The output of this phase is an Action Plan for e-business implementation. The process gate at the end of this phase is a hard gate. Three activities included in this phase: define actions, timescales, and responsibilities, obtain buy-in essential support from collaborators for implementing the Action Plan, and allocate resources for implementing the Action Plan.

13.3.3.5 Implement action plan

Implement Action Plan refers to executing all the planned actions to achieve the strategic objectives. Implementing a solution is the crucial part of strategies planning and execution (Johnson *et al.*, 2008). This phase is a very difficult one to implement. At the Implement Action Plan Phase, Senior Management Board is not directly involved and other members in the organisation, such as Middle Level Management staff, and Lower Level Management staff, and other End-users, carry out all the actions. The e-business champion is crucial at this phase as a leader for the implementation of the Action Plan. The champion must ensure the execution of all the defined actions and seek a suitable management approach to monitor the roll out of the Plan and address the problems that may arise during or after the execution of the various actions. This means that the e-business champion may have to carry out some corrective actions (e.g. the defined actions are not correctly executed). When problems arise, the e-business champion must consider the answers to the following key questions:

- *Evaluate the current situation:* What will happen if things continue as they are?
- *Consider various corrective solutions:* Are there any measures that could be applied and to assess the pros and cons of adopting each alternative course of action?
- *Select and implement one of the course actions:* What should be done to solve the problem?
- *Link back into the monitoring process:* Has the corrective action had the desired effect?

The output of this phase is the implementation of an Action Plan. The process gate at the end of this phase is a soft gate, which means that the activities in this phase and their related activities in the following phase can be executed sequentially or concurrently. For example, organisations can review the effectiveness of their e-business strategies against the CSFs of people, either during or after executing the defined actions related to people. Four activities included in this phase: attend the defined skill training and execute the identified actions, execute the identified actions and rationalise the key business processes, update the identified technological systems and criteria, and get essential supports from collaborators when implementing the Action Plan.

13.3.3.6 Review strategy

Review Strategy refers to the process for evaluating the adopted strategies after they had been implemented, and determining lessons learnt from the review, which may include reshaping the vision and objectives, and modifying the CSFs. To achieve both the tangible and intangible benefits of e-business, organisations need to consider units of measurement to connect their organisational performance and their strategic actions. Moreover, organisations need to consider how to connect the critical success factors with their organisational-specific e-business capabilities when attempting to identify units of measurement for reviewing e-business strategies.

The Review Strategy Phase provides an effective way of reviewing e-business strategy, and more importantly, it enables companies to learn from their previous e-business implementation experience. The outputs of this phase include two reports: (1) a report giving feedback to the personnel who identified the CSFs, and (2) a report to personnel who analysed the company's e-business situation. The process gate at the end of this phase is a soft gate, which means that the activities in this phase and their related activities in the following phase can be executed sequentially. At this phase, the feed-back loop is adopted to connect the Review Strategy Phase to the Analyse Situation Phase. This enables the Strategic e-Business Framework to act as a cycle for carrying out all the phases once again. Three activities included in this phase: review the effectiveness of the e-business strategy against CSFs, share the success of e-business strategy implementation across the organization, and review the e-business situation if the review against CSFs is a failure.

13.3.4 The evaluation of the framework

The main purpose of the framework evaluation process was to gauge the appropriateness and effectiveness of the Strategic e-Business Framework from the industry perspective. The evaluation involved three steps: (1) preparing a questionnaire for conducting structured interviews; (2) carrying out the structured interviews with the industry practitioners; and (3) analysing the interview results and presenting the findings. Altogether, six evaluations were completed. Four evaluations were undertaken by the industry practitioners who took part in the multiple-case studies described in Section 2, which aided in the development of the framework. The involvement of the same personnel sought to inspect the internal validity and consistency of the current research (Creswell, 2003). Industry practitioners who were new to the study carried out another two evaluations, which offered different perspectives and tested the applicability of the framework in a wider scope (Wellington, 2000). Table 13.1 displays the industry practitioners that participated in the framework evaluation processes.

Table 13.1 Industry practitioners participating in the framework evaluations

			Experience	
Evaluator	*Organisation discipline*	*Role of interviewee*	*Industry*	*IT*
Evaluator 1	Contractor, construction and engineering	System and Technology Director	38	22
Evaluator 2	Contractor, construction and development	Technical Service Director	25	20
Evaluator 3	Contractor, construction and engineering	Senior IT Manager	25	20
Evaluator 4	Consultant, construction, engineering and technical service	Information System Director	20	20
Evaluator 5	Consultant, architectural and engineering	Senior IT Manager	22	22
Evaluator 6	Consultant, construction and assess management	Technical Service Director	34	25

Overall, the findings of the evaluation revealed that the evaluators gave positive about the ease with which the framework could be implemented. The evaluators rated the framework as highly appropriate in general. Moreover, the evaluators were highly satisfied with the provision of phases, included activities in each phase, and the factors considered for categorising the activities. The evaluators also reported that the applicability of the main activities was high and confirmed that those activities were appropriate in delivering the final outcomes of the associated phases. Furthermore, the evaluators reported that the framework could benefit their organisations in the following areas:

- Helping them to recognise the attributes of their organisations in e-business practices, specifically by means of the strategic scanning or positioning at the Analyse Situation Phase;
- Providing a useful business case for organisations to evaluate the risks and requirements of e-business;
- Assessing the awareness and commitment of the Senior Management Board in e-business strategies and implementation;
- Helping them to establish a culture that is conducive to e-business implementation by including People as one main factor in the framework and identifying associated activities;
- Guiding them to perform decision-making on technological issues by including Technology as one main factor in the framework and identifying associated activities; and
- Helping them carefully consider the e-business practices of external business partners by including 'external environment' as one main factor in the framework and identifying associated activities.

13.4 Recommendations and conclusions

This section discusses the future implementation consideration of the Strategic e-Business Framework.

This chapter presented a Strategic e-Business Framework that organisations in the construction industry can use to define, execute and review their e-business strategies.

The framework was developed from an IT director's and/or senior IT manager's perspective based on a concept of harnessing the abilities and commitment of other staff within the organisation. The framework addresses all factors necessary for developing and implementing an e-business strategy, such as advising organisations to review e-business strategies after execution, linking technological systems, people and processes within the strategy, and including external organisations in the strategy. The framework enables IT directors or senior IT managers to include multiple elements in their e-business strategies and carry out consistent e-business planning.

The framework can also assist organisations in the construction industry to better utilise their available IT resources and maximise the benefits of e-business through strategic practices. Six phases and phase gates (hard gates or soft gates) defined in the framework ensure a sequential approach for defining, managing and reviewing e-business strategies, and a feedback loop was defined to enable the framework to act as a cycle for going through repeatedly, which makes the continuous improvement of e-business implementation possible within organisations. The outputs of each phase were defined to assist IT leaders in the review of procedures during their e-business implementation. Five factors defined in the framework ensure organisations to have the opportunities to improve their capabilities in these areas, and to make organisations ready for immediate e-business implementation. Five groups of people or teams assigned to the activities of the framework provided a basis for organisations to utilise their resources, specifically human resources. Three groups of Critical Success Factors (CSFs), people, process and technology, defined in the framework, ensure that organisations can improve, execute and review their e-business implementation effectively and purposefully. The analysis of the evaluation of the framework by practitioners have demonstrated the intense interest of industry practitioners in implementing the framework to improve their current strategic practices in e-business and remain diligent in e-business strategy development.

The following recommendations are made on strategic implementation of e-business in the construction industry:

- The industry should consider a comprehensive business solution addressing both the organisation's current needs and its future emerging needs in order to make full utilisation of their existing investment in e-business including effective benefits realisation planning.
- The industry should consider their e-business solutions in a collaborative environment. Organisations must give serious thought to the engagement and support of external organisations. This includes linking e-business systems with the information set that is collected from the clients and submitted by suppliers. This is crucial because secure, transparent information exchanges and transfers along the whole supply chain are necessary to ensure the effective implementation of e-business and unlock the full substantial benefits that e-business has to offer.
- Organisations in the construction industry should consider the appropriate and effective staff training and skill assessment programmes in order to motivate employees to use e-business tools and applications, build their confidence in using new e-business technologies, and create a good culture to e-business practices.
- The industry should revise the measures and methods used for evaluating their level of success in implementing e-business. The current measures are limited and ineffective. The measures for evaluating e-business

implementation should include performance metrics for implementation at both the organisational level and the project level. The measures used to evaluate e-business strategies should include criteria that ensure the success of strategy implementation. The evaluation of e-business strategies should define consistent methods for pursuing those success criteria.

In conclusion, the framework is a positive step towards e-business management. Organisations in the industry can enhance their e-business development and plan proactively when new technologies emerge.

Acknowledgement

The content of the chapter has been previously published in the *Journal of Information Technology in Construction* (ITcon) in Chen *et al.*, (2013). We are thankful to ITcon to give us the permission to re-publish the content as a book chapter.

References

Al-Mashari, M. (2002) Enterprise resource planning (ERP) systems: a research agenda. *Industrial Management & Data Systems*, 102(3), 165–170.

Alshawi, M. (2007) *Rethinking IT in Construction and Engineering—Organisational Readiness*. Abington, PA: Taylor and Francis.

Alshawi, M., Goulding, J., Khosrowshahi, F., Lou, E. and Underwood, J. (2008) Strategic positioning of IT in construction: an industry leaders' perspective. *A Report for BERR by Construct IT for Business and the Research Institute for the Built and Human Environment*. [Accessed March 2016]. Available from: http://www.salford.ac.uk/built-environment/research/research-centres/construct-it/publications.

Anice, I.A., Strat, D.L. and Moor, W.C. (2001) Building blocks of a successful e-business strategy. In: *Proceedings of Portland International Conference on Management of Engineering and Technology*, Portland, OR, 29 July–2 August 2001. IEEE Conference Publications, 1, p. 144.

Billingham, V. (2008) *Project Management: How to Plan and Deliver a Successful Project*. Abergele, UK: Studymates.

Business Dictionary Online (2016) The definition of action plan. [Accessed March 2016]. Available from: http://www.businessdictionary.com/definition/action-plan.html.

Chaffey, D. (2009) *E-business and E-commerce Management: Strategy, Implementation and Practice,* 4th Edition. Harlow: Pearson Education Limited.

Chaffey, D. (2011) *E-business and E-commerce Management: Strategy, Implementation and Practice*. 5th Edition. Harlow, UK: Pearson Education.

Charlesworth, A. (2016) Is the definition of e-commerce the same as the definition of e-business. In: AlanCharlesworth.edu—A Market View of Marketing on the Internet. [Accessed March 2016]. Available from: http://www.alancharlesworth.eu/alans-musings/e-commerce-or-e-business.html#.

Chen, Y., Ruikar, K.D. and Carrillo, P.M. (2013) Strategic e-business framework: a holistic approach for organisations in the construction industry. *Journal of Information Technology in Construction*, 18, 306–320, http://www.itcon.org/2013/15.

Cooper, R., Kagioglou, M., Aouad, G., Hinks, J., Sexton, M. and Sheath, D. (1998) The construction process protocol. Engineering and Physical Science Research Council. The Generic Design and Construction Process Protocol (GDCPP) Project Report.

Creswell, W.J. (2003) *Research Design: Qualitative, Quantitative, and Mixed Methods Approaches,* 3rd Edition. London: SAGE Publications Ltd.

Creswell, W.J. (2009) *Research Design: Qualitative, Quantitative, and Mixed Methods Approaches,* 3rd Edition. Thousand Oaks: SAGE Publications Ltd.

Creswell, W.J. (2013) *Research Design: Qualitative, Quantitative, and Mixed Methods Approaches*, 4th Edition. London: Sage Publications.

Craig, T. (2004) *Three Issues to Supply Chain Management Success: Process, People, Technology.* [Online]. LTD Management: Supply Chain Management Consulting. Lehigh Valley. [Viewed March 2016]. Available from: http://www.ltdmgmt.com/072503.php.

Daghfous, A. and Al-Nahas, N. (2006) The role of knowledge and capability evaluation in e-business strategy: an integrative approach and case illustration. *SAM Advanced Management Journal*, 71(2), 11–20.

Deise, M., King, P., Nowikow, C. and Wright, A. (2000) *Executive's Guide to E-Business: From Tactics to Strategy.* New York: John Wiley & Sons.

European Commission. (2006) ICT and electronic business in the construction industry: ICT adoption and e-business activities in 2006. *The European E-business Market Watch Sector Report No.7–2006*. [Accessed March 2016]. Available from: http://www.umic.pt/images/stories/publicacoes/BR06.pdf.

Fink, D. (2006) Value decomposition of e-commerce performance. *Benchmarking: An International Journal*, 13(1/2), 81–92.

Griffin, R.W. (2013) *Management*. 8th Edition. Andover, MA: Cengage Learning.

Jarvenpaa, S.L. and Tiller, E.H. (2001) The new frontier in e-business: integrated Internet strategy. IBM Faculty Partnership Award, 2001. [Accessed March 2016]. Available from: https://www.yumpu.com/en/document/view/35810437/the-new-frontier-in-e-business-integrated-internet-strategy.

Jelassi, T. and Enders, A. (2009) *Strategies for E-business: Creating Value Through Electronic and Mobile Commerce,* 2nd Edition. Harlow: Pearson Education Ltd.

Jelassi, T., Enders, A. and Martinez-Lopez, F. J. (2014) *Strategies for E-business: Creating Value Through Electronic and Mobile Commerce,* 3rd Edition. Harlow: Pearson Education Ltd.

Johnson, G. and Scholes, K. (2003). *Exploring Corporate Strategy: Text and Cases*, 6th Edition. Harlow: Pearson Education Ltd.

Johnson, G., Scholes, K. and Whittington, R. (2008) *Exploring Corporate Strategy.* 8th Edition. Harlow, UK: Pearson Education.

Jelassi, T. and Enders, A. (2009) *Strategies for E-business: Creating Value Through Electronic and Mobile Commerce,* 2nd Edition. Harlow: Pearson Education Ltd.

Jelassi, T., Enders, A. and Martinez-Lopez, F. J. (2014) *Strategies for E-business: Creating Value Through Electronic and Mobile Commerce,* 3rd Edition. Harlow: Pearson Education Ltd.

Johnson, G. and Scholes, K. (2003). *Exploring Corporate Strategy: Text and Cases,* 6th Edition. Harlow: Pearson Education Ltd.

Johnson, G., Whittington, R. and Scholes, K. (2011) *Exploring Corporate Strategy: Text and Cases*. 9th Edition. Harlow, UK: Prentice Hall.

Jutla, D.N., Craig, J. and Bodorik, P. (2001) A methodology for creating e-business strategy. In: *Proceedings of the 34th Hawaii International Conference on System Sciences*, Hawaii, January 2001, IEEE Conference Publications, pp. 1–10.

Kalakota, R. and Robinson, M. (2004) *E-business 2.0: Roadmap for Success*. 2nd Edition. Reading, UK: Addison-Wesley Professional

Laudon, K.C. and Laudon, J.P. (2014) *Essentials of Management Information Systems*. 11th Edition. Upper Saddle River, NJ: Prentice Hall.

Levenburg, N.M. and Magal, S.R. (2004) Applying importance performance analysis to evaluate e-business strategies among small firms. *e-Service Journal*, 3(3), 29–48.

McDonald, M. (2008) *Malcolm McDonald on Marketing Planning: Understanding Marketing Plans and Strategy*. London: Kogan Page.

Moen, R. and Norman, C. (2009) Evolution of the PDCA cycle. [Accessed March 2016]. Available from: http://www.westga.edu/~dturner/PDCA.pdf.

Mongollon, M. and Raisinghani, M. (2003). Measuring ROI in E-business: A Practical Approach. *Information Systems Management*, 20(2), 63–81.

Mohammadian, A., Pursultani, H. and Akhgar, B. (2010) An integrative framework of e-business strategy building blocks for knowledge based and intelligent systems. In: *Proceedings of IEEE 2010 International Conference on Computing and Automation Engineering*, Singapore, February, 2010, IEEE Conference Publications, 5, pp. 593–597.

Norton, J. (2001) *Winning in the Race for E-business*. London: The Royal Academy of Engineering.

Pai, J.C. and Yeh, C.H. (2008) Factors affecting the implementation of e-business strategies: an empirical study in Taiwan. *Management Decision*, 46(5), 681–690.

Porter, M.E. (1985) *Competitive Advantage*. New York: The Free Press.

Process Protocol. (2016) Process protocol: background. Salford University. [Accessed March 2016]. Available from: http://citeseerx.ist.psu.edu/viewdoc/download?doi=10.1.1.503.2035&rep=rep1&type=pdf.

Ruikar, K.D. (2004) Business process implications of e-commerce in construction organisations. EngD Thesis. Loughborough University.

Ruikar, K.D., Anumba, C.J. and Carrillo, P.M. (2006) VERDICT—an e-readiness assessment application for construction companies. *Automation in Engineering*, 15(1), 98–110.

Ruikar, K.D., Anumba, C.J. and Carrillo, P.M. (2008) Organisational readiness for e-business. In: Anumba, C.J. and Ruikar, K.D (eds). *E-business in Construction*. Oxford: Blackwell Publishing.

Shelbourn, M., Bouchlaghem, D., Anumba, C. and Carrillo, P. (2006) The PIECC decision-making framework. *The Planning & Implementing Effective Collaboration in Construction Industry (PIECC) Project Report*. Loughborough University.

Smith, P.R. and Taylor, J. (2004) *Marketing Communications: An Integrated Approach*. London: Kogan Page.

Sultan, F. and Hussain, H.A. (2001) Design a trust-based e-business strategy. *Marketing Management*, 10(4), 40–45.

Tashakkori, A. and Teddlie, C. (2010) *Handbook of Mixed Methods in Social & Behavioral Research*. 2nd Edition. Thousand Oaks, CA: Sage Publications.

Traynor, V. (2004) Mixed methods presentation by Vanessa Traynor. Research Methodology Resources for Beginners. [Accessed March 2016]. Available from: http://www.scoop.it/t/research-methods.

Wellington, J.J. (2000) *Education Research: Contemporary Issues and Practical Approaches.* London: Continuum International Publishing Group.

Zeng, Q. and Li, X. (2008) Evolution of e-business transformation strategy: a four dimension model. In: *Proceedings of IEEE 2008 International Conference on Service Systems and Service Management*, Melbourne, VIC, 30 June–2 July 2008, IEEE Conference Publications, pp. 1–5.

14 Innovation in e-business

Issues related to adoption for micro and SME organisations

Eric Adzroe and Bingunath Ingirige

14.1 Introduction

According to Kenny (2007), construction is a $1.7 trillion industry world-wide, amounting to between 5% and 7% of GDP in most countries and accounts for a significant part of the global gross capital formation a little under one-third. As noted across the globe the sector's role in economic development is undeniable. Despite the significant contribution that the construction industy makes towards a country's economy, the international construction industry is dominated by small businesses. Therefore, in both developed and developing countries, small businesses collectively create the major impact on the economy. As the small business impact is made collectively, it is often the case that the turnover gap between the few large firms and the smaller firms is very wide. This makes innovation in small businesses very challenging as the uptake should be significant in number to make an impact. Collaborative practice enhanced by extensive Information and Commmunication Technology (ICT) use (Fulford and Standing, 2014) is seen as a mainstay of efficiency improvements as it enables integration and automation of processes. E-business from both a method and the process point of view can influence this uptake so that the appropriate use can possibly trigger an increase that is needed in construction.

Ashrafi *et al.* (2014) based on an EU wide methodology, classify the scale of businesses based on the number of employees. Enterprises with less than 10 employees are regarded as Micro, between 11 and 50 employees as Small, between 51 and 250 employees as Medium and over 250 employees as Large organisations. SMEs constitute more than 95% of of firms in many countries. However, wide the turnover margins between the large and the small firms are, SMEs in construction can achieve dominant positions in the industry and as a whole the community of SMEs command a major impact in the construction industries of both developed and developing countries. E-business use with their ability to enhance communication and collaboration could, therefore, create a major impact in SMEs. However, harnessing the technology use and e-business within the smaller and micro businesses does not come without its disadvantages and barriers. The objective of this chapter is to present some of the highlights of e-bsiness use in this context and argue both the case and the potential, for smaller businesses to take a more leading role in this area.

Although there are disparities of size of organisations in construction, it is important to achieve some degree of consistency between the players in terms of processes and their e-business use as the construction industry is a prime example of a project based industry (Buvik and Rolfsen, 2015) and a lot of SMEs operate within global supply chains (Parmigiani and Rivera-Santos, 2015). Increasingly, there are several developing countries that are spearheading the developments of e-business within a framework of small and medium scale businesses and within a context of a lower scale of technology adoption (Tran *et al.*, 2015). This chapter examines e-business technologies that are appropriate for SME and Micro organisations in construction. The chapter also reports from a developing country case of the Ghananian construction industry to demonstrate the importance of e-business within their industry and for populating its use among small businesses.

14.2 Why ICT in small businesses?

Acar *et al.* (2005) found a connection between firm size and ICT use, whereby the larger the firm size, more extensive the ICT use becomes. Quoting previous research, Acar *et al.* (2005) also shows that SMEs' current use of ICTs in construction firms continues to be 'piecemeal' with few contractors fully integrate ICTs with their business processes. SMEs more often use ICT for limited applications such as Book-keeping and invoicing even in the construction industries' of relatively advanced countries (Samuelson, 2002). SMEs usually do not have any long term plans and in most instances do not have the spare capacity to engage in activites of long term exploitation for growth. Their considerations are mostly of a short term orientation.

Despite the apparent advantages ICTs offer, construction firms in general are slow to exploit their potential benefits (Egbu and Botterill, 2002). The building construction industry invests little in ICTs compared with the other sectors such as financial services and manufacturing (Egan, 1998: 27–28). Although SMEs form a substantial constituent of the global economy, there is limited knowledge available surrounding the adoption of information and communication technologies (ICTs) by SMEs (Shiels *et al.*, 2003). Use of ICT for collaborative practice (Fulford and Standing, 2014) by SMEs is even smaller. From the point of view of Malaysia, Chong *et al.* (2014) states that despite the contribution that SMEs make towards the economy, just over 30% of the SMEs actually use e-business in the country (not just construction SMEs but the overall SMEs within the economy). According to the above authors, lower e-business use is a common occurrence within the developing world. ICT use offer numerous benefits across a wide range of intra- and inter-firm business operations and transactions

(OECD, 2004). ICT uses improve information and knowledge management inside the firm and can reduce transaction costs and increase the speed and reliability of transactions activities for both business-to-business (B2B) or business-to-consumer (B2C) transactions (Damanpour and Damanpour, 2001; OECD, 2004; Ruikar and Anumba, 2008). Additionally, ICT is in fact an effective tool for improving external communications and quality of services for established and new customers. These advangages provided a compelling reason for SMEs to adopt ICT models to support their specific needs in terms of business transactions.

ICT is defined as *"technologies dedicated to information storage, processing, and communications"* (Ang and Koh, 1997). According to Björk (1999) Information Technology (IT) include all kinds of technologies used for the storage, transfer and manipulation of information, thus, also including devices such as copying machines, faxes and mobile phones. El-Ghandour and Al-Hussein (2004) consider ICT as a collective reference to the integration of computing technology and information processing. ICT hold within its domain many supporting technologies such as computers, software, network, telephone and fax machines and combination of these different items transform raw or semi raw data into useful information. The ultimate purpose of ICT is to facilitate the exchange and management of information and has a lot of potentials for information process component of the construction industry (Rivard, 2000). The immense contribution of ICT in business development has proven to be a crucial factor for all sectors both Large and SMEs. According to Ekholm and Molnár (2009) *"modern object oriented ICT with well-defined information structures and efficient communication interfaces in the manufacturing industry shown to be an efficient tool and supports the integration of processes for product development, production, materials supply and maintenance"*. It is noted that ICT at this level of development is however missing in today's business processes (Oliveira and Martins, 2011). ICT plays a vital strategic role in the sustained growth of business organisations incluing small ones (Seyal *et al.,* 2000; Underwood and Khosrowshahi, 2012). It is in this vein that there have been several attempts by the developed economies to leverage the application of ICT in the industrial sector, which for several years accepted the use of ICT in a fragmented and sporadic manner than a coordinated and integrated approach. There have been numerous strategic national and international initiatives to address the application of ICT within the industrial sector, such as Department of Trade and Industry (DTI, 2001). ICT have played a major role in business processes and the impact is evident in many industrial activities most especially in the developed economies. Therefore, any discussion about the impact of ICT within businesses should be built on the role it might play in facilitating the industry towards meeting its goals successfully (Hosseini *et al.,* 2012). ICT has become, an increasingly important part of today's global economy. Businesses and corporations including construction have moved beyond their borders to the international arena in order to find progressively more efficient techniques to run their operations (Bukartek *et al.*, 2007). Advances in ICT application in diverse ways in recent times have affected business culture and all industries have been greatly influenced. The effect of ICT is far reaching when utilised within the domain of Internet technology. Using ICT within the domain of the Internet is referred to as e-business (Amit and Zott, 2001; Oliveira and Martins, 2011). At this point, companies can conduct business over the Internet utilising the opportunities provided by the Internet. Businesses, especially the established ones including creating new businesses which are being facilitated by the use of ICT and the Internet to create online business activities, further, ICT and the Internet have supported the exploitation of new business ventures, and this is most evident in developed economies. As suggest by Hosseini *et al.* (2012) the relevance of ICT and the Internet can no longer be ignored by business organisations, especially smaller ones as the benefits are available in addition to pressure from globalisation where businesses are moving beyond their geographical locations.

Although many have argued that ICT is common within large organisations some of which have collaborated with universities and research organisations in several ICT research activities and initiatives, this is in no means to suggest that small businesses have not contributed to ICT development and worth creation within the industrial sector (Combe, 2012). For example, small construction businesses in the UK have been identified as major contributors to the national economy through employment generation and worth creation accounting for approximately 49%, or £51 billion of annual construction output (Griffith, 2011).

14.3 The opportunities and barriers of mainstreaming e-business within SMEs

This section explored the opportunities and barriers of mainstreaming e-business technologies within SMEs.

14.3.1 e-Business opportunities within SMEs

In recent times despite the slow pace of e-business adoption by SMEs, they have gradually recognised the positive impact that e-business techonlogies, such as computer facilities, e-mail and the Internet and their applications can have on their business outlook (OECD, 2004; Combe, 2012). There are evidence to suggest that in developed countries, most SMEs, including micro-enterprises with fewer than ten employees, now have at least one computer facility, usually with Internet access. This also is a sign that most SMEs are beginning to realise the opportunities offered by main-streaming e-business in their business operations. A wide range of business software can improve information and knowledge management within SMEs, leading to more efficient business processes and better firm performance. Communication via e-mail utilising the Internet can help to improve external communication, in either B2C or B2B contexts, and may reduce transaction costs, increase transaction speed and reliability, and extract maximum value from each transaction in the value chain (OECD, 2004; Molla and Heeks, 2007).

E-business as earlier noted offer benefits for a wide range of business processes and operation. Starting from the firm level, e-business uses can make communication within the firm faster and make the management of the firm's resources more efficient. OECD (2004), noted that seamless transfer of information through shared electronic files and networked computer system improves the efficiency of business processes such as documentation, data processing and other back-office functions (e.g. organising incoming orders and preparing invoices). Several opportunities exist for SMEs to become progressively mature in e-business technology. For example, sophisticated e-business technology application such as Knowledge Management System (KMS) and Enterprise Resource Planning (ERP) improves organisation's capacity to store, share and use their knowledge and know-how capabilities. Discussing e-business from inter-firm level, ICT propelled by the Internet produces e-business which has a great potential to reducing business transaction costs as well as increasing the speed and trustworthiness of transactions. As noted earlier, e-business in SMEs can also reduce inefficiencies resulting from lack of co-ordination between firms in the value chain. Typical example could be cited of B2B interaction and real-time communication can reduce information irregularities between buyers and suppliers and build closer relationships among trading partners (Moodley, 2003). The discussion shows that, adopters of e-business technology can benefit from reduced transaction costs, increase transaction speed and reliability, and extract maximum value from transactions in their value chains (OECD, 2004; Perera *et al.*, 2012).

14.3.2 Barriers to e-Buisness within SMEs

Relocating the discussion within the construction industry, Eadie *et al.* (2007, 2010a,b) undertook a study in Northern Ireland primarily to determine barriers to the adoption of e-business among construction SMEs, the study identified infrastructure, culture, security, and legal as some of the issues that are of concern or things to consider critically, as these issues have the tendency to derail implementation of e-business within construction. Love *et al.* (2001) also discovered, among other things; technical, financial, organisational and behavioural as major barriers to e-business implementation in construction. However, in a more general view Ruikar and Anumba (2008) pointed out that general barriers to e-business mainly fall into three categories, namely infrastructure, trust and reliability, and regulatory issues. Additionally, Ismail and Kamat (2008), Issa *et al.* (2008), Wilkinson (2008) and Ruikar *et al.* (2008) conducted scholarly studies on different aspects of e-business in construction from both developed and developing country perspectives. On the contrary, the small construction organisations have been completely left out in most of these referenced studies, nonetheless, evidence from those studies provides a theoretical framework and platform and the opportunity for the commencement of research into the implementation of e-business within the small construction organisations. It is worth noting that small construction organisation stand to gain from the opportunity provided by main-streaming e-business within their operations. When these technologies are applied to business processes it can increase organisation's efficiency, productivity, reduce costs and expand market reach. It is further, argued that e-business has the potential to improve performance through an integrated environment where communication and information flow remains an essential component (Acar *et al.*, 2005). The benefits of e-business are not confined to only big organisations, small organization can also equally benefit from e-business. Even in developing countries small businesses operating in the construction industry can also gain benefits appropriate to their scale and operations (e-Business W@tch, 2006). It is important to understand the capabilities and roles of these technologies and their tangible effects on construction industry activities especially when considering from a developing country's perspective where these technologies have understandably not fully attained maturity.

14.4 e-Business in SMEs: developed country perspective

Under this section, the concept of e-business technology was explored and key elements of e-business technology identified. The identified elements have been discussed and general benefits of construction have been pointed out.

14.4.1 Concept of e-business technology and application

In the recent years, the use of the Internet technology for business has been on the increase mostly across manufacturing, retail, banking and many other business sectors. The benefits of using the Internet technology to conduct business has been well noted and researched, the emergence of the Internet technology has far-reaching ramifications on the way business is conducted (Gunasekaran and Ngai, 2008). This act of conducting business using the Internet technology in conjunction with ICT infrastructure can be referred to as electronic business (e-business) and in some research documentations it also referred to as electronic commerce (e-commerce). Many businesses including construction believed that the adoption of e-business provides the opportunity to improve operational efficiency, profitability and strengthen the completive position (Beheshti and Salehi-Sangari, 2007; Sabri *et al.*, 2014). Ahmed *et al.* (2005) believe that e-business is an umbrella terminology that encompasses e-commerce and e-procurement activities and refer to the utilisation of network computing and the Internet to transform a firm's value chain (i.e. internal processes, suppliers and partner interaction, and customer relationships with the prime goal of creating value and competitive edge). To gain understanding into e-business or e-commerce, this research attempt to identify definitions of these terminologies and then relate them to construction business process.

There are ranges of definitions for e-business and e-commerce. From the work of Damanpour and Damanpour (2001), they are of the view that e-business and e-commerce is any "net" business activity that transforms internal and external relationships to create value and exploit market opportunities driven by new rules of the connected economy. Similarly, e-commerce is referred to as business transaction by electronic means through the Internet and/or dedicated networks (Anumba and Ruikar, 2002; Ahmed *et al.*, 2005). Accordingly, Damanpour and Damanpour (2001), described e-business in terms of a quantity rather than an absolute state of a company. They consider a business an e-business to the degree that it targets the market opportunities of conducting business under new electronic channels, which revolve around the Internet. This is an acknowledgement that e-business comes in many forms and can be implemented to a very small or large degree. It is also an acknowledgement that the Internet and the Web are essential components of e-business and e-commerce strategies. Fundamentally, e-business can be defined as the interchange of goods, services, property, ideas or communications through an electronic medium for purposes of facilitating or conducting business (Cheng *et al.*, 2001; Xiao *et al.*, 2015).

Ruikar and Anumba (2008) define e-business as the use of the Internet and other digital technology for organisational communication, coordination and the management of the firm. On the other hand, Wamelink and Teunissen (2003) defined e-business as the use of information and communications technology to change and improve business relationships. In the simplistic possible term, however, e-business is an electronic way of doing business (Anumba and Ruikar, 2008). Therefore, companies must participate in external business relationships by using computer interactions (i.e. transactions, support, marketing, communication and collaboration) by either business-to-business (B2B) or business-to-consumer (B2C), if it is to be considered an e-business (Damanpour and Damanpour, 2001). In relation to this Cheng *et al.* (2001), argued that e-business infrastructure is used to improve communication and coordination, and encourage the mutual sharing of inter-organisational resources and competencies. This was further corroborated in a general perspective by Muffatto and Payaro (2004), arguing that e-business is the process whereby Internet technology is used to simplify certain company processes, improve productivity, and increase efficiency. It allows companies to easily communicate with their suppliers, buyers, and customers, to integrate "back-office" systems with those used for transactions, to accurately transmit information, and to carry out data analysis in order to increase their competitiveness. To support the inter-organisational sharing of resources and competencies in a network structure, communication and co-ordination need to be maintained (Cheng *et al.*, 2001; Kauffman and Tallon, 2014).

From the above definitions Kalakota and Whinston (1996) are of the view that the original meaning of e-business is attached to the establishment of computer network system to search and retrieve information in support of business decision making and inter-organisational co-operation. To further understand e-business it is important to gain an understanding in general terms of the various forms and application of e-business in both private and public sectors.

14.4.2 Forms of e-business

Ruikar and Anumba (2008) discussed broadly six (6) forms of e-business functionalities as depicted in Figure 14.1. These functionalities are briefly described below:

14.4.2.1 Business-to-business (B2B)

Business-to-Business (or B2B as it is commonly referred to) is an electronic means of carrying out business transactions between two or more businesses. It incorporates from manufacturing to service providers. For

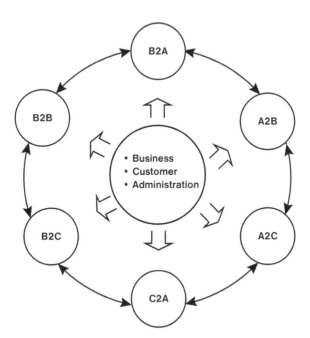

Figure 14.1 Six forms of e-business activities.
Source: Adapted from Ruikar and Anumba (2008).

example, using B2B a company can leverage the Internet to place orders electronically, receive electronic invoices and make electronic payments.

14.4.2.2 Consumer-to-consumer (C2C)

Examples of C2C business models include consumer e-auction and blogs. Although, there may be no financial transaction in C2C business models there is still an exchange of value and these are economic activities and could be referred to as peer-to-peer. Blogs, for example, have led to the development of new C2B and C2C applications by presenting the opportunity and tools for virtually anyone to express their views easily and to communicate these globally and inexpensively.

14.4.2.3 Administration-to-administration (A2A)

Using A2A business model, government departments and agencies can nationally and or internationally communicate and exchange classified information through dedicated portals. Typical examples in many advanced jurisdictions include the national DNA database and other policing information.

14.4.2.4 Business-to-consumer (B2C) or consumer-to-business (C2B)

The B2C model refers to commercial transactions between an organisation and customer or between customers and an organisation. For instance, when applied to the retail industry, a B2C process will be similar to the traditional method of retailing, the main difference is the medium used to carry out the business, the Internet. This method requires the consumer to have access to the Internet. This as explained by Laudon and Laudon (2000) cited in Ruikar and Anumba (2008) by selling direct to customers or reducing the number of intermediaries, companies can achieve higher profits whilst charging lower prices. Some typical examples of the B2C category include Amazon.co.uk and eBay.co.uk. C2B on the other hand, is a business model in which consumers offer products and services to companies at a cost. This business model is a reversal of the traditional business model, where companies offer goods and services to consumers. Typical examples of C2B model are online surveys.com, and survey monkey, where individuals offer the service to reply to a company's survey and in return the company pays the individual for their service.

14.4.2.5 Business-to-administration (B2A) or administration-to-business (A2B)

The B2A category covers all transactions that are carried out between businesses and government bodies using the Internet as a medium. An example is that of Accela.com, a software company that provides round the clock public access to government services for asset management, emergency response, permitting, planning, licensing, public health and public works. A2B on the other hand, is an electronic means of providing business-specific information such as policies, regulations directly to business. A typical example of the A2B category in construction is e-tendering solutions that enable potential construction stakeholders to bid for government-led projects using online tendering tools (Eadie *et al.*, 2012).

14.4.2.6 Consumer-to-administration (C2A) or administration-to-consumer (A2C)

Examples of C2A include applications such as e-democracy, e-voting, information about public services and e-health. Using such services consumers can post concerns, request feedback, or information (on planning application progress) directly from their local governments/authorities. On the other hand, A2C provides a direct communication link between governments (e.g. local authority) and consumers.

14.4.3 Faces of e-Business

Damanpour and Damanpour (2001) discussed four faces of e-business (see Figure 14.2), which originally was proposed by the Gartner Advisory Group.

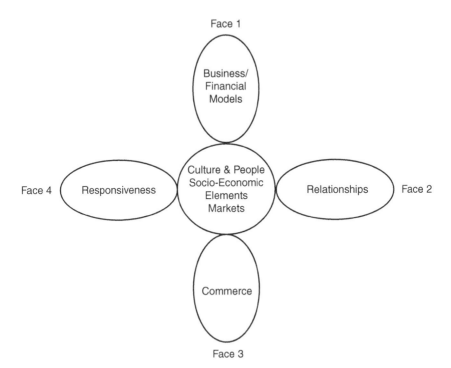

Figure 14.2 Four different faces of e-business models.
Source: Adapted from Damanpour and Damanpour (2001).

Each of this faces looks at e-business from different perspective (Ruikar and Anumba, 2008). The four faces of e-business are discussed below:

14.4.3.1 Face 1: business and financial models

This phase focuses on the business model and the opportunities that operate as an electronic entity. Financial considerations such as reduced costs and operating efficiency are primary considerations. This model regards technology as an enabler of business opportunity and requires strict changes in the corporate culture, image, and accounting guidelines. The model can be used for an existing company (brick-and-mortar), new spin-off form brick-and-mortar or a small, unknown company (like amazon.com).

14.4.3.2 Face 2: relationships

This face looks at e-business from a relationship perspective as new relationships and collaborations are created and forged in e-business to enter new markets or enhance customer, supplier and business interactions (Qin, 2010). Customer relationship management, supply chain management and technology infrastructure management can be created by e-commerce change. For example, the traditional order and invoice process can be reported and transmitted electronically. Electronic marketplaces, catalogues and bidding systems, and Internet search can revolutionise business conduct, accelerate business activities, increase global competition, create global logistics networks, provide better customer relationships and cheaper and better services, and speed up goods and information along the entire supply chain.

14.4.3.3 Face 3: commerce

This face focuses on electronic buying and selling. This requires the building of systems, services, models, and relationships to support the most effective buying and selling mechanism. This particular face overlaps the other three faces. It emphasises the importance of technology to business success, and customer demands (Damanpour and Damanpour, 2001; Barnes and Hunt, 2013). Central to the opportunity of leveraging the Internet and the Web for e-business is the ability to use this medium to reach buyers throughout the buying process at all times, including those that might otherwise be inopportune (Qin, 2010).

14.4.3.4 Face 4: responsiveness

This face deals with efficiency and timing of business transactions. Responsiveness, in e-business terms mean reducing the time between a business request and its fulfilment. It is also about increasing the efficiency of the computing systems (operating systems and their support services) that provide fulfilment. This will help a company to complete a business transaction electronically, without resorting to hand-carried or faxed information. For example, the direct connection of a rent-a-car automobile request system to insurance companies results in improved efficiency, reduction of errors, and hence customer satisfaction.

14.4.4 Synthesis of forms and faces of e-business with respect to SMEs

Considering the various definitions of e-business, and other fundamentals of e-business it is, therefore, imperative to conclude that e-business technology is a process improvement led technology in supporting business process and inter-organisational relationships and coordination using the Internet technology platform, a medium through which a functioning e-business can be achieved. For example, from e-business forms viewpoint; construction firms can use the Internet to place orders electronically, receive invoices electronically and possibly make electronic payments and can support project extranets. Whilst e-business faces determines the focus of a business in terms of what e-business intends to achieve for the firm, they neither limit the usage of any model or idea, nor are all faces applicable to all companies. It is up to the construction firm to adopt and apply any of the four faces to their business.

However, from this discussion perspective, relationships and responsiveness faces fall within the confines of performance improvement. The relationship face seeks to promote collaboration between construction teams and that has to do with supply chain management and technology (Onetti *et al.*, 2012). The responsiveness face on the other hand, deals with issues concerning efficiency and timing of business transaction. In effect, achieving a reduction in the time a request is made and its fulfilment. In essence, the responsiveness face can as well promote improvement within SMEs construction organisations. It is, therefore, important to carry forward and contextualise these faces (relationship and responsiveness) in the context of this discussion and by extension within SMEs construction organisations which is the primary beneficiary of this research.

14.4.5 Key e-business application from developed countries perspective

This section illustrates key uses of e-business from developed country perspective. There is evidence within construction that there are impacting trends of e-business in construction. For example, Ruikar and Anumba (2008), noted that reliance on third parties such as courier services, can sometimes lead to delays and also high amount of added expenses incurred in the delivery of project documents to project members who are geographically distributed. e-Business has the potential to overcome some of the process and communication inefficiencies (London and Bavinton, 2006). Some of the common construction e-business trends (see Figure 14.3) include:

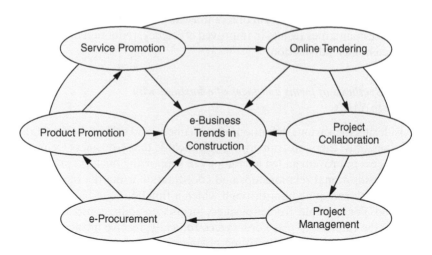

Figure 14.3 e-Business trends in construction.
Source: Adapted from the discussions in Issa *et al.* (2003) Ruikar and Anumba (2008).

14.4.5.1 Service promotion

The Internet facility is being used to promote *"companies by the dissemination of company service information"* principally to potential clients of their services (Ruikar and Anumba, 2008; Chang *et al.*, 2013). Architect, designers, contractors and others within the construction sector are using the Internet for promotion.

14.4.5.2 Product promotion

The Internet is used by construction related firms for the purpose of product sales through online promotion. This is done either through an independent website or through an online vendor. Such promotion site displays all product and material specifications many include manufacturer and supplier details, product availability, quality assurance, cost and mode of delivery.

14.4.5.3 e-Procurement through web directorate and search engines

The Web can be used as a tool to procure information about construction related suppliers and their products (Chang and Wong, 2010). Several websites provide a search tool for the user to access varied information about the construction industry. Information about the industry may be varied and can range from items such as jobs and products to specific information about bidding process (Chang *et al.*, 2013).

14.4.5.4 Project management

Some websites are designed to streamline the construction business process. These services can be referred to as web-enabled project management (Alshawi and Ingirige, 2003). Principally, the Internet can be used to improve and integrate the process of design and management of construction projects (Barnes and Hunt, 2013). This can yield several benefits. It can result in speeding up the process of communication between different parties involved in the project and avoid unnecessary delays.

14.4.5.5 Project collaboration

The web can be used to facilitate online collaboration for project partners, which allows project partners to collaborate and communicate with each other in real time (Galliers and Leidner, 2014). The concept of online collaboration defies the boundaries of time and geography and allows construction stakeholders to, among other things, exchange ideas and make submissions and comments no matter where they are located (Teece, 2010; Oliveira and Martins, 2010).

14.4.5.6 Online tendering

According to Ruikar and Anumba (2008), the Internet has revolutionised many aspect of the construction business and also made it possible to have online tendering services. This service allows for tendering information online along with project specification (Ruikar, 2004). e-Business has emerged to become an important asset in many economic sectors as noted from literature (Weill and Vitale, 2013). Even though the uptake appears slow in the construction industry, efforts must be made to promote e-business within the construction business. *This research attempts to highlight the key role that e-business is perceived to play in assisting construction organisations to use the Internet to carry out their procurement related activities in the context of the Ghanaian construction industry.* As pointed out by Issa *et al.* (2003), the *"Internet is changing the way business is done in construction"*. Despite the immaturity of the technology and its short history, e-business initiatives are already transforming industries and becoming a key component (Issa *et al.*, 2003; Aboelmaged, 2010). Further, the Web has become a source for information, goods, and services, and a means of communication (Alshawi and Ingirige, 2003; Issa *et al.*, 2003).

In the light of this, the next section looks into the application of e-business in construction.

14.4.6 Extension of e-business concept in construction

As indicated in literature e-business potentially can be deployed and applied across all economic sectors and non-economic activities. According to Hashim and Said (2011) few writers define e-business in a broader context; that is 'the facilitation and integration of business process'. However, in construction industry specifically, London *et al.* (2006) provide an extensive definition of e-business in the context of the construction industry as reported by Hashim and Said (2011):

"E-business in construction involves any electronic exchanges of information in relation to the various stages of the design, construction and operation asset life cycle which includes:

1   *Internal organisational driven activity for firm core and support business including industry specific and generic business software applications, websites, email and electronic banking.*
2   *Externally linked online web based portals involving:*

   • Design collaboration and document management
   • Online tendering
   • Procurement, purchasing and invoicing
   • Information

3   *Online or internal organisational facility management systems"*.

It can be deduced from the discussions that there is no conclusive definition for e-business. It showed that definitions are adopted based on the particular sector in question or where a particular research is being carried out. In the context of this research, the definition provided by London *et al.* (2006) is relevant as it encompasses all the component identified in the discussion presented in Section 14.4.4.

14.5 E-business in SMEs: developing country perspective

The adoption of e-business in developing countries differs greatly from one country to the other (WTO, 2013). This notwithstanding many face a number of similar barriers to e-business whilst at the same time encountering similar opportunities. The implementation of various types of e-business models in developing countries depends on several factors. For example, the existing structure of an industrial sector as well as how it complements into a given sectoral value chain. Besides that, the significant difference of cultures as well as business philosophies across developing countries has additionally been seen to limit the applicability and transferability of the e-business models originated by some developed countries.

Most studies on e-business technology implementation in construction organisations have been conducted within the confines of developed economies. However, in recent times there have been attempts to undertake similar studies within developing countries' economies. For example, Hinson and Sorensen (2006) conducted a study into the application of e-business within the non-traditional export sector, principally arguing that the adoption of e-business practice has benefits for small Ghanaian exporters' organisational improvement. A framework for small firm exporter, e-business development was developed. The framework identified four main activities, these activities include international triggers, macro triggers and micro trigger has been grouped and mapped to e-business technologies. The third section, e-business, organisational transformation centres, mentioned among other things, finance, marketing, strategy leverage capabilities and human resources management, this is then mapped to e-business value delivery section and finally to the output section namely, enhanced export performance.

Arguably, the study by Hinson and Sorensen (2006) can be confirmed to be among the first in the context of e-business from developing countries perspective. For example, Sørensen and Buatsi (2002) assessed the use of the Internet within the export business in Ghana. Further, Hinson *et al.* (2007) focused on the Internet usage patterns amongst internationalising Ghanaian exporters and this gradually enter construction. For a better understanding of the tremendous benefits of e-business/e-commerce, Iddris (2012) examined the need to identify and measure the perceived importance of the driving forces and barriers in the adoption of e-business solutions in small and medium-sized enterprises in developing countries. Although, there is an acknowledgement that some amount of work has been done on e-business technology transfer to the construction industries in developing countries as indicated previously, and some different sector(s), the body of knowledge did not appear to have supported the fundamental necessities of e-business technology transfer within developing countries construction industries.

14.6 How does Ebuisness use in SMEs and MOs spawn possible innovations?

The discussions in the previous sections of this chapter provided the background for e-business in Small and Medium Enterprises (SMEs) and Micro Organisations (MOs) with emphasis on opportunities offered by e-business adoption within SMEs and MOs. Additionally, barriers to e-business adoption and implementation within SMEs and MOs has been outlined principally to throw light on key areas that should be of concern to SMEs and MOs as barriers to e-business may vary from one SME and MO to the other. It is also important to note that SMEs and MOs that have sophisticated e-business applications may face different barriers from those that are just at the beginning of the adoption ladder. For the opportunities offered by e-business to SMEs the discussion in Section 14.3.1 shows that despite the slow pace of e-business adoption by SMEs, they have gradually recognised the positive impact that e-business technologies, such as computer facilities, e-mail and the Internet and their applications can have on their business outlook. Further, there are evidences to suggest that in developed countries, most SMEs, including MOs with fewer than ten employees, now have at least one computer facility, usually with Internet access from which there are able to communicate via emails etc. the discussion also argued that this is a sign that most SMEs and MOs are beginning to realise the opportunities offered by main-streaming e-business in their business operations. The success of adopting and migrating business processes onto an e-business platforms within SMEs and MOs has to revolve around both hardware and software facilities. Section 14.3.1 further shows that a wide range of business software can improve information and knowledge management within SMEs and MOs, leading to more efficient business processes and better firm performance. Communication via e-mail utilising the Internet can help to improve external communication, in either B2C or B2B contexts, and may reduce transaction costs, increase transaction speed and reliability, and extract maximum value from each transaction in the value chain.

E-business as earlier discussed offer benefits and opportunities for a wide range of business processes and operations within SMEs. The discussions show that from the firm level, e-business uses can make communication within the firm faster and make the management of the firm's resources more efficiently, thereby improving operational processes within the organisation. Whilst seamless transfer of information through shared electronic files and networked computer system improves the efficiency of business processes such as documentation and data processing. Again, several opportunities exist for SMEs and MOs to become progressively mature in e-business technology. For example, sophisticated e-business technology applications such as Knowledge Management System (KMS) and Enterprise Resource Planning (ERP) improves an organisation's capacity to store, share and use their knowledge and know-how capabilities. Discussing e-business from inter-firm level, ICT systems with the use of the availability of the Internet produces e-business which has a great potential for reducing business transaction costs as well as increasing the speed and trustworthiness of transactions. Specifically, in developing countries such as Ghana, private sector SMEs and e-business together is likely to improve its record in terms of transparency when performing transactions electronically with the Government sector counterparts specifically in terms of public procurement. As noted earlier, e-business in SMEs and MOs can also reduce inefficiencies resulting from lack of co-ordination between firms in the value chain. Typical example, could be cited of B2B interaction and real-time communication can reduce information, irregularities between buyers and suppliers and build closer relationships among trading partners. The discussion shows that, adopters of e-business technology can benefit from reduced transaction costs, increase transaction speed and reliability, and extract maximum value from transactions in their value chains.

Despite the highlights of e-business technology, the discussions in Section 14.3.2 presented barriers to e-business adoption and implementation within SMEs both within developed and developing countries. Eadie *et al.*

(2007, 2010a,b) studying the Northern Ireland construction industry identified barriers to the adoption of e-business among construction SMEs. They found that factors such as infrastructure, culture, security and legal as major issues that might seeminly derail implementation of e-business within construction. Hence the different combination of the factors could impede innovation for the SMEs and MOs. Considering the cumulative effect that SMEs and MOs can have on an economy of a country, Governments and policy makers would strive to gain at least smaller improvements, to benefit from the commulative effect that these smaller improvements might have on the overall economy. On the other hand, Love *et al.* (2001) identified; technical, financial, organisational and behavioural factors as major barriers to e-business implementation in construction. However, in a more general view Ruikar and Anumba (2008) pointed out that general barriers to e-business mainly fall into three categories, namely infrastructure, trust and reliability, and regulatory issues. These barriers can be grouped under three key elements, namely: infrastructure, training and legal and security issues. These issues, as shown in Figure 14.4 can form a framework for an innovative deployment of e-business technology within SMEs. Infrastructure in this context refers to network infrastructure at the SME firm level together with the availability of a wide range of high quality Internet and communication services at competitive prices is particularly important. Firms can then choose services that are appropriate to their needs and operations. Particularly, broadband, characterised by high-speed. It is important to also focus on the wireless network infrastructure and cloud computing also important. The second element of Figure 14.4 is training. In this context, training for the purposes of this discussion focuses on competence factors, which include internal ICT knowledge and e-business management capabilities. These skill sets are considered crucial for successful e-business adoption, and training programmes for SMEs and MOs are among major policy targets in many countries. In many cases, ICT skills have been addressed in a broader context of education and training, and initiatives for reducing the digital divide (OECD, 2004). For example, in developing countries, governments support industries through the provision of ICT training or training support. Some of these trainings focus on the basic ICT user. The third and last element as shown in Figure 14.4 is a legal and security system. The argument that is put forward for this element is that business and consumer confidence in the security and trustworthiness of online transactions is essential to the implementation of e-business technology (OECD, 2004; Ruikar and Anumba, 2008). It is of importance to assure both businesses and consumers that their use of online services is secure, reliable and verifiable. Businesses need a legal framework that is predictable and practical for domestic and cross-border transactions. Creating the appropriate level of confidence in e-business requires a mixture of trustworthy technologies and regulatory and self-regulatory arrangements (OECD, 2004).

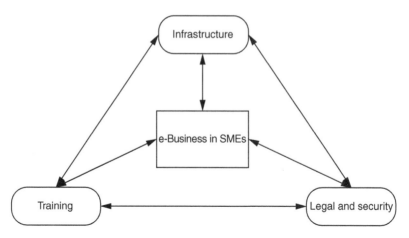

Figure 14.4 Framework for e-Business development strategy.

14.7 Recommendations and the way forward

General ICT and policies have an important role in enhancing the conditions for SMEs to adopt and exploit e-business technologies. In addition, specific organisation policies are important for initial uptake (e.g. management and ICT skills, technology and security). To improve the uptake of e-business technologies and thereby step up new innovations, it is argued that the following overall policy approaches need to be supported and extended:

*   Encourage business and sector associations to provide tools to assess e-business opportunities, benefits and costs, and the development of niche products and services
*   Training programmes for SME managers and employees focusing on both technical and managerial skills need to improve abilities to benefit from e-business technology. Some of these trainings can be provided in cooperation with business sector organisations, training institution and government agencies. In this context, human resource development is a priority
*   A broad policy geared towards addressing infrastructure and security, can provide trust and confidence for businesses and customers using the e-business platform to transact business.
*   In the context of developing countries, use e-government initiatives as in the case of Ghana (Adzroe and Ingirige, 2014) and to provide incentives for SMEs to go on line by simplifying administrative procedures, reducing costs and allowing them to enter new markets.
*   SMEs are usually part of a supply chain, both within the context of developing and developed countries. Hence large companies and Government procurement systems should favour a more SME centric approach in setting up the lowest common denominator.

This chapter supported the view that e-business technologies are not just for large companies, but SMEs in the construction industry can also adopt them instead of their often traditional labour intensive methods. SMEs can reduce their duplicate work and inefficient management practices and improve their ability to develop innovative new methods and systems and processes that are transparent and trustworthy. SMEs can improve their skills through continuous training schemes which are sympathetic to SME needs. Governments in both developed and developing countries need to kick start their planning processes to achieve this new trend as part of their long term growth strategy for the construction sector. A quick reality check on their economy will validate their vision and goals in this direction.

References

Aboelmaged, M. G. 2010. Predicting e-procurement adoption in a developing country. *Industrial Management and Data Systems*, 110, pp. 392–414.

Acar, E., Koçak, I., Sey, Y. and Arditi, D. 2005. Use of information and communication technologies by small and medium-sized enterprises (SMEs) in building construction. *Construction Management and Economics*, 23, pp. 713–722.

Aduwo, E. B., Ibem, E. O., Uwakonye, O., Tunji-Olayeni, P. and Ayo-Vuaghan, E. K. 2016. Barries to the uptake of e-procurement in the Nigerian building industry. *Journal of Theoretical and Applied Information Technology,* 89, pp. 113–147.

Adzroe, E. K. A. and Ingirige, B. 2014. Improving the technological capacity of the local contractors through e-business technology transfer – the case of the local Ghanaian contractors. *CIB W55/65/89/92/96/102/117 and TG72/81/83 International Conference on Construction in a Changing World Heritance Kandalama, Sri Lanka,* 4th–7th May 2014.

Ahmed, M. S., Ahmad, I., Azhar, S. and Arunkumar, S. 2005. Current State and Trends of E-Commerce in the Construction Industry: Analysis of a Questionnaire Survey. Retrieved 07 February 2013 from www.cteseerx.ist.psu.edu/viewdoc/download, pp. 1–21.

Alshawi, M. A. and Ingirige, B. 2003. Web-enabled project management: an emerging paradigm in construction. *Automation in Construction*, 12, pp. 349–364.

Amit, R. and Zott, C. 2001. Value creation in e-business. *Strategic Management Journal*, 22, pp. 493–520.

Ang, J. A. and Koh, S. 1997. Exploring the relationships between user information satisfaction and job satisfaction. *International Journal of Information Management*, 17, pp. 169–177.

Anumba, C. J. and Ruikar, K. 2008. *e-Business in Construction*, Wiley-Blackwell, London.

Anumba, C. J. and Ruikar, K. 2002. Electronic commerce in construction – trends and prospects. *Automation in Construction*, 11, pp. 265–275.

Ashrafi, G., Schlehe, J. S., Lavoie, M. J. and Schwarz, T. L. 2014. Mitophagy of damaged mitochondria occurs locally in distal neuronal axons and requires PINK1 and Parkin. *JCB*, 206, pp. 655–670.

Barnes, S. A. and Hunt, B. 2013. *E-Commerce and V-Business*, London: Taylor & Francis.

Beheshti, H. M. A. and Salehi-Sangari, E. 2007. The benefits of e-business adoption: an empirical study of Swedish SMEs. *Service Business*. Retrieved 10 July 2013 from http://www.diva-portal.org/smash/get/diva2:474933/FULLTEXT01.pdf, 1, pp. 233–245.

Björk, B.-C. 1999. Information technology in construction: domain definition and research issues. *International Journal of Computer Integrated Design And Construction, SETO, London*. Retrieved 08 August 2012 from https://helda.helsinki.fi/bitstream/handle/10227/617/bjork.pdf?sequence=2, 1, pp. 1–16 (The journal is no longer published).

Buvik, M. P. and Rolfsen, M. 2015. Prior ties and trust development in project teams – a case study from the construction industry. *International Journal of Project Management*, 33, pp. 1484–1494.

Chang, H. H., Tsai Y.-C. and Hsu, C.-H. 2013. E-procurement and supply chain performance. *Supply Chain Management: An International Journal*, 18, pp. 34–51.

Chang, H. H. and Wong, K. H. 2010. Adoption of e-procurement and participation of e-marketplace on firm performance: trust as a moderator. *Information & Management*, 47, pp. 262–270.

Cheng, E. W. L., Li, H., Love, P. E. D. A. and Irani, Z. 2001. An e-business model to support supply chain activities in construction. *Logistics Information Management*, 14, pp. 68–78.

Chong, A. Y.-L., Ooi, K.-B., Bao, H. and Lin, B. 2014. Can e-business adoption be influenced by knowledge management? An empirical analysis of Malaysian SMEs. *Journal of Knowledge Management*, 18, pp. 121–136.

Combe, C. 2012. *Introduction to E-Business*, Routledge, New York.

Damanpour, F. and Damanpour, J. A. 2001. E-business e-commerce evolution: perspective and strategy. *Managerial Finance*, 27, pp. 16–33.

e-Business W@tch. 2006. ICT and e-Business in the Construction Industry: ICT adoption and e-business activity in 2006. Retrieved 19 December 2013 from http://www.umic.pt/images/stories/publicacoes/BR06.pdf.

Eadie, R., Millar, P., Perera, S., Heaney, G. A. and Barton, G. 2012. E-readiness of construction contract forms and e-tendering software. *International Journal of Procurement Management*, 5, pp. 1–26.

Eadie, R., Perera, S. A. and Heaney, G. 2007. Drivers and barriers to public sector e-procurement within Northern Ireland's construction industry. *ITcon*, 12, pp. 103–119.

Eadie, R., Perera, S. A. and Heaney, G. 2010a. A cross discipline comparison of rankings for e-procurement drivers and barriers within UK construction organisations. *ITcon*, 15, pp. 217–233.

Eadie, R., Perera, S. A. and Heaney, G. 2010b. Identification of e-procurement drivers and barriers for UK construction organisations and ranking of these from the perspective of quantity surveyors. *ITcon*, 15, pp. 23–43.

Egan, J. 1998. Rethinking construction: The report of the Construction Task Force to the Deputy Prime Minister, John Prescott, on the scope for improving the quality and efficiency of UK construction. Retrieved 8 September 2014 from http://constructingexcellence.org.uk/wp-content/uploads/2014/10/rethinking_construction_report.pdf.

Egbu, C. O. and Botterill, K. 2002. Information technologies for knowledge management: their usage and effectiveness. *ITcon*, 7, pp. 125–136.

Ekholm, A. A. and Molnár, M. 2009. ICT development strategies for industrialisation of the building sector. *ITcon*, 14, pp. 429–444.

El-Ghandour, W. A. and Al-Hussein, M. 2004. Survey of information technology applications in construction. *Construction Innovation: Information, Process, Management*, 4, pp. 83–98.

Fulford, R. and Standing, C. 2014. Construction industry productivity and the potential for collaborative practice. *International Journal of Project Management*, 32, pp. 315–326.

Galliers, R. D. A. and Leidner, D. E. 2014. *Strategic Information Management: Challenges and Strategies in Managing Information Systems*, London: Routledge.

Griffith, A. 2011. Delivering best value in the small works portfolio of public sector organizations when using preferred contractors. *Construction Management and Economics*, 29, pp. 891–900.

Gunasekaran, A. A. and Ngai, E. W. T. 2008. Adoption of e-procurement in Hong Kong: an empirical research. *International Journal of Production Economics*, 113, pp. 159–175.

Hashim, N. A. and Said, I. 2011. Exploring e-business applications in the construction industry: issues and challenges. *Annual Summit on Business and Entrepreneurial Studies (ASBES 2011) Proceedings.* Available at http://econpapers.repec.org/paper/cmslasb11/2011-014-134.htm.

Hinson, R., Sorensen, O. A. and Buatsi, S. 2007. Internet use patterns amongst internationalizing Ghanaian exporters. *The Electronic Journal of Information Systems in Developing Countries*, 29(3), pp. 1–14.

Hinson, R. A. and Sorensen, O. 2006. E-business and small Ghanaian exporters: preliminary micro firm explorations in the light of a digital divide. *Online Information Review*, 30, pp. 116–138.

Hosseini, M. R., Chileshe, N., Zuo, J. and Baroudi, B. 2012. Approaches for implementing ICT technologies within construction industry. *Australasian Journal of Construction Economics and Building, Conference Series*, 1, pp. 1–12.

Ibem, E. O., Aduwo, E. B., Tunji-Olayeni, P., Adekunle Ayo-Vaughan, E. A. and Uwakonye, U. O. 2016. Factors influencing e-procurement adoption in the Nigerian building industry. *Construction Economics and Building* 16, pp. 54–67.

Iddris, F. 2012. Adoption of E-commerce solutions in small and medium-sized enterprises in Ghana. *European Journal of Business and Management*, 4(10), pp. 48–57. Available online at www.iiste.org.

Ismail, I. A. A. and Kamat, V. R. 2008. Integrated multi-disciplinary e-business infrastructure framework. In: Anumba, C. J. and Ruikar, K. (Eds) *e-Business in Construction*, Wiley-Blackwell, Oxford, pp. 65–78.

Issa, R., Flood, I. A. and Caglasin, G. 2003. A survey of e-business implementation in the US construction industry. *Journal of Information Technology in Construction*, 8, pp. 15–28.

Issa, R. R. A., Flood, I. A. and Treffinger, B. 2008. Assessment of e-business implementation in the US construction industry. In: Anumba, C. J. and Ruikar, K. (Eds) *e-Business in Construction*, Wiley-Blackwell, Oxford, pp. 248–264.

Kalakota, R. A. and Whinston, A. B. 1996. *Frontiers of Electronic Commerce*, Addison Wesley Longman Publishing Co., Inc., Reading, MA.

Kauffman, R. J. and Tallon, P. P. 2014. *Economics, Information Systems, and Electronic Commerce: Empirical Research*, Taylor & Francis.

Kenny, C. 2007. Construction, Corruption, and Developing Countries. World Bank Policy Research Working Paper 4271, June 2007. Available at http://www2. globalclearinghouse.org/Infradev/assets%5C10/documents/WB%20(Kenny)%20-%20Construction,%20Corruption%20and%20Developing%20Countries%20 (2007).pdf.

London, K., Nathaniel, B., Jonathan, M., Benjamin, E., Ron, W. A. and Guillermo, A.-M. 2006. *E-business Adoption in Construction Industry.* Queensland University of Technology.

Love, P. E. D., Irani, Z., Li, H., Cheng, E. W. L. A. and Tse, R. Y. C. 2001. An empirical analysis of the barriers to implementing e-commerce in small-medium sized construction contractors in the state of Victoria, Australia. *Construction Innovation: Information, Process, Management*, 1, pp. 31–41.

Molla, A. A. and Heeks, R. 2007. Exploring e-commerce benefits for businesses in a developing country. *Information Society*, 23, pp. 95–108.

Moodley, S. 2003. The challenge of e-business for the South African apparel sector. *Technovation*, 23, pp. 557–570.

Muffatto, M. A. and Payaro, A. 2004. Implementation of e-procurement and e-fulfillment processes: acomparison of cases in the motorcycle industry. *International Journal of Production Economics*, 89, pp. 339–351.

OECD 2004. ICT, E-Business and Small and Medium Enterprises. OECD Digital Economy Paper, No. 86. Retrieved 12 July 2015 from http://www.oecd-ilibrary.org/science-and-technology/ict-e-business-and-small-and-medium-enterprises_232556551425.

Ofori, G., Ai Lin, T. E. and Tjandra, K. I. 2011. Developing the Construction Industry: A Decade of Change in Four Countries. In Laryea, S., Leiringer, R. and Hughes, W. (Eds) *Procs West Africa Built Environment Research (WABER) Conference*, 19–21 July 2011, Accra Ghana.

Oliveira, T. and Martins, M. F. 2010. Understanding e-business adoption across industries in European countries. *Industrial Management and Data Systems*, 110, pp. 1337–1354.

Oliveira, T. A. and Martins, M. F. 2011. Literature review of information technology adoption models at firm level. *Electronic Journal of Information Systems Evaluation*, 14, pp. 110–121.

Onetti, A., Zucchella, A., Jones, M. A. and Mcdougall-Covin, P. 2012. Internationalization, innovation and entrepreneurship: business models for new technology-based firms. *Journal of Management and Governance*, 16, pp. 337–368.

Parmigiani, A. and Rivera-Santos, M. 2015. The influence of institutions on trust and governance. *Strategic Management Society Conference*, Denver CO, October 2015.

Perera, S., Udeaja, C., Zhou, L., Rodrigo, A. A. and Park, R. 2012. Mapping the E-business Profile and Trends in Cost Management in the UK Construction Industry. Retrieved 7 February 2013 from http://nrl.northumbria.ac.uk/11823/

Qin, Z. 2010. *Introduction to E-commerce*, Springer, Berlin and Heidelberg.

Rivard, H. 2000. A survey on the impact of information technology on the Canadian architecture, engineering and construction industry. *Electronic Journal of Information Technology in Construction*, 5, pp. 37–56.

Ruikar, K. 2004. *Business Process Implications of E-commerce in Construction Organisations.* Doctor of Engineering (EngD) thesis, EngD, Loughborough.

Ruikar, K., Anumba, C. J. and Carrillo, P. 2008. e-Business: the construction context. In: Anumba, C. J. and Ruikar, K. (Eds) *e-Business in Contruction.* Wiley-Blackwell, Oxford, pp. 6–21.

Ruikar, K. and Anumba, C. J. 2008. Fundamentals of e-Business. In: Anumba, C. J. and Ruikar, K. (Eds) *e-Business in Construction,* Wiley-Blackwell, Oxford, pp. 1–22.

Sabri, S. M., Sulaiman, R., Ahmad, A. and Tang, A. 2014. A review on IT outsourcing practices for e-business transformation among SMEs in Malaysia. *2014 International Conference on Information Technology and Multimedia (ICIMU),* 18–20 November, pp. 124–129.

Samuelson, O. 2002. IT-Barometer 2000-the use of IT in the Nordic construction industry. *ITcon,* 7, pp. 1–26.

Seyal, A. H., Rahim, M. M. A. and Rahman, M. N. A. 2000. An empirical investigation of use of information technology among small and medium business organizations: a Bruneian scenario. *The Electronic Journal of Information Systems in Developing Countries,* 2, pp. 1–17.

Shiels, H., Mcivor, R. and O'Reilly, D. 2003. Understanding the implications of ICT adoption: insights from SMEs. *Logistics Information Management,* 16, pp. 312–326.

Sørensen, O. J. A. and Buatsi, S. 2002. Internet and exporting: the case of Ghana. *Journal of Business and Industrial Marketing,* 17, pp. 481–500.

Teece, D. J. 2010. Business models, business strategy and innovation. *Long Range Planning,* 43, pp. 172–194.

Tran, Q., Zhang, C., Sun, H. and Huang, D. 2015. Initial adoption versus institutionalization of e-procurement in construction firms: an empirical investigation in Vietnam. *Journal of Global Information Technology Management,* 17, pp. 91–116.

Underwood, J. A. and Khosrowshahi, F. 2012. ICT expenditure and trends in the UK construction industry in facing the challenges of the global economic crisis. *ITcon,* 17, pp. 25–42.

Wamelink, H. A. and Teunissen, W. 2003. E-Business in the construction industry: a search for practical applications using the Internet. *International Association for Automation and Robotics in Construction.* Available at http://www.iaarc.org/publications/fulltext/isarc2003-93.pdf, pp. 543–547.

Weill, P. A. and Vitale, M. 2013. *Place to Space: Migrating to Ebusiness Models,* Harvard Business Review Press, Boston, MA.

Wilkinson, P. 2008. The role of extranet in construction e-Business. In: Anumba, C. J. and Ruikar, K. (Eds) *e-Business in Construction,* Wiley-Blackwell, Oxford, pp. 81–102.

WTO. 2013. e-Commerce in developing countries: opportunities and challenges for small and medium-sized enterprises. *World Trade Organization.* Retrieved 11 July 2015 from https://www.wto.org/english/res_e/booksp_e/ecom_brochure_e.pdf.

Xiao, Z., Zhang, W., Li, Q., Liu, L. and Cui, L. 2015. A method of e-commerce trading process construction based on PaaS platform. e-Business Engineering (ICEBE), *2015 IEEE 12th International Conference on e-Business Engineering (ICEBE),* 23–25 October, 2015. Retrieved (04/12/2015) from https://www.computer.org/csdl/proceedings/icebe/2015/8002/00/8002a220.pdf 2015. pp. 220–227.

15 Application of social media in the construction industry

Paul Wilkinson

15.1 Introduction

This chapter provides a brief introduction to and history of the emergence of social media, highlighting the evolution from the first generation 'Web 1.0' technologies of the early 1990s to more user-friendly and interactive 'Web 2.0' Internet platforms. This shift was facilitated by technological changes such as digital bandwidth improvements and advances in mobile telephony, but – despite their enthusiastic adoption of some other technologies – take-up of social media by most construction organisations was characteristically cautious and slow. However, the need to embrace platforms widely used by new generations of employees (and so help fill skills shortages) has accelerated adoption. Additionally, the integration of social media into mainstream business communication, marketing and PR has meant it complements and sometimes replaces previously used channels. The industry currently known as construction is constantly adopting new technologies; social media is a shift in communications – however, its use will eventually become as normal as email.

15.2 Background

What is today commonly understood as social media has been in development since at least the 1980s, but its emergence into the mainstream of media, business and commerce is predominantly an early 21st century phenomenon. As a sector, Architecture, Engineering, Construction (AEC) and its related built environment disciplines,[1] with some occasional exceptions, is not generally in the vanguard when it comes to adopting new technology (Sepasgozar and Bernold, 2012; Singh and Holmstorm, 2015), and social media has posed particular challenges to an often conservative industry.

As this chapter will outline, social media is not a single technology, but a sprawling and constantly evolving, inter-dependent set of platforms, spread across different devices, allowing millions of users to connect and create, share and combine different types of digital content in countless ways (and given the many 1000s of sites and applications created, this chapter can only 'scratch the surface'). Compared to relatively straightforward concepts such as mobile phones, email, computer-aided design or even the worldwide Web, the nebulous nature of social media means it can be difficult for construction business managers to identify its relevance to their work (Figure 15.1).

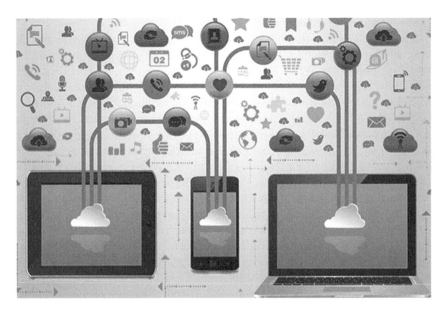

Figure 15.1 Device, service, content and connectivity = complexity.

This chapter looks at the emergence of social media and how various platforms have emerged on a scale rivalling traditional communication channels and offering new ways for users to interact with content and with each other. It is mainly focused on experiences in the UK and US, where there is a high degree of similarity in social media adoption, and aims to help the reader understand how social media tools and techniques can be applied in the construction sector to support core communication functions (from recruitment to marketing and customer service), share knowledge and even open up new business opportunities. It will become clear that social media is not a short-lived 'fad', but part of an ongoing evolution of Internet-enabled communications that will eventually become part of 'business as usual'.

15.3 What is social media?

> a group of Internet-based applications that ... allow the creation and exchange of user-generated content.
>
> (Kaplan and Haenlein, 2010)

Definitions of social media vary (some focus on the content exchange, some on the interpersonal networking they enable, or are based on descriptions of the technologies or platforms deployed) and have evolved over time as the hardware and software applications have developed, but the core elements tend to include use of computer tools that allow people to create, share or exchange information, ideas, pictures, videos, audio or other content in virtual communities and networks.

There is a useful distinction to be made here between social media and the first generation of Web tools, and to see social media as part of the cumulative evolution of the Web. During the 1990s, the first stage – so-called Web 1.0 – saw Websites that were developed by relatively few content creators (creating and maintaining a website generally demanded good knowledge of HTML and other technologies) and they published content to multiple users who were broadly passive consumers of that content. As such Web 1.0 tools largely replicated existing one-way marketing communications such as brochures; during the 1990s, many corporate Websites were largely static 'brochure-ware'; onsite interactivity between the creator and the consumer was limited (some sites had guest-books, for example, but otherwise any interaction tended to involve separate channels: email, telephone, fax). Server capacity and consumer reliance on slow Internet connections meant that content had to be optimised to speed up information flow (few photos, used small and at low resolution, for example). And Web 1.0 was also more controllable by those publishing content.

Web 2.0, by contrast, shifts the emphasis from the previous 'one to many' or broadcast mode, to a more 'many to many' or conversation mode, changing the balance between publishers and consumers of content, and dramatically reducing any notion of 'control' – the notion, instead, is one of 'influence'. Rather than being passive, users can now participate and interact, and can become publishers of content at little or no cost to them. Detailed technical knowledge and programming skills are unnecessary with most of today's social media platforms, and the reduction of bandwidth capacity constraints has opened up new opportunities to share a wider variety of richer media content (e.g. video) and to create new interactive options (real-time chat, online meetings, etc.). User-friendly services have also been developed that allow users to create their own Websites, set up their own online communities, to create and share content almost continuously and across multiple platforms, and – importantly – to use these platforms to freely interact with friends, colleagues and other users, and for such communities to co-create, discuss and modify user-generated content. (Levine *et al.*, 2000)

15.4 Social media: a brief early history

It is tempting to regard the emergence of computer-mediated social media as a 21st century phenomenon, but people began interacting through computer networks almost as soon as they were first connected – and some years before the inception of the worldwide Web (c. 1993). The first public bulletin board services were created in the late 1970s, for example; USENET was established in 1980; and the first IRC (Internet relay chat) services began in 1988. In the US, CompuServe began offering its users file exchange services in 1981, and, overshadowing competing systems such as Prodigy and GE-nie, started marketing Internet connectivity services in 1989 with email and forums proving hugely popular. CompuServe's main rival AOL (with whom CompuServe later merged, in 1998) started as a games-related service during the 1980s and started to compete with CompuServe's online service in 1989. As they expanded internationally, common features of both CompuServe and AOL were services to share files and access news and events, but they also offered interaction. Forums, arguably, were the precursor of today's social networks; AOL also had member-created communities complete with searchable member profiles.

Less proprietary, more open social networks, typically built upon relational databases and delivered via the Web, soon followed (a useful history and timeline for social networking sites is given by Boyd and Ellison, 2008). Personal networking site Classmates.com launched in the US in 1995, SixDegrees.com followed in 1997 but lasted just 4 years, folding in 2001. Friendster started up in 2002, business working site LinkedIn and MySpace in 2003, and Bebo and Piczo in 2005. Facebook was launched in 2004 as a US university-focused network, allowing users to create profiles, post photographs, status updates and other content. It became a public service in 2006, the same year that messaging service Twitter was launched – both also enabling open application programming interfaces, APIs, which allowed thousands of third party developers to create applications that worked with these platforms (Digital Trends, 2014). And Internet search giant Google launched its own social network, Google+, in June 2011.[2]

Alongside these social networks, other more focused content-sharing platforms developed, allowing people to create and share different types of 'social objects' or user-generated content – for example:

- **Blogs** – short for 'weblogs', a term coined in 1997; blogging platforms soon followed (e.g. LiveJournal and Blogger in 1999, TypePad and the open-source WordPress in 2003, the micro-blogging platform Tumblr in 2007)
- **Wiki articles** (in its most basic form, a wiki is a Web application that allows users to collaborate and modify, extend or even delete content; the most well-known example, Wikipedia, 'the free encyclopedia that anyone can edit', was launched in January 2001)
- **Instant messages** (e.g. WhatsApp, launched in 2009; Viber, 2010, Facebook Messenger, 2011)
- **Photographs** (e.g. Picasa, founded in 2002 and purchased by Google in 2004; Flickr, founded in 2004, acquired by Yahoo in 2005; and later mobile-oriented apps such as Instagram and Snapchat, launched in 2010 and 2011 respectively)
- **Videos** (e.g. Vimeo, established in 2004, and YouTube, founded in 2005 and acquired by Google in 2006; similarly, mobile-oriented apps followed later: Vine was founded in 2012 and promptly acquired by Twitter;

live video-streaming apps Periscope and Meerkat both launched in 2015)
- **Music and other audio** content such as podcasts (a term coined in 2004; networks in this space included the shortlived Napster (1999–2002), Last.fm and Spotify – founded in 2002 and 2007 respectively)
- **Location** (e.g. Foursquare (later Swarm), launched in 2009)

As well as the emergence of APIs to allow different platforms to share and reuse content, some common Web 2.0 conventions helped provide links. For example, RSS (rich site summary, or really simple syndication) allows blog posts and other social content to be syndicated and shared via 'feeds', while tagging means users can categorise, index and link content by use of common keywords or searchable hashtags. And some applications (e.g. IFTTT – If This Then That – Zapier) have been created to help users connect actions between devices and platforms.

15.5 Social media today

A few of the afore-mentioned leading proprietary platforms – Facebook, LinkedIn, Twitter, YouTube and Wikipedia – have achieved widespread international familiarity and adoption, being translated into different languages and their reach extended across multiple platforms. Some types of social media – blogs, for example – are less associated with particular platforms, being supported by several different providers, ranging from open-source networks and content management systems to proprietary solutions – but have also achieved widespread familiarity, and will share many familiar social media concepts such as tagging and sharing.

The scale of adoption of individual platforms is also significant insofar as total use of some platforms today exceeds what a few years ago amounted to total worldwide access to the Internet:

- Facebook in particular has started to dominate the rest of the Web as a platform for sharing information; in March 2015, it had 1.44 billion registered users – not far short of half of all 3 billion global users of the Internet reported by the International Telecommunications Union (Dent, 2014).
- Not far behind in terms of domination relating to consumption of a particular form of content (video) is YouTube, which in March 2013, announced it had over a billion unique users every month ('Nearly one out of every two people on the Internet visits YouTube').
- Wikipedia may not have the same active user base (some 19 million registered users, and 69,000 active editors in November 2014) but was delivering 18 billion page views to a user base of nearly 500 million unique visitors each month.

Table 15.1 Social networking registrations and active use, global

Network use (Q1 2015)	Total registered users	Monthly active users
Facebook	1,440 million	1,250 million
YouTube	–	1 billion+
LinkedIn	364 million	347 million
Twitter	302 million	236 million

Source: Statista.com.

Many other individual social media platforms have in excess of 100 million monthly active users (in addition to those shown in Table 15.1, among those familiar in English-speaking markets, Statista lists WeChat, Skype, Google+, Instagram, Viber, Tumblr, Snapchat and LINE).

Illustrating the scale of the blog sector is more difficult, but the aggregate worldwide blogging community is on par with the major networks. No single blogging platform dominates. Some blogs are supported by generic content management systems (CMSs), while some blog platforms (most notably WordPress) have become Web CMSs widely used for Websites as well as blogs, ahead of more Website-specific CMS platforms such as Joomla, Drupal and Magento; in 2012, *Forbes* reported:

> WordPress powers one of every 6 Websites on the Internet, nearly 60 million in all, with 100,000 more popping up each day. Those run through its cloud-hosted service, which lets anybody create a free Website online, attract 330 million visitors who view 3.4 billion pages every month.
>
> (Colao, 2012)

In 2014, WordPress alone was supporting around 75 million blogs (the other major blogging platform, Blogger, owned by Google, does not release statistics about its use; Livejournal has around 50 million sites). WordPress conservatively[3] claimed its users produced about 58.7 million new posts each month; in terms of readership, over 409 million people viewed more than 20.1 billion pages each month (WordPress, 2015). Add micro-blogging to the picture, and the scale of blogging expands considerably; on 6 June 2015, Tumblr, the leading platform reported it had 239.6 million blogs, whose users to date had published 112.6 billion posts, with a typical single day's output shown as over 72 million posts.

Clearly, the range of social media platforms and types of content and interactions have expanded dramatically since the 1990s, and it has also been a highly volatile market (which has sometimes deterred businesses, concerned that they may choose a platform that then discontinues). Some of the successful social networks of the early 2000s have disappeared, numerous mergers and acquisitions have occurred, and, partly reflecting the inter-dependence of content, platforms, hardware and connectivity, new social media services – both niche and generic – are launched almost constantly.

15.6 Social media – fast-changing and constantly evolving

Various efforts have been made to categorise or to describe the scope of the various tools, but they remain at best snapshots, and can be influenced by the audience for which the description was developed (construction industry adoption will be explored in detail later in this chapter). For example, as the UK construction industry began to take interest in social media, a report for the Royal Institution of Chartered Surveyors (Waller and Thompson, 2009) defined the main areas of social media in 12 sectors. Another illustration, the Conversation Prism, created for marketing and PR professionals, maps the leading platforms and applications in the social media universe across 26 sectors (the first edition was developed by Brian Solis in 2008; reflecting the volatility of the social media world, it has been regularly updated and the fourth edition was published in 2013) (Figures 15.2 and 15.3).

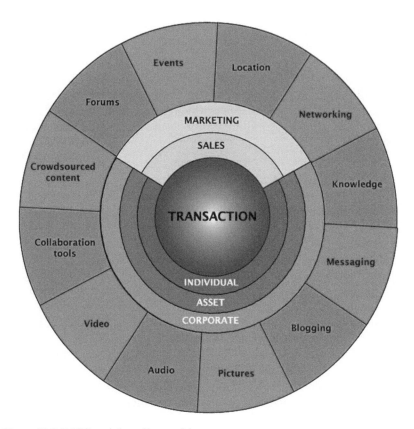

Figure 15.2 RICS social media graphic.

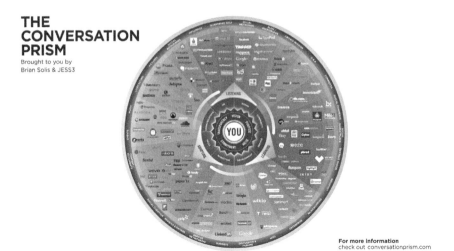

Figure 15.3 The Conversation Prism, Brian Solis.

If we attempt to reconcile the two diagrams (or even the different iterations of the Prism alone), it is clear that the social media world moved on in 4 years, with some new types of services emerging (for example, live-streaming video and audio services, taking advantage of increased availability of high bandwidth connectivity), and others diminishing in importance. Solis's latest iteration included five categories not included in the third edition in 2011; 122 services were removed and 111 added; and he reviewed but decided against the addition of five other categories including virtual worlds, social CRM, attention/communication dashboards, and DIY/custom social networks (Table 15.2).

The emergence and eventual adoption of these different types of social media has been particularly influenced by factors associated with hardware. Three areas stand out:

Connectivity – The roll-out and increased consumer availability of broadband Internet connections capable of handling 'richer' media: commonly imagery and audio data captured in larger file sizes than the text-led and low-resolution, often static graphics characteristic of early Web 1.0 Websites.

Digital media – The growing availability of digital still and video cameras, computer webcams, and the digitisation of images, text and other content such as music.

Mobile telephony – The launch of new mobile devices, taking advantage of the wider wireless connectivity, and incorporating an increasing range of functions, including email, cameras, voice recorders, Internet browsers, and GPS (global positioning system).

Table 15.2 The changing social media world

Royal Institution of Chartered Surveyors (Waller and Thompson 2009)	Conversation Prism (Solis 2013)
Location	Location
Networking	Social networks
	Business networks
	Service networking
Knowledge	Wiki
	Q&A
	Social curation
	Content/documents
	Social bookmarks
Messaging	Social streams
Blogging	Blogs and Microblogs
	Comments
Pictures	Pictures
Audio	Music
Video	Video
Collaboration tools	Nicheworking
	Enterprise
Crowdsourced content	Crowd wisdom
Forums	Discussion and forums
Events	Events
	Live-casting
	Social commerce
	Social marketplace
	Reviews and ratings
	Influence
	Quantified Self

These developments (briefly considered in the next three sections) have not been the only factors, of course. As technology became cheaper, more widely available and simpler to use it was also rapidly adopted by younger people, who, as they entered the workplace, sought to use their favoured applications alongside – and sometimes instead of – established corporate communication tools.[4] And, with user-generated content threatening to swamp content created by existing media outlets, publishers and broadcasters also began to embrace social media, sharing their own content on common platforms. (Social, commercial and economic factors influencing social media adoption, particularly as they apply in the construction and built environment sectors, will be discussed in more detail later in this chapter).

15.6.1 Socially connected

By 2005, one in two of the UK households had Internet access; in 2015, the figure was 84% (Ofcom, 2015). The explosive growth in access to, and use of, the Internet both stimulated, and was stimulated by, advances with respect to telecommunications technology. Consumers moved from reliance on – by today's standards – slow external modems (e.g. 9.6Kbps) to, initially, ISDN, ADSL/SDSL and fibre-optic cables. Over time, connectivity no longer demanded a physical connection to a network; users could access data wirelessly.

Digital mobile phone systems first emerged during the early 1990s, adopting the European GSM standard. The first GSM network was launched in Finland in 1991, with UK networks following soon afterwards. This technology also enabled a, then, new form of communication: text messaging (the first person-to-person SMS was sent in 1993). This proved popular, as did the ability to access and download digital content, such as ringtones, introduced in 1998. Mobile phones were now genuinely pocket-sized, less expensive, and had also become increasingly ubiquitous in the AEC industry, not least because they aided communication as workplaces were often geographically dispersed and required frequent travel, and because they enabled SMEs (a third of the UK's construction workforce are employed in businesses employing under ten people) to keep in contact with existing and potential customers, and project team colleagues.

The mobile phone market exploded during the early 2000s as mobile providers moved from GSM through GPRS and EDGE, to 3G and now 4G – telecommunications standards that expanded use of mobile devices beyond voice communication, facilitated by the dramatic increases in bandwidth availability and therefore speed. Increasingly, phones were used more for viewing and moving data – texts, photographs, emails, file attachments, Web pages and other digital media – than they were used for voice conversations. 4G also supported bandwidth-intensive applications like streaming media, and also allowed voice calls to be treated like any other type of streaming audio media.

Users also become less reliant on telecoms providers' mobile networks. Wi-Fi allowed smartphone and other device users to access Internet-based applications and services via locally broadcast wireless networks, bypassing mobile networks and often getting a faster and so better user experience. And if Wi-Fi was unavailable, good quality 3G helped create a market for specialised devices to access the mobile Internet: first, USB plug-in 'dongles', then wireless routers that made 3G, and later 4G, Internet connectivity available to multiple computers (phones, laptops, tablets) simultaneously over Wi-Fi, rather than just to a single computer via a USB plug-in.

Individuals and businesses also started to embrace 'the cloud'. Information was no longer mainly held on a company's own servers behind a firewall, with access limited to authorised users. New services offered similar access (anytime, anywhere, from any device) often via standard Web-browsers (and/or 'apps') with shared hosting infrastructures and password-protected profiles governing security. Social media providers are nowadays, therefore, among the leading exponents of cloud-based computing, storage and software-as-a-service.

15.6.2 *Visually social*

Echoing the shift from analogue to digital telecommunications seen in the UK mobile sector in the 1990s and in the transition of recorded music from vinyl or type to CDs and later MP3s, etc, photography and videography also began (a more gradual) transition to become digital media during the 1990s. Replacing traditional film cameras, chemical processing and printing, film reels and video cassettes with devices that could capture and store images as digital files, however, took some years. But by the early 2000s, digital cameras were outselling film cameras, traditional photography-related companies such as Kodak and Nikon were frantically trying to reshape their businesses, and Webcams were being incorporated into laptop and desktop computers. By 2015, near-professional grade photography and video could be achieved by users of consumer-level devices including high-end smartphones.

15.6.3 *Socially mobile*

The first 'smartphone', the IBM Simon, was launched in 1993, combining mobile phone, pager, fax machine, calendar, address book, clock, calculator, notepad, email, and a stylus-activated touch-screen with a QWERTY keyboard (the first phone incorporating a camera appeared in 2000). Personal digital assistants, PDAs, followed: pocket-sized computers from companies such as Psion and Palm. Web connection usually involved placing the PDA in a cradle connected to a PC to synchronise data held on the two devices – but eventually it became a more portable alternative to a laptop or desktop. Other operating systems, such as Windows CE (later Pocket PC and Windows Mobile), also improved users' experience and synchronisation capabilities, while RIM's Blackberry helped grow the smartphone market. By 2006, half of all mobile phones sold featured a camera, with an increasing proportion able to capture low-resolution video as well as still images.

However, the smartphone market began a dramatic change in June 2007 when Apple launched the iPhone, consciously taking powerful computer processing onto mobile devices, and providing mobile-oriented applications – mobile 'apps' – to help users undertake tasks; the App Store opened in 2008. HTC's first phone to use Google's open source Android operating system, was launched in 2008 when Google also opened its own mobile application market, today Google Play. The iPhone was a huge success, as was its sister product, the iPad tablet, launched in April 2010, but Android smartphone and tablet adoption also expanded dramatically.[5] Hardware manufacturers have since released new smartphones and tablets on these two platforms regularly, expanding processing capacity, increasing storage, enhancing the cameras and adding GPS, compasses, accelerometers and other functions.

Quickly, Psion's Symbian, Windows and Blackberry became minor players in the mobile OS market. However, the launch of Microsoft's Windows 8 and Windows Phone 8 in late 2012 (coupled with easy integration with existing Microsoft-based back-office IT infrastructures), made it attractive to some corporate IT directors. As a result, Microsoft retained a small foothold in the mobile computing market.

In line with most other developed economies, by 2015, two thirds of UK adults owned a smartphone, with penetration higher in younger age groups – 9 out of 10 adults under 35 had smartphones – and 37% of adults used tablet devices to go online (Ofcom, 2015, pp. 28–29).

Once online, use of social media has become an increasingly widespread activity. Until 2007, Ofcom did not ask about social media use; that year, 22% of online users said they had profiles on social media sites. In 2015, 72% of respondents said they had at least one social media profile (Facebook being the most common), with four out of five saying they visited their sites at least once a day. The impact of smartphones was also evident:

> 46% of those who use social media say the device they use most often for it is a smartphone (laptop/netbook is the second most popular device at 29%).

> (Ofcom, 2015, p. 32)

15.7 Construction and social media

The rise of social media was monitored by many people in the architecture, engineering, construction and real estate/property industry, but adoption was initially cautious. The AEC sector has occasionally been enthusiastic about new technologies – as already mentioned, many construction people were keen early adopters of mobile telephones, for example. But, in common with some other areas of business and commerce (particularly business-to-business, B2B), social media attracted less enthusiasm. This has been a common trait of construction.

Construction businesses, by and large, are not technophobic, but in a fragmented, highly mobile, conservative, risk-averse, project-oriented, information-centric, cost-conscious and sometimes litigious industry, companies have often adopted a slower and more measured approach to technology adoption particularly if it might also be perceived as consumer-rather than business-oriented. Often they will seek to ensure they are implementing reliable solutions – ideally already with a proven track record of use by other industry organisations or on previous projects (Wilkinson, 2005, p. 19), and they will be keen to see a tangible return on investment (reduced costs, saved time, improved quality).

Mobile phone adoption in the 1990s had had a readily identifiable ROI, as had the replacement of manual approaches to information creation and dissemination – replacing type-writers with word-processors, moving from manual draughting to computer-aided design, and switching from paper and postage to email, for example. And, in parallel with the emergence of social media, the launch of new Web-based tools (c. 2000) to support construction project document and drawing collaboration also proved attractive; start-ups such as 4Projects, Aconex, Asite and BIW (today known as Conject)[6] and eBuilder developed into successful Web-based businesses (Wilkinson, 2005, pp. 23–31).

However, social media applications did not fit easily into such adoption strategies, and many construction businesses tried to prevent social media use. A 2009 US survey of architectural, engineering and construction firms by the Society for Marketing Professional Services (Connell *et al.*, 2009) found that many firms maintained IT and personnel policies actively discouraging social networking; 67% of the firms surveyed blocked employees from viewing social networking sites while at work (p. 14).[7] This was not unique to construction. Across many business sectors, organisations frequently saw social media tools as irrelevant or viewed them with suspicion and so blocked employee access to them in the workplace. Reasons for blocking social media often included fears relating to:

- Computer viruses
- Loss of confidential data through employee carelessness
- Hacking
- Impacts on productivity
- Reputational damage if used inappropriately
(ClearSwift Research cited by Leach, 2011)[8]

Social media was also often initially seen as more personal or consumer-oriented, and so less relevant in the workplace or to construction. From a UK property professional's perspective, for example, Waller and Thompson (2009) pointed out:

> ... most of the uses to which it [social media] has been put are Business to Consumer (B2C), rather than Business to Business (B2B). (p. 8)

As a result, while mobile phones, new hard/software and new cloud-based applications that improved traditional construction processes gradually found favour among construction businesses, social media applications took far longer to achieve even grudging acceptance within companies. Even if social media tools were essentially 'free', many company managers were reluctant to allow their people to invest in the time needed to select and learn their efficient use; they feared productivity losses as employees 'wasted time' using new social tools alongside conventional ones; and there were also common misconceptions that social media was mainly for young people, an ephemeral 'fad', or just something to be used for marketing purposes.

15.7.1 Economic factors

In this context, it is striking, then, that attitudes towards social media began to change as an indirect consequence of the late 2000s great recession. But before this is examined in more detail, it is worth noting considering the AEC industry's reluctance to invest in IT – at least at similar levels to other industry sectors.

In 2012, *Engineering News Record* reported Gartner Research findings indicating that the construction industry was 'dead last in IT spending compared to 14 other industries' (Sacolick, 2012). Even on a wider definition including materials and natural resources, this low spend was later re-confirmed:

> In 2013, IT spend was 1% as a percent of revenue, 1.2% as a percent of operating expenses, and last on both measures compared to the other 19 industries surveyed. Gartner also put Construction in one of 12 industries where revenue is growing faster than IT spend, implying that it is unlikely Construction as a sector will break out of its underspending trend anytime soon.
>
> (Sacolick, 2014)

Construction industry spend on IT was about one-third the average across all industries. Sacolick continued:

> That's astonishing – especially at a time when construction leaders recognize that recruiting a new generation of architects and engineers is critical to sustain and grow the industry. When college graduates expect to work with sophisticated collaboration tools, the latest BIM technologies or BYOD, it will be hard to compete for this talent if construction companies are significantly underspending in technology for its employees.

15.7.2 Demographic factors

This new generation of potential industry recruits is the so-called Generation Z, and yet many senior construction business managers are two generations removed: Generation X – commonly understood as the population cohort born in the 1960s and 1970s, and who were educated and entered work at a time before information technology was common in everyday business life. The following generation – Gen Y (also known as the 'Millennials', born in the 1980s and 1990s) – grew up in a world where technology was becoming increasingly important and mobile, and where the Internet and worldwide Web were rapidly expanding. The latest cohort – Gen Z – is maturing in a technology environment where (at least so far as the developed world is concerned) Internet access is almost as commonly available as water or electricity, and where mobile communication and social media are, for them, simply a normal part of everyday life.

These rapid changes have also led writers to spawn new terms to attempt to explain the differing digital competencies of the generations. Prensky (2001), for example, famously distinguished between 'digital natives' and 'digital immigrants', defining the former as the generation of 'native speakers' of the digital language of computers, video games, social media and other Internet tools. The 'immigrants', on the other hand, date from a pre-digital time, and may struggle to interact intuitively with digital devices and media.

The global construction industry was ravaged by the great recession around the end of the 2000s. Output in the UK construction sector, for example, fell faster than the whole economy in 2008, staged partial recoveries and then suffered further dips, only emerging slowly from recession in 2013. Over this 5-year period, gross value added dropped 11.2%, putting output in real terms around the 2003 level, and over a quarter of a million jobs were lost (Rhodes, 2014). As the industry has started rebuilding its capacity, in common with other developed economies, the UK has found it increasingly difficult to gain new recruits to its trades and professions, with some sectors facing particular skills challenges – both short- and long-term. As industry pundit Brian Green wrote in *Building* magazine in November 2014:

> There has been, however, massive but less immediately evident damage caused by half a decade in an economic quagmire. The construction industry's reputation, never that splendid, as a career choice for young folk has once again been tarnished.

The poor image of the industry has long been cited as a problem, but the skills shortages have focused minds like never before on the need to engage better with young people. The UK Government's 2013 industrial strategy, *Construction 2025*, for example, mentions the industry's image 14 times and its reputation a further four times, stressing: 'fundamental change is required in how the construction industry is perceived by the general public' (p. 40). It recommended four areas where action was needed to reform the image of the industry, starting with 'engaging young people and society at large'.

15.8 Social media in construction – recruitment and retention

The value of social media to help such engagement has been recognised for some years. In the US, for instance, the SMPS research (Connell *et al.*, 2009, which additionally cited a 2008 *Economist* survey on professional services) identified some strong justifications for enterprise or business-to-business use of social media, including recruiting talent and retention (p. 17). At the same time, as previously noted, the same study had identified two-thirds of AEC businesses attempted to stifle social media activity. Recruitment and retention objectives will not be achieved if professional services firms block use of social media, as Waller and Thompson identified:

> In much of the world of property, status is measured by the ability to attract work and successful projects from a network of contacts. We, as an industry, encourage graduates to 'get out into the market' to make contacts because we know that improving their personal brand will enhance the chances of them earning revenue for the firm. We then set up our computer systems to stop them accessing Facebook, MySpace and Twitter – the methods they have been using at college to network.
>
> (2009, p. 4)

15.8.1 Recruitment and retention case study: HOK

Connell *et al.*'s 2009 report included a case study of international planning and design practice HOK, which had consciously embraced Web 2.0 technologies as part of a strategy to recruit new joiners by showcasing what it was like to work at HOK. A blog, *Life at HOK*, launched in 2008, let employees talk about their work:

> On this blog we want to introduce some of the HOK people and personalities behind these projects. ... HOK is made up of teams of creative individuals who specialize in many different planning and design disciplines and work together in locations all over the world. ... Here we are unleashing the voices of a small group of talented HOK people who live and breathe [our] values every day. These are the diverse voices – our tribal storytellers—who are eager to share with the world their Lives at HOK.
>
> (quoted in Wilkinson, 2009)

John Gilmore, one of HOK's corporate communications team, explained about the wish to connect with its target audience:

> Our hope was that blog would help change the external perception from 'HOK the big company' to 'HOK the creative people'. We thought that if we could set up a talented team of bloggers with a good blog tool, the pieces would fall in place in terms of helping with recruitment and retainment, appealing to future business partners and clients, connecting with traditional and new media members (everybody Googles!), and even strengthening our internal design culture. As you know well, today's Gen-Yers are living their lives online and we needed to get in on these conversations. Life at HOK also serves a practical purpose as a launching point for all HOK's Web 2.0 properties, which we call the 'HOK Network'.
>
> (Gilmore, 2009)

The HOK team recruited existing staff members with experience in and/
or enthusiasm for using blogging and other social media, and used them
to help create guidelines and to train on appropriate use of the different
platforms ahead of the launch, and then put in place structures for ongoing
support:

> HOK's Communications team joined with our HR group to recruit 20
> bloggers representing offices in the US, Canada, London, Hong Kong
> and Singapore. We wrote a 'Blogger Manifesto' outlining the creative
> vision and goals for the blog. We wrote an HOK Blog Policy covering
> topics like etiquette and confidentiality. Although bloggers always de-
> cide what to post, they committed themselves to client confidentiality,
> professionalism, mutual respect and good taste. The Corporate team
> hosted a two-day, in-person training session in St. Louis for bloggers
> to learn about the concept, to receive training on how to use the soft-
> ware, and to get to know each other. A half-day team-building session
> included visits to several local HOK-designed projects.
>   The team developed a Help Blog that houses troubleshooting tips and
> suggestions and that provides a resource that helps newly added bog
> team members get up to speed. To keep the blog front of mind with
> bloggers, the Communications team sends a Weekly Blog Update to all
> bloggers. This message includes pats on the back, blog traffic reports
> and content suggestions. The Communications team edits posts (after
> they are published) for grammar, punctuation and spelling – never for
> content. The team will hold semi-annual WebEx meetings with all blog-
> gers worldwide to continue the conversation begun at the launch and
> to get their ideas on what is working well, what could be improved and
> what could be the 'next big thing'.
>
> (Gilmore, 2009)

The benefits of this bottom-up approach resulted in an HOK social media
system that met both individual and corporate objectives, and which also
proved self-sustaining.

A small 'Opportunities search' link on the blog home page clearly showed
that one purpose of the site was for recruitment. In the first six months (to
March 2009), the blog accumulated over 500 posts from 30 contributors,
and had more than 43,000 visits and 120,000 page views from people in
over 60 countries, with occasional spikes of more than 1,000 daily unique
visitors, and over 1,000 comments were posted to the blog (Connell *et al.*,
2009, p. 20).

It also helped users find other HOK resources, including HOK presences
on YouTube, Flickr, Twitter, bookmarking site Delicious, Facebook, LinkedIn, VisualCV, and SlideShare, plus a handful of other HOK blogs. The
initial selection of social media platforms played to HOK's strengths and to
its recruitment objectives:

- As a design practice it wanted to display its design work and projects alongside its people, so had both formal and informal videos displayed in a corporate HOK YouTube channel and in a HOKlife channel
- Similarly, it used two Flickr accounts – one corporate and HOKLife – to display photographs of HOK people and projects
- A corporate @HOKNetwork Twitter account (then with two hundred followers) provided firm-wide updates and news information about the firm, and were fed into a live Twitter feed on the blog (individual HOK employees also had their own Twitter accounts)
- HOK also established a robust presence on social bookmarking site, Delicious, tagging two hundred fifty items related to news about HOK or the design profession.
- An HOK Careers page on Facebook (popular with HOK's recruitment target audience of university and design school students) gained nearly five hundred fans in the first six months; having been designed to give potential recruits to ask questions about internships and jobs at the firm, it received nearly fifty comments and inquiries during the same period.
- LinkedIn was HOK's secondary careers page, giving a brief snapshot of the firm, and 1,200 of its employees were on the network.
- VisualCV allowed recruits to submit interactive multimedia resumes and view more information about the company, including videos and office information. In five months, fifty VisualCV resumes were submitted.

According to Gilmore and to Connell *et al.* (2009, p. 22), the blog also received dozens of inquiries from people looking for jobs, and helped the firm's communications team work better with the group's human resources function. The initiative also extended to working with other internal teams:

> We are beginning to work with designers and practice groups on incorporating these Web 2.0 tool into their marketing and project work. PlanningNet, for example, is a blog-powered internal workspace for our global Planning team.
>
> (Gilmore, 2009)

Underlining that use of social media is not just a short-term strategy but a sustained part of its communications programme, the HOKLife blog (Figure 15.4) and most of its related social media channels were still going strong in 2015. A review of the updated platforms in June 2015 showed:

- A further 1,289 blog posts had been added to the HOKLife blog since March 2009;
- The HOK Network YouTube channel hosted 154 videos (excluding liked and favourites), had 861 subscribers, and 239,088 views;

Figure 15.4 Image of HOKLife.

- The HOK Network Flickr account displayed 4,258 photographs;
- The @HOKNetwork Twitter account had tweeted 2,277 times and had 31,867 followers. Further corporate Twitter accounts are linked from the blog, including one for @HOKLondon, added in January 2011 and gathering 1,484 followers, and a technology-oriented account, @HOKBuildingSMART, added in December 2012 with 675 followers;
- On Delicious, HOK Network had shared 4,622 public links;
- With fan pages now longer provided, HOK's company Facebook page, created in 2011, had 9,181 'likes'; and
- HOK's LinkedIn profile was linked to 2,007 employees, with 57,351 people following its updates.

Some platforms are no longer prominent (VisualCV has gone), but new platforms have been added to the blog home page links, reflecting the emergence of new media:

- Following the launch of Google+ in 2011, HOK created a +HOK Design profile, gaining 468 followers, and 152,414 page views; and
- 26 HOK publications are displayed via Issuu.com, and its page has 243 followers.

Monitoring the effectiveness of media channels has long been a key role for marketing and PR departments, and over the past decade they have had to develop new knowledge and expertise in social media, and help their businesses maintain effective use of their different channels. As communication specialists they have had to become increasingly tech-savvy and able to influence their business's adoption and appropriate use of new media. For example, the emergence of websites and of self-published online content has diminished the importance of printed brochures, while social networking platforms can be used instead of, or alongside, face-to-face meetings for community engagement. Traditional outlets have also been affected; trade print publications, for instance, have seen their recruitment and display advertising dwindle to the extent that some titles have become web-only, with corresponding changes in how they interact with information providers.

15.9 Social media for construction marketing and public relations (PR)

Marketing an AEC (Architectural Engineering and Construction) organisation to potential future recruits through online conversations[9] will overlap considerably with promoting that business to existing and potential customers and to other stakeholders, so it is perhaps surprising that construction businesses have not been quick to seize such opportunities. After all, in the construction industry, conversations are the lifeblood of business. Conversations are used to ask questions, to listen to feedback, to learn about colleagues, to collaborate on projects, to share news and ideas, to pick up information, to share knowledge, to recommend (and get recommended), to make new contacts and so on. If people are taking these conversations online, organisations which don't participate or limit interaction could be falling behind their competitors. For example, it is common to see Twitter used to ask questions about a company or its products – not responding to such queries may open a sales or recruitment opportunity to a competitor; Twitter has also been used very powerfully by the UK building information modelling (BIM) community to share information using the hashtag #ukbimcrew (Butcher, 2014); and LinkedIn discussions are used extensively by businesses to position staff as 'thought leaders' in their fields.

Connell *et al.* (2009) suggested social networking ought to be a natural tool for marketing AEC professional services firms, where reputation and referrals are vital, but firms were often initially reluctant to deploy social media for marketing purposes. Echoing arguments still heard at construction conferences years later, Connell *et al.* (2009) cited concerns about:

- The 'fit' with audience, brand, organisational capability and goals;
- Social networking being the province of teens, not business people (they felt that it is simply 'not professional');
- Social networking being the province of individual rather than enterprise communications;
- Social networking reducing control of message, brand ambassadors and image; and
- Additional work needed to manage social networking initiatives (pp. 7–8).

As a result, and setting aside examples such as HOK's use of social media for recruitment and retention, US AEC professional services were then mainly using social media and networking for:

- Research and business development (using LinkedIn, for example);
- Awareness and brand building through blogging; and
- Individual (but not enterprise) social networking (i.e. employees marketing themselves as experts in their discipline).

The firms surveyed tended to extend conventional individual networking (participation in organisations and conferences) to professional online places like LinkedIn, private social networks (alumni and trade associations) and specialist industry or interest groups. Just 37% were beginning to use social media for marketing purposes (p. 12). Connell *et al.* argued that AEC firms should not ignore social networking:

> Marketers and others should explore its use as key components of the firm's efforts to build relationships with a wide range of constituencies. As it becomes the norm in other marketing communications (particularly for consumer products companies) we believe that clients, prospects, and employees will expect to be able to engage in social networking with their professional services providers (p. 14).

They laid out some considerations, made recommendations to help firms start to use social media more strategically, and included some examples of how firms were already using social networks in their businesses (pp. 15–19). Such examples, coupled with increased integration by mainstream media channels, have started to alter perceptions of social media as a 'province for teens'. With newspapers, magazines, television and radio all incorporating social media alongside conventional channels such as letters to the editor or phone-ins, a growing number of commercial organisations are understanding its strategic importance and now routinely including corporate social media channels alongside email links, for instance.

Since 2009, adoption has continued to grow. A further SMPS (Society for Marketing Professional Services) publication in 2011 (Bolton *et al*. 2011; see also Butcher, 2015) showed that, when used appropriately and strategically, social media helped build and foster relationships through marketing a firm's employees and their expertise, knowledge, and creativity. Looking specifically at clients' use of social media, it identified owners who used social media in their RFP (Request for Proposal - tendering) and selection processes, or who partnered with their design consultants to promote project messages to stakeholders via social media. In 2015, Bolton *et al*. said social media was more accepted, regarded less as a fad, and its use had become more widespread in the industry, but it was not necessarily being used effectively. Research team member Dana Galvin felt progress was needed in three areas:

> First, we are still figuring out how to sell and market our services. ... Second, as soon as we feel like we have made some ground, the social media landscape changes or we perceive it to change. ... I think some are still waiting for the social media landscape to solidify.[10] Third, we are still catching up from the recession. Marketing budgets and staffs were cut heavily so as we pull our way out of that, new marketing initiatives like social media can begin to thrive at firms where they could not before.
>
> (Butcher, 2015)

Social media can also help clients find and research firms that they want to work with, continued Bolton:

> Social media platforms play an important role in the communication of thought leadership – which can be delivered through content marketing tactics like blogs, white papers, Webinars, etc. – by helping drive traffic to those online destinations. As decision makers increasingly look only for valuable content and resources that help them innovate and solve problems, firms using social media and content marketing effectively should gain a foothold through their visibility.
>
> (Butcher, 2015)

15.9.1 *From broadcasting to conversation: integrating social media with conventional construction marketing, PR and customer service*

Visibility is perhaps a key word here, particularly when it comes to promoting construction businesses online. While word-of-mouth recommendations and referrals do carry weight, many users will follow-up by researching the individual or company online, and the extensive inclusion of social media links in Search Engine Results Pages (SERPs) can quickly either confirm or undermine the reputation of the subject. As a result, construction marketing and PR people are deploying Search Engine Optimisation (SEO) techniques, including use of social media platforms, to increase their businesses' and people's prominence in online searches. For example:

Videos and images within results increase the likelihood of users clicking on a particular item; the Google search algorithm gives added prominence to online sources, often individuals, perceived as authoritative;

- Items from Facebook, LinkedIn, and Google+ are frequently displayed in SERPs; and
- Wikipedia articles about companies or their products often appear prominently in SERPs.[11]

Accordingly, and as previously mentioned, marketing and PR professionals are increasingly expected to have a good knowledge of digital techniques that extends beyond conventional Web 1.0 era expertise in advertising, printed collateral, events, Websites, direct marketing and media relations. Fundamentally, this may also involve a move away from old broadcasting-type approaches (an approach sometimes denigrated as 'spray and pray'), but there are still risks that firms simply treat mainstream Web 2.0 and social media platforms in the same way they treated Web 1.0 media, as a SMPS contributor Holly Bolton explained:

> The key to success in social media is engaging with your connections and providing them with valuable content. What I would call an unsuccessful application is when a company starts up a Twitter account and uses it to merely Tweet out links to press releases and features of their product or services rather than using it to provide valuable content and interact with others.
>
> (Butcher, 2015)

Accordingly, organisations which regard their communications as mainly one-way, one-to-many broadcasts will be less likely to hear what's being said about them, their projects, products, services and communications (at least, not until it's probably too late).

15.10 Use of social media in construction project delivery

It would, however, be wrong to consider social media as something mainly to be used for recruitment or for communications by the marketing and PR team. In a construction business, the latter should also act as advisers and facilitators to enable their colleagues to make appropriate use of the various channels to support project delivery. As should now be apparent, businesses now have a rich array of tools that they can use to share content and interact with other project stakeholders in the context of construction projects.

Alongside generic social media platforms (Facebook, Twitter, LinkedIn, Flickr, YouTube, blogs, etc), entrepreneurs have developed industry-specific applications that deploy social media approaches to fulfil construction, property and other built-asset related functions. This might best be illustrated by showing how these tools might be deployed during the project lifecycle: the planning, design, construction, handover and future operation of built assets.

- When organisations are contemplating a new development, for example, they might consult with local residents using a consultative tool designed to capture people's views on the existing environment. Stickyworld is one example of a socially-enabled platform that enables local authorities to collate multi-media information (e.g geo-located Tweets, photographs, video as well as online comments) from local people about their locality; the same business also developed an interactive tool, You-CanPlan, that used an interactive gaming engine to display early masterplanning proposals and solicit feedback. Conventionally, this might be undertaken through a series of 'town hall' meetings, planning consultations and 'charrette' workshops, but an interactive online process could get feedback from locals daunted by, or unable to attend, such conventional interactions. Augmented reality tools such as Layar have also been deployed to help people visualise the impacts of proposed new developments.
- Design is an intensely collaborative process often spanning multiple disciplines. Collaboration (both during design and during the construction phase) might be enabled through information sharing on project extranets or BIM Common Data Environments, but social media paradigms are also being deployed. Some collaboration platforms – particularly those launched as mobile-first tools relevant to social-savvy professionals (e.g. US-based FieldLens) – have adopted Facebook-like 'wall' interfaces to share ideas or comments rather than rely on one-to-one email-type feedback loops. Wikis are being used for internal knowledge management (UK architect Feilden Clegg Bradley was one of the first practices to pioneer such approaches, capturing project-related information – suppliers, products, specifications, etc – for potential future re-use), and to provide industry information sources (e.g. DesigningBuildings.com, OpenBuildings.com). And TripAdvisor-style construction product review portals are also being developed (e.g. SpecifiedBy).
- Use of mobile devices on active construction sites is often banned (usually for health and safety reasons), but where their safe use might accelerate project processes, they, and relevant apps, are increasingly used for managing both discrete processes (e.g. defects management) or for

providing a wider site team collaboration platform (e.g. US-based Jobsite Unite, Australia's Envision, and Denmark-based GenieBelt). For communications outside the core team, generic platforms such as blogs are used in community relations; a project blog on a Dorset, UK highway project, for example, welcomed questions from local schoolchildren and residents, and explained different stages of the project from pre-project archaeological explorations to road white-lining (Dorset County Council, 2010–2012)

- And, post-construction, social media tools are also being developed for facility management and maintenance. For example, with the mobile social app CrowdComfort, building occupants can report site-specific building issues, keeping workplaces running smoothly and everyone safe, happy and productive.

Unsurprisingly, given the explosion in social media use over the past decade or so, and the high market capitalisations of the some of the leading platform providers, entrepreneurs are constantly looking for the 'next big thing', whether that is something addressing a global or regional need or something specific to a particular industry, profession or even set of tasks.

Some construction businesses and individual professionals have established new businesses which have developed new social media technologies specifically for the construction industry. For example:

- The aforementioned social platforms YouCanPlan and StickyWorld were devised by London-based architectural practice Slider Studio to support consultation processes relating to the built environment.
- In 2011, a Belfast-based groundworks engineer created Senubo as a 'private social network for the construction industry' aiming to help companies create internal LinkedIn-type networks where the expertise, experience and project-related knowledge of individuals could be researched and shared via a handheld application.
- Similarly, but more targeted at tradespeople, in 2015 US startup JobsiteUnite.com aimed to simplify, streamline, and record critical jobsite communication, capitalising upon the increasingly common use of mobile devices and social media platforms by site-based workers.

And some construction-oriented platforms have attempted to emulate the purpose of existing generic platforms:

- New York, US-based mobile construction collaboration application FieldLens was dubbed 'The Facebook of Construction' by *Engineering News Record* (Sawyer and Abaffy, 2013);
- Construction startup ProTenders' founder told *VentureBeat* it wanted to be 'The LinkedIn of Construction' (Grant, 2013); and

- UK-based DesigningBuildings.com applies the same technology used by Wikipedia to provide a wiki for construction professionals, Real-EstateWiki.com does the same for US-based property professionals, and OpenBuildings.com is a wiki (with supporting mobile apps) about buildings.

15.11 Communication in the Web 2.0 era

As has been described, Web 2.0 technologies require a more wide-ranging approach, recognising that communication must be more interactive and that barriers between companies and stakeholders have diminished. Numerous tools have been added to the communications toolbox; within organisations, old 'command and control' communication strategies (where information was cascaded from management downwards) need to be revised to reflect more horizontal and bottom-up communication in real-time; and external audiences can now directly engage with businesses via social media, putting a new onus on customer service. Responding to this interactive reality will require both reactive and proactive approaches. Across businesses (not just among the PR or marketing staff), managers will need to have working knowledge of social media tools and their use so that they can monitor what is being said about the organisation and respond accordingly. Where they discern that key audiences are best reached via social media, they will need to understand how to develop appropriate content, distribute it and then respond to feedback.

Today, user-generated content can be published to the Web in seconds – by anyone, anywhere, anytime – by individuals asking a question on a discussion forum, sending a Tweet, writing a blog post, commenting on Facebook or LinkedIn, uploading a video to YouTube or a photograph to Flickr, checking in on Swarm, or editing a Wikipedia page. Social media platforms allow anyone to talk about a company, its projects, people or products – and, critically, are totally outside the direct control of that organisation. Today's businesses need to monitor such conversations, and, when appropriate, engage with those people and seek to influence that online conversation.

If negative statements about a person, organisaiton or project, particularly from influential individuals, are allowed to go unchallenged, reputations can be seriously damaged. For example, in July 2005, influential blogger Jeff Jarvis wrote a scathing review of a computer company, 'Dell Hell: Seller Beware', lamenting its poor customer service. Dell responded and not just through social media – it also set about improving its customer experiences through better services and products (Solis, 2012, pp. 257–258). It created a company command centre as its operational hub for listening and engaging across all social media globally (Solis, 2012, p. 263). Similarly, telecommunications giant BT has a team of more than 30 full-time employees dedicated to monitoring what is being said about BT online, including on social media (Smith, 2012, p. 159). Dedicated departments are not always the answer, of course. In smaller organisations, communication staff and business managers can use desktop tools (e.g. Hootsuite) to search for mentions of their company, products or projects, and then engage with content-generators as necessary.

Aware that staff may already be using various social networks, construction businesses should follow the examples set by enlightened organisations (including local authorities, the Civil Service, the BBC and BT), and develop clear and consistent policies or guidelines covering appropriate and responsible use of social media. And – as design firm HOK did – and, as social media should not be solely the responsibility of the marketing or PR team, they should include a much wider range of staff in devising and implementing those policies.

Particularly given the extent of mobile phone possession today, it is almost impossible to ban employees from using social media (and would raise ethical issues about 'freedom of speech'). However, in addition to monitoring content, organisations can also monitor quantitative usage of platforms via their corporate networks: they could limit use of social media so as not to impair individual productivity, while recognising that appropriate use may, in some cases, replace laborious or time-consuming alternatives). Organisations may also legitimately block access to certain sites or platforms – access to YouTube may be constrained, for example, for the impact that video streaming has on Internet bandwidth.

Some policy content may already have been created by HR and legal teams and be implicitly covered in employee handbooks or contracts regarding, for example, appropriate use of company IT, safeguarding commercially sensitive information, protecting customer confidentiality and respecting copyright; this should be explicitly extended to cover social media. Organisations might also extend their guidelines to clarify if and how the company name or brand identity might be used (some businesses require staff to include disclaimers in their social media profiles – 'these are personal views and do not represent the views of [company]'), appropriate user-names, and who takes leadership on managing particular platforms, particular types of audience, or particular subjects (bearing in mind that organisations may have subject matter experts who can act as 'thought leaders' in their field). And to support the policies and to encourage good practice, firms should also train employees in how to use their chosen social media platforms, both technically, and in terms of sharing content and responding appropriately to critical comments.

With individual staff contributing to LinkedIn discussions, blogging and tweeting, interesting and useful contributions to online conversations can quickly reap benefits.

- Many construction businesses are now incorporating blogs into their corporate Websites, and, as previously mentioned, a few even encourage project-related blogs so that, for example, landmarks in project progress can be shared (eg: Dorset County Council, 2010–2012).

- Twitter is widely used by many construction people, and some highly constructive [*sic*] conversations have taken place, some indexed by the use of hashtags; since 2011 #ukbimcrew, for example, has facilitated numerous conversations about building information modelling in the UK (Butcher, 2014), while Leeds Beckett University's ThinkBIM events have used Twitter, live-blogging, live video-streaming and content capture tools such as Storify to share its BIM events online (Leeds Beckett University, 2016).

Organisations also need to be responsive: social media exchanges happen in real-time – so organisations must be regularly monitoring and be prepared to respond quickly, particularly if someone identifies a customer service problem, as unanswered criticism can leave an indelible negative 'footprint' on the Web.

Social media is also highly measurable (important in businesses which demand metrics showing how well the organisation, its projects, services and people are performing). Most of the major platforms come with extensive inbuilt analytics capabilities; further third-party tools provide additional insights, including areas such as sentiment analysis; and the impacts of social media on traffic to corporate Websites can also be measured in fine detail by tools such as Google Analytics. This allows communicators, managers and their colleagues to monitor the impacts their social media interaction is delivering, and to adjust their tactics if necessary. For example, if social media campaigns are developed to generate sales leads, businesses should be monitoring the sources of incoming enquiries so that the return on investment (ROI) of the campaigns can be assessed.

Creative use of social media to augment traditional communications can yield measurable differences for a business, generating sales leads, increasing brand awareness, improving customer service, differentiating it from competitors and driving traffic to its Website – among other benefits. Wall-tie manufacturer Helifix, for example, used a blog to increase its Website visibility by 45% and boosted search traffic by 40%.

Even if companies decide not to actively engage online, knowledge of social media and the various platforms, along with some monitoring, is still desirable from a risk management point of view. By setting up Google alerts or employing someone to monitor online mentions, for example, a company can quickly see when its name, brands or people are being discussed online and assess whether any response might be appropriate. If a business suddenly found itself embroiled in a crisis, social media channels allow a company to rapidly respond to both online comments and coverage in conventional print and broadcast media. Prompt response requires some experienced users, but the impact of these responses may not be widespread unless they or the company have accumulated some 'social capital' to help them deal with criticism – an engaged community of Twitter followers or other online brand advocates can help mitigate negative comments, particularly if there is ample online evidence of positive sentiment towards the company, product, project or person involved.

Such responses are necessary due to sheer persistence of online content on the Web. Whether on mobile devices or on conventional computers, the advent of social media has increased the volume of recorded interactions via the Web, creating content – text, photographs, images, video, audio, etc – that will often remain searchable and available online for years to come. This not only affects organisations but also their employees, who will need to be conscious that what they say and do online is widely discoverable (corporate policies and guidelines on social media use not only protect the business, but can protect individuals from posting inappropriate information that might jeopardise their careers.

This persistent 'digital footprint' for an organisation or individual can also, of course, present new business or career opportunities. For example, a lot of social media content is structured information that can be rapidly processed and analysed alongside other 'big data' to help businesses gain insights into industry trends or to identify issues, problems or exceptions. At a more personal level, some organisations increasingly value employees with proven online networks – influence assessment tools such as Klout have been used in the US by Silicon Valley employers to ensure new recruits join them with already-established social kudos (Poeter, 2012).

15.12 Looking forward

Social media tools and philosophies are now moving increasingly from the periphery of construction industry communications towards becoming an intrinsic part of everyday working and professional interaction. This is only to be expected given the pace of online adoption in society at large. Ofcom (2015, p. 28), for example, identified that UK residents now spend an average of 20.5 h a week on the Internet, more than double the figure in 2005, when the average was 9.9 h. Social media use during this period has played a big part in this increase. In 2007 22% of online users said they social media profiles; in 2015 the proportion with at least one social media profile stood at 72%, with 81% of these using social media at least once a day (p. 32). According to GlobalWebIndex research covering the third quarter of 2015, the typical global Internet user is now spending 1.77 h per day on social networking; however, Britons are slightly less digitally obsessed, spending around 1 h and 20 min each day (ie: over 9 h a week), managing an average of four social networks (GWI, 2015).

Against this backdrop and regardless of the social media readiness (or not) of individual construction businesses, social media has also begun to move into the mainstream of government and construction industry policymaking. The 2015 *Digital Built Britain* strategy made several references to social media. First, it identified the ease-of-use and simple clear interfaces of social media and said these should be the model for software used in engineering and especially BIM:

> Normal lay users are mostly conversant with applications such as email and social media, both of which perform complex processes, yet manage to present the user with clear simple interfaces. Our aim must be to present the day to day user with useful easy to consume and interact with information and knowledge (p. 25).

Second, the document lamented:

> With an industry so keen to enable collaboration of diverse people, the uptake of social media in the supply chain has been relatively slow. Where uptake has taken place it has been with tools such as LinkedIn which have a more business focus. Lessons should be learnt from this and the patterns of social media uptake to create an appropriate toolkit to encourage very wide adoption and usage (p. 25).

Third, it anticipated that delivery of the *Digital Built Britain* strategy (which draws together strands from four separate industrial strategies: *Construction 2025*, Information Economy, Smart Cities, Business and Professional Services) would bring about a paradigm shift in the interactions between people, technology and the environments they work in, continuing:

> They will enable collaborative working in virtual, mobile workplaces and combine a high degree of self-regulated autonomy with decentralized leadership and an enabling rather than directive management style. Fortunately most of the young people who will be entering the industry over the next decade have grown up with social media and have already acquired many of the skills needed to work in this environment (p. 29).

Clearly, even the entry of large numbers of web-savvy young people into the construction industry will not result in a dramatic change to how the industry communicates. For a start, their industry education and training should ensure that new entrants understand and respect standards relating to quality, health and safety and other aspects of project and corporate governance and compliance. And once junior staff start employment they will be working alongside incumbent professionals, many of them well versed in the sector's necessarily conservative and risk-averse culture, and who will provide a buffer against rapid and/or inappropriate change (this was one reason why social media policies, mentioned earlier, need to be compiled by a cross-section of managers and staff, and related to existing approaches to project delivery, data and intellectual property protection, risk management, client confidentiality, etc).

Parallel industry changes such as the introduction of building information modelling (BIM) – the focus of the *Digital Built Britain* strategy – have necessarily taken several years to move from initial government statement of intention in 2011 towards a Level 2 BIM deadline in 2016, and industry organisation are still grappling with the implication of BIM for industry processes long focused on 2D information. Similarly, social media will not lead to an overnight change. Just as businesses gradually replaced hard copy correspondence with email, and hand-drawing with CAD, so the introduction of BIM processes and of social media approaches will also take several years to be introduced and incorporated into industry organisations' and project teams' normal working methods. While social media may currently be regarded as novel, it is worth remembering that many in the industry felt the same way about email and CAD, but as we saw with these technologies and associated processes, social media will eventually simply be normal – assimilated into 21st century digital working in the reshaped industry currently known as construction.

Notes

1 The abbreviation AEC is often used in the context of construction to cover all the relevant disciplines, sometimes adding 'O' for owner/operator (AECO) to include clients and fields such as facilities management.
2 This outline, and the following section, focuses mainly on social networks with extensive international reach. Other services have substantial user bases but these are predominantly located in one country or region. For example, most users of Tencent QQ, QZone, WeChat, Baidu Tieba and Sina Weibo live in China; LINE's user-base is located mainly in Japan and south-east Asia; and VK (originally VKontakte) users are mainly in Russia and Eastern Europe.
3 The numbers are likely an under-estimate. WordPress says its statistics 'are for blogs we host here on WordPress.com, both on subdomains and their own domains, or externally-hosted blogs that use our Jetpack plugin and are part of our network.' Some blogs will not use this plugin, so will be excluded from the figures.
4 And supporting internal collaboration, various private social media products also emerged. For example, Yammer.com (now part of Microsoft), for example, was used internally several construction businesses.
5 Android tablet devices had been launched in 2009, but the release of the 'Honeycomb' operating system in February 2011, hastened a new generation of Android tablets.
6 Conject was acquired by Aconex in March 2016.
7 The effectiveness of a workplace ban is debatable, of course. Most employees now have access to the Web through mobile devices, and given that most households in developed economies also have Internet access, workplace bans are not going to stop people accessing social networks from home and talking—as many do—about their work, their colleagues, their employer, customers, etc.

8 While such fears are not irrational, they should not be related only to social media, as the impacts could just as easily be felt through careless or inappropriate use of email or other digital media. Many organisations have employee contracts, staff handbooks and policies and procedures covering appropriate use of email, phones and other corporate IT, and such provisions can be easily updated to explicitly cover appropriate use of social media.

9 Traditional print advertising of vacancies has been largely superseded by web-based jobs portals and use of social networking platforms, particularly those, like LinkedIn, that are geared towards business use. From the recruiter's point of view, conversations can be instigated in seconds, while the applicant can often mine social media sources to access a wealth of online information prospective employer and its people.

10 Of course, it is debatable whether the social media landscape will ever solidify: new hardware, better connectivity and other technological advances are constantly opening up new avenues for people to apply Web 2.0 approaches, and the scope of services offered by different platforms changes frequently.

11 Businesses should exercise great caution regarding Wikipedia, particularly when they might wish to create or edit articles about their company, its product(s), projects or people. Wikipedia has strong guidelines on neutrality and conflicts of interest, and it advisable not to edit articles if the user has any other involvement that might be perceived to colour his/her judgement.

References

Bolton, Holly, Galvin, Dana, and Kilbourne, Adam (2011), *The Clients Use of Social Media and Social Networking* (white paper), Society for Marketing Professional Services, Alexandria, VA.

Boyd, Danah M, and Ellison, Nicole B. (2008), Social network sites: Definition, history, and scholarship. *Journal of Computer Mediated Communication*, 13 (1), 201–230. doi:10.1111/j.1083-6101.2007.00393.x. https://www.researchgate.net/profile/Nicole_Ellison/publication/259823204_Social_network_sites_Definition_history_and_scholarship/links/541354060cf2bb7347db216a.pdf (accessed: 12 January 2016).

Butcher, Scott (2015), 'The State of Social Media in the A/E/C Industry (aka, The Three SoMegos Ride Again!)', Engineering News Record, (Marketropolis blog, 27 January 2015). http://enr.construction.com/opinions/blogs/butcher.asp?plckController=Blog&plckBlogPage=BlogViewPost&newspaperUserId=623ad72a-6766-4594-8d42-07bfdc0821a1&plckPostId=Blog%3a623ad72a-6766-4594-8d42-07bfdc0821a1Post%3aae7de5c6-22c3-47e3-806f-52ad9a22117b&plckScript=blogScript&plckElementId=blogDest (accessed: 5 June 2015).

Butcher, Su (2014), 'The #ukbimcrew is Not a Clique; it's for Everyone', Just Practising blog, 7 May 2014. http://www.justpractising.com/social-tools/networking/ukbimcrew-clique-everyone/ (accessed: 16 January 2016).

Colao, J.J. (2012), 'With 60 Million Websites, WordPress Rules The Web. So Where's The Money?', *Forbes* (24 September 2012; web article dated 5 September 2012). http://www.forbes.com/sites/jjcolao/2012/09/05/the-internets-mother-tongue/ (accessed: 6 June 2015).

Connell, Regina M., Shuck, Barbara D., and Thatch, Marion (2009), *Social Networking for Competitive Advantage* (white paper, July 2009), Society for Marketing Professional Services, Alexandria, VA.

Dent, Steve (2014), 'There are Now 3 Billion Internet Users, Mostly in Rich Countries', Engadget (25 November 2014). http://www.engadget.com/2014/11/25/3-billion-internet-users/ (accessed: 6 June 2015).

Digital Built Britain/UCL (2015), *Digital Built Britain: Level 3 Building Information Modelling—Strategic Plan*, BIS, London.

Digital Trends (2014), 'The History of Social Networking' (5 August 2014). http://www.digitaltrends.com/features/the-history-of-social-networking/ (accessed: 5 June 2014).

Dorset County Council (2010–2012), Weymouth Relief Road Blog. https://weymou-threliefroad.wordpress.com/ (accessed: 14 January 2016).

Gilmore, John (2009), 'Life at HOK—the HOK perspective', *pwcom.co.uk* blog post (19 March 2009), online at http://blog.pwcom.co.uk/2009/03/19/live-from- hok-guest-post/ (accessed: 7 June 2015).

Global Web Index (2015), GWI Social Summary, Q3, 2015—download from http://www.globalwebindex.net (accessed: 16 January 2016).

Grant, Rebecca (2013), 'Startups from Around the World take on America at Plug and Play Expo 2013', VentureBeat, 13 June 2013. http://venturebeat.com/2013/06/13/startups-from-around-the-world-take-on-america-at-plug-and-play-expo-2013/ (accessed: 5 June 2014).

Green, Brian (2014), 'A £20 Billion Repair Bill to Fix the UK Construction Industry After the Recession', *Building*, 19 November 2014. http://brickonomics.building.co.uk/2014/11/20-billion-repair-bill-fix-uk-construction-industry-recession/ (accessed: 4 June 2014).

Kaplan, Andreas M., and Haenlein, Michael (2010), Users of the world, unite! The challenges and opportunities of social media. *Business Horizons*, 53 (1), 61. http://www.sciencedirect.com/science/article/pii/S0007681309001232 (accessed: 5 June 2014).

Leach, Anna (2011), 'Most bosses monitor or block social-network use at work', *The Register*, 7 September 2011. http://www.theregister.co.uk/2011/09/07/fear_of_social_media_holds_back_tech_adoption_survey/ (accessed: 5 June 2015).

Leeds Beckett University Centre for Knowledge Exchange (2016), ThinkBIM blog. http://ckegroup.org/thinkbimblog/ (accessed: 16 January 2016).

Levine, Rick, Locke, Christopher, Searls, Doc, and Weinberger, David (2000), *The Cluetrain Manifesto: The End of Business as Usual*, Perseus, Cambridge, MA.

Ofcom (2015), *Adults Media Use and Attitudes*, Ofcom, London. http://stakeholders.ofcom.org.uk/binaries/research/media-literacy/media-lit-10years/2015_Adults_media_use_and_attitudes_report.pdf (accessed: 5 June 2014).

Poeter, Damon (2012), 'Tweet or Die: Employers Hiring Based on Applicants' Klout Scores?', PC Mag (26 April 2012). http://uk.pcmag.com/internet-products/64362/news/tweet-or-die-employers-hiring-based-on-applicants (accessed: 16 January 2016).

Prensky, Marc (2001), Digital natives, digital immigrants. *On the Horizon*, 9 (5), 1–6.

Rhodes, Chris (2014), 'The construction industry: statistics and policy' (Standard Note SN/EP/1432), House of Commons Library.

Sacolick, Isaac. (2012), 'Construction Industry Dead Last in IT Spend', *Engineering News Record*, 28 November 2012. http://enr.construction.com/technology/construction_technology/2012/1203-gartner-stats-aec-dead-last-in-it-spend.asp (accessed: 4 June 2015).

Sacolick, Isaac. (2014), 'Construction Industry Continues to Underspend in Technology', *Engineering News Record*, 5 May 2014. http://enr.construction.com/technology/information_technology/2014/0303-construction-industry-continues-to-underspend-in-technology.asp (accessed: 4 June 2015).

Sawyer, Tom, and Abaffy, Luke (2013), 'Facebook of Construction' Uses Timeline to Manage Projects', *Engineering News Record*, 9 October 2013. http://enr.construction.com/technology/information_technology/2013/1014-facebook-of-construction-uses-8216timeline8217-to-manage-projects.asp (accessed: 5 June 2015).

Sepasgozar, Samad M. E., and Leonhard, E. Bernold. (2012), Factors influencing the decision of technology adoption in construction. *ICSDEC 2012*, pp. 654–661, doi:10.1061/9780784412688.078.

Singh, Vishal, and Holmstrom, Jan (2015), Needs and technology adoption: observation from BIM experience. *Engineering, Construction and Architectural Management*, 22 (2), 128–150. doi:10.1108/ECAM-09-2014-0124.

Smith, Andrew (2012), Social media monitoring in Waddington, Stephen (ed), *Share This: The Social Media Handbook for PR Professionals*, John Wiley, London, pp. 157–162.

Solis, Brian (2012), *The End of Business as Usual*, John Wiley & Sons, Hoboken, NJ.

——— (2013), *The Conversation Prism* online at https://conversationprism.com/ (accessed: 5 June 2015).

Tumblr.com (2015), *Activity statistics* online at https://www.tumblr.com/about (accessed: 6 June 2015).

UK Government (2013), *Construction 2025: Industrial Strategy: Government and Industry in Partnership*, July 2013. https://www.gov.uk/government/uploads/system/uploads/attachment_data/file/210099/bis-13-955-construction-2025-industrial-strategy.pdf (accessed: 4 June 2015).

Waller, Andrew, and Bob Thompson (2009), *The Role of Social Media in Commercial Property*, Remit Consulting/RICS, London.

Wilkinson, Paul (2005), *Construction Collaboration Technologies: The Extranet Evolution*, Taylor & Francis, London.

——— (2009), 'Life at HOK', *pwcom.co.uk blog*—blog post, 16 March 2009, online at http://blog.pwcom.co.uk/2009/03/16/life-at-hok/ (accessed: 7 June 2015).

WordPress.com (2015), *Activity statistics* online at https://wordpress.com/activity/ (accessed: 6 June 2015).

Appendix: List of Social Networks Tools

Name	Description	Web Address
Baidu Tieba	Chinese communication platform	tieba.baidu. com
Blogger	Google's weblog publishing tool	blogger.com
Classmates	Social networking to reconnect with classmates	classmates. com
CrowdComfort	Mobile facility management app	crowdcomfort. com
Delicious	Social bookmarking service	delicious.com
Envision	Project delivery platform for infrastructure, energy and resources, and building projects	envisionapp. com
Facebook	Social networking site	facebook.com
FieldLens	Construction communication tool	fieldLens.com
Flickr	Yahoo's online photo management and sharing application	flickr.com
FourSquare	Location intelligence to build meaningful consumer experiences and business solutions	FourSquare. com
Friendster	Social gaming site	friendster.com
GenieBelt	Construction project management software and mobile app	GenieBelt.com
Google+	Google's social networking site	plus.google. com/
Hootsuite	Social media management system for brand management	Hootsuite.com
IFTTT	Connects apps	ifttt.com
Instagram	Online mobile photo-sharing, video-sharing, and social networking service	instagram. com
Issuu, Inc.	Free electronic publishing platform for magazines, catalogs, newspapers	Issuu.com
Jobsite Unite	Social collaboration tool for construction jobsites	jobsiteunite. com
Klout	A website and mobile app that uses social media analytics to rank its users according to online social influence	Klout.com
Last.fm	Online music catalogue with free music streaming, videos, photos, lyrics, charts, artist biographies, concerts and Internet radio	Last.fm
Line Corporation	A communication app for users to make free voice calls and send messages	LINE.me
LinkedIn	Business-oriented social networking service	LinkedIn.com
LiveJournal Inc.	Social networking service where Internet users can keep a blog	LiveJournal. com
Meerkat	Mobile app that enables users to broadcast live video streaming through their mobile device	Meerkatapp. co
MySpace	Social networking website offering an interactive, user-submitted network of friends, personal profiles, blogs, groups, photos, music, and videos	MySpace.com
Napster	Online music store	Napster.com
Periscope	Live video streaming app	Periscope.tv
Picasa	An integrated photo sharing tool to organize and edit digital photos	picasa.google. com

Name	Description	Web Address
Piczo	Social networking and blogging website for teens	Piczo.com
ProTenders	Platform to manage construction tendering activities	ProTenders.com
QZone	Social networking site that allows users to write blogs, keep diaries, send photos, listen to music, and watch videos	qzone.qq.com
Sina Weibo	Chinese microblogging (weibo) website	weibo.com
Sixdegrees.com	Social networking website. Discontinued in 2001	SixDegrees.com
SlideShare	Online slide sharing tool	SlideShare.com
Snapchat	Video messaging application	Snapchat.com
SpecifiedBy	Catalogues building products and materials through modern search, comparison and information management capability	SpecifiedBy.com
Spotify	A music streaming, podcast and video service	Spotify.com
Storify	Social media storytelling platform	Storify.com
Swarm	A location sharing app within social networks	Swarm.com
Stickyworld	Enables involvement of wider groups of people in projects	Stickyworld.com
Tencent QQ	Internet-based instant messaging (IM) software service from China	imqq.com
Tumblr	Yahoo's microblogging platform and social networking website	Tumblr.com
Typepad	Service for hosting and publishing weblogs and photo albums	TypePad.com
Viber	Instant messaging and VoIP app for various mobile operating systems	Viber.com
Vimeo	Video sharing website	Vimeo.com
Vine	Vine is a short-form video sharing service where users can share six-second-long looping video clips	Vine.co
VKontakte	One of the largest European online social networking service	VK.com
VisualCV	An online tool to create professional CVs	VisualCV.com
WeChat	A free messaging and calling app that allows users to connect with family and friends across countries	WeChat.com
Wikipedia	A free encyclopaedia built collaboratively using wiki software	Wikipedia.com
WhatsApp Inc.	A cross-platform instant messaging app	Whatsapp.com
WordPress	An open-source content management system	WordPress.com
Yammer	An enterprise social network for businesses	Yammer.com
YouCanPlan	Social networking for project reviews and consultation	youcanplan.stickyworld.com
YouTube	A video sharing website	YouTube.com
Zapier	A data integration and sharing website	Zapier.com

Social networks mentioned in this chapter (note: some older sites or apps have been discontinued).

16 Social media in construction: an exploratory case study

Srinath Perera, Michele Victoria and Samuel Brand

16.1 Introduction

Social media are considered as powerful tools to influence people as well as businesses in a short time span. Therefore, systematic use of social media in a business environment can bring greater benefits in no time. However, there are opposing views and criticisms around social media implementation within businesses. Nevertheless, many businesses are now active in social media platforms as means of marketing, recruiting people, improving brand image and so on. Yet, there are issues in implementing social media in a business environment as it involves devising proper guidelines and protocols for its usage and effective communication to the employees of the organisation. This becomes more complicated in the case of construction industry as it is fragmented and the industry is distinctly different compared to other industries. Therefore, this paper presents an exploratory study of a construction organisation in term of its social media implementation and use. It mainly focuses on the types of social media platforms used for personal, business and career development purposes; usage policy of the organisation; level of integration with business goals; significance of social media on various branches of business and barriers in implementation.

Social media can take different forms, each with different users, intended uses and unique features. It leaves the users with a specific or wholly personal experience. A simple definition of social media is, 'web-based sites where a conversation can take place between two or more people where, in short, there is the opportunity to exchange information' (Andrews, 2012).

There are multiple views on social media and how it should be defined. One of these is that all media is social, based on an information model. However, Fuchs (2014) confronts this definition and suggests that not all media are social, based on a communication model. The idea is that media becomes social only when communication takes place.

Furthermore, Ahlqvist *et al.* (2008) presents a distinct definition of social media as follows:

> Social media is built on three key elements: content, communities and Web 2.0. Content refers to user created content which may be of very different types; it may be photos, pictures or videos, but also presents information, tags, reviews and play-lists to mention some examples of this wide choice of input that people may create and publish on the Web.
>
> (Ahlqvist *et al.*, 2008)

The technology used to create and share the content within the network and allow people to participate is called the Web 2.0 technology which serves as the foundation of the whole social media platform development (see, O'Reilly, 2005).

16.2 Classifications of social media platforms

There are various forms of classifications of social media platforms. In an ownership perspective, Golden (2010) classifies social media platforms into two types such as: firm sponsored or individual publications (e.g. blogs) and third-party forums (e.g. Facebook, Twitter and LinkedIn).

Nevertheless, six types based on activities that are commonly agreed by most of the writers are as follows (Scott, 2014; Grahl, 2015; SEOPressor, 2015):

1   Social networks: Facebook (http://www.facebook.com), LinkedIn (http://www.linkedin.com), Google+ (http://www. plus.google.com)
2   Media sharing: YouTube (http://www.youtube.com), Instagram (http://www.instagram.com), Pinterest (http://www.pinterest.com), Flickr (http://www.flickr.com), Spotify (http://www.spotify.com)
3   Microblogging: Twitter (http://www.twitter.com), Tumblr (http://www.tumblr.com)
4   Blog Comments and Forums: Blogger (http://www.blogger.com)
5   Social news: Reddit (http://www.reddit.com)
6   Bookmarking site: Delicious (http://www.delicious.com), StumbleUpon (http://www.stumbleupon.com)

Social networking platforms allow users to connect to people all over the world and share messages and media files with personalised settings to each account. Media sharing platforms enable distribution of various media types including videos, audios and images. Microblogging platforms designed for concise posts where the user can update short messages and link with other sites. Blog Comments and Forum platforms engage users in a public conversation. Social news sites allow registered members to submit content from external sources. Finally, bookmarking sites enable users to save links, add bookmarks or tag webpages (Scott, 2014; Grahl, 2015; SEOPressor, 2015).

In addition, Myers (2012) identified following types where some examples might overlap with another category as well:

1   Publishing tools: WordPress (http://www.wordpress.com), Squarespace (http://www.squarespace.com)
2   Collaboration tools: Wikipedia (http://www.wikipedia.org), WikiTravel (http://www.Wikitravel.org), WikiBooks (http://www.wikibooks.org)
3   Rating/Review sites: Amazon ratings (http://www.amazon.com), Angie's List (http://www.angieslist.com), Trip Advisor (https://www.tripadvisor.com)
4   Personal broadcasting tools: Blog Talk radio (http://www.blogtalkradio.com), Ustream (http://www.ustream.tv), Livestream (http://www.livestream.com)
5   Virtual worlds: Second Life (http://www.secondlife.com), World of Warcraft (http://www.warcraft.com), Farmville (http://www.farmville.com)
6   Location based services: Check-ins, Facebook Places (http://www.facebook.com/places), Foursquare (http://www.foursquare.com), Yelp (http://www.yelp.com)
7   Widgets: Profile badges, Like buttons
8   Group buying: Groupon (http://www.groupon.com), Living Social (http://www.livingsocial.com), BoomStreet (http://www.boomstreet.com)

Out of several types identified above, most commonly used types in construction industry are social networks, media sharing and micro blogging platforms (Michaelidou *et al.*, 2011; Pauley, 2014; Whiston Solutions, 2015) mainly for business purposes. More details on social media usage in construction organisations are discussed later in this chapter.

16.3 Popular social media platforms and their origins

It can be said that millennium gave birth to many social media platforms. Figure 16.1 illustrates the timeline of the birth of various social media platforms. Among which Facebook, LinkedIn, Twitter and YouTube are being most popular sites accessed by almost all around the world (Bennett, 2014; Moreau, 2014; eBiz, 2015). Therefore, this section discusses the origins and techniques used in the development of these four popular platforms.

16.3.1 Facebook

Facebook is a social networking site founded by Mark Zuckerberg and Eduardo Saverin, in 2004. Initially, the platform was developed for the use of Harvard University students. Later the user circle was expanded step by

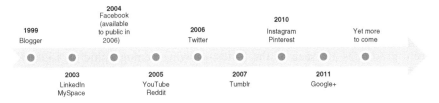

Figure 16.1 Timeline of birth of various social media platforms.

step. Finally, it was made available to everyone who is older than 13 years and has a valid email address by September 2006 (Wikipedia, 2015a). It allows users to connect with people all over the world by creating their own profiles to share messages, links, photos and also videos. It is being upgraded from time to time revising the outlook of the website and introducing new features to improve the user experience. Further, Facebook is recognised as the most popular social networking platform with over 1.44 billion monthly active users at present (Statista, 2015). The functionality of Facebook is based primarily on relationships where identity, presence, reputation and conversations are secondary criteria (Kietzmann *et al.*, 2011).

HipHop for PHP is a custom made source code transformer developed to build the platform. This was reported to reduce average CPU consumption by 50% on Facebook servers (Wikipedia, 2015a). However, this source code transformer was discontinued in 2013 and replaced by HipHop Virtual Machine (HHVM) which is a just-in-time compilation-based execution engine for PHP (Wikipedia, 2015b).

16.3.2 LinkedIn

LinkedIn is also a social networking site founded by Reid Hoffman, Allen Blue, Konstantin Guericke, Eric Ly and Jean-Luc Vaillant in 2002 and launched in 2003 specifically for developing professional networks (LinkedIn, 2015). LinkedIn profile allows the users to demonstrate their professional skills, abilities, achievements, publications and work experience. A well-presented profile has the potential to attract more connections and thereby brings a great deal of new opportunities leading to a stronger professional career. Earlier LinkedIn provided 'InMail' option where users can send messages to other users within the platform who are not connected with them. However, the option is now available only for premium accounts which require a monthly subscription for upgrading the ordinary account to a premium account. Kietzmann *et al.* (2011) identifies LinkedIn functionality is primarily based on Identity followed by relationships and reputation as performance criteria.

Currently, LinkedIn is reported to have over 364 million users out of which over 115 million is from United States of America (USA). It is also reported that in the first quarter of 2015 more than 75% of new members registered in LinkedIn from outside USA. Furthermore, Students and recent graduates are seen as the fastest-growing demographic on LinkedIn (LinkedIn, 2015).

16.3.3 Twitter

Twitter is a microblogging social media platform launched in 2006 by Jack Dorsey, Evan Williams, Biz Stone and Noah Glass. It allows users to share short messages (140 characters long) which are known as a 'tweets'. Therefore, it is known as 'the SMS of the Internet'. Distinctly different from Facebook and LinkedIn it allows unregistered users to also read others' tweets while only registered users can comment on tweets. Further, it is reported that Twitter currently has over 302 million active users out of over 500 million users (Wikipedia, 2015c). Its functionality is primarily focused on conversations followed by sharing, presence and relationships.

Initially, the Twitter web interface used Ruby on Rails framework which was then replaced by a Java server in 2011 due to high volume of activity. The tweets are stored in a MySQL database and then sent to the search engines via the Firehose API where the whole process said to be taking about 350 ms (Wikipedia, 2015c).

16.3.4 YouTube

YouTube is a media sharing website created in 2005 and acquired by Google in 2006. It enables users to view, upload and share video files. Registration is not mandatory as unregistered users can view videos except for the flagged videos and comments. However, to upload videos and comment on videos, registration is necessary (Wikipedia, 2015d). YouTube is ranked to be the most popular media sharing platform with over 1 billion unique visitors a month (Peansupap and Walker, 2006). Its functionality is based primarily on sharing videos followed by conversations, groups and reputation (Kietzmann et al., 2011).

Previously, YouTube required the Adobe Flash Player plug-in to be installed in a computer in order to watch videos. However, from 2010 YouTube capitalised on built-in multimedia features in HTML5 standard supported web browsers which does not require Adobe Flash Player (Wikipedia, 2015d)

Facebook, LinkedIn, Twitter and YouTube are accessible through a website interface as well as mobile applications. In addition, Facebook and Twitter provides SMS accessibility. Further, all these platforms are available in multilingual versions which allow the users to customise their profile.

16.4 Benefits of social media

Social media can provide a wide range of benefits as follows (Patten, 2007; Bradwell and Reeves, 2008; Salcido, 2011; Brown, 2012; Pauley, 2014):

- Are easily accessible and free to use
- Enable fast specialist feedback/better communication
- Have the potential to reach to a global audience
- Help improve networks
- Improve awareness of a brand
- Enhance knowledge on current issues and initiatives/innovations
- Facilitate knowledge management
- Have simplified functions and features and are user friendly.
- Drive traffic towards website
- Raise company profile
- Improve the online visibility of the organisation, the products and the services
- Allow consumers to engage with companies/organisations
- Improve work efficiency
- Increase PR opportunities with industry publications
- Monitor competitors and help to act on negative comments

Nevertheless, impact of social media on different business functions hugely varies. For instance, Smith *et al.* (2011) state that information technology, sales, marketing, services and human resources are highly affected; research and development and supply chain are moderately affected; and legal and finance are less affected by social media which is illustrated in Figure 16.2. The list of benefits presented above supports the figure.

Specialist on social media suggests that employees must be encouraged to use social media in work in order to gain the best output from it (Broughton *et al.*, 2010). However, this has to be done in a systematic way by devising online policy protocols. In addition, it is also believed that a successful social business strategy will integrate social media into business goals and

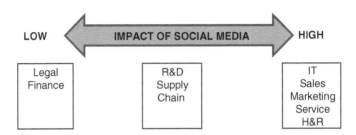

Figure 16.2 Impact of social media on business functions.
Source: Smith *et al.* (2011).

objectives (Myers, 2012; Altimeter, 2015; Li and Solis, 2015). This calls for sound usage guidelines, protocols and awareness among the employees.

16.4.1 Drivers of social media

Drivers of social media are identified and short listed by Ahlqvist *et al.* (2008) through a workshop and 12 key drivers in the order of significance according to the study are:

- Changing cultural environment – cultural changes influence work life and drive social media usage.
- Commercial and customer orientation – customer requiring transparency from businesses, client participate in product developments, quick feedback and the like are facilitated through social media.
- Networking – businesses are evolving from individual to relationship base intimacy which demands communication based tools.
- Human resources – fragmentary working life, attitude of youngsters, work not being tied to place or time, entrepreneurship at some level all drives the use of social media.
- Accessibility of technology – accessible cost and increased technology literacy are major reasons for the reach of social media.
- Environmental challenges – reduced travelling (reduced emissions and energy consumption) and connectivity through social media.
- Social responsibility and sustainable development – social media as a new bottom-up communication channel in all fields of sustainable development.
- International markets and competition – dynamics, remigration and outsourcing and the like demands presence in social media.
- Segmented needs of diverse customer sectors – where all customers cannot be served individually social media comes into play.
- Aging – extended working age with virtual communications requires the technology.
- Outsourcing (subcontracting) – social media facilitates scattered working models and ad-hoc working.
- Development of technology – technology development serves as an enabler of wide social media usage.

These factors are further grouped into major six categories such as societal and cultural change (changing cultural environment, networking and aging); business and clients (commercial and customer orientation and segmented needs of diverse customers); science, technology and innovations (accessibility of technology, development of technology); sustainable development (environmental challenges, social responsibility and sustainable development); work life (human resources and outsourcing); and markets (international markets and competition).

Table 16.1 Drivers and barriers of social media implementation in business environment

Drivers	Barriers
Efficiency requirements	Information security
Hectic working culture	Possibilities of headhunting
Rising demands to combine work and leisure	Perceived as waste of working time

16.5 Social media in different industries

16.5.1 Drivers and barriers of social media usage in business environments

In addition to the drivers identified by Ahlqvist *et al.* (2008) above, drivers and barriers in a work related environment are also identified as listed in Table 16.1. Accordingly, findings suggest that tightening of efficiency requirements and hectic working culture drives multiple parallel meetings and communication channels to be used in the work place. However, organisation culture is often considered to be a barrier due to concerns about information security, possibilities of headhunting and regarding it as a waste of time. These resist social media usage and implementation in a working environment.

Nevertheless, following case studies present positive outcomes of successful social media implementation within different industries.

16.5.2 Health and beauty industry

Famous beauty company L'Oreal has been reported to be using all forms of social media to the attract job seekers. Further, company's vacancies webpage is linked directly to the company profiles in Facebook, LinkedIn and twitter. In addition to that L'Oreal provides a link in their Facebook profile for an internship programme which is linked to an application called 'Work for Us'. This targets candidates and allows collation of important information. Through this method L'Oreal was able to reach inactive candidates as well as improved their brand. It was also reported that L'Oreal Australia saved nearly 20,000 Australian dollars in recruitment process (LinkedIn, 2012; Szkolar, 2012).

The above example of L'Oreal (LinkedIn, 2012; Szkolar, 2012) showcases effective recruitment process through social media platforms which helps to identify potential candidates at a lower cost and time than the conventional method of recruitment.

16.5.3 Food and drink industry

Coca-Cola is a world famous drink company and well-known for its brand. Ahuja *et al.* (2009) reported that Coca-Cola to have over 63 million fans to its main Facebook page, over 700,000 followers in Twitter and also seemed to be active in Pinterest and Google+. This also reveals that Coca-Cola is able to attract more fans and followers and improve their brand image by being active in social media.

Coca-Cola example conveys the message that social media serve as good marketing tools and help improve brand image of businesses. However, Myers (2012) points out that small businesses feel overwhelmed by various social media platforms and as a result, there is a tendency of being active in all platforms which might not be beneficial (Myers, 2012).

16.5.4 Academia

Now social media are also popular among scholars. The identified key benefits of social media to scholars includes: keeping up-to-date with topics, following other researchers' work, discovering new ideas or publications, promoting current work/research, making new research contacts, personal management and impact measurement (Szkolar, 2012; Coleman, 2013; Vasquez *et al.*, 2015). Platforms like Google Scholar, Academia.edu, Research Gate, Mendeley and ImpactStory are a few examples of academic social media platforms. However, due to the credibility of the information present in social media sometimes users refrain from approaching it for academic purposes. Nevertheless, Szkolar (2012) highlights that social media have a role to play within the academic world though it is too early to predict whether scholarly social media sites will become a norm.

16.6 Social media in the construction industry

Construction industry is a distinctly different industry compared to manufacturing and other similar industries. There are no consumers as such but building users and building clients who ensure the construction achieves required specification and standards. There is a heavy competition in the industry with greater price competition, lower profits and possible inflated prices. As a result, construction companies strive for greater competitive advantage than ever before.

The Kotrlik and Higgins (2001) reported that up to 77% of companies in information and communication sector are leaders in social media adoption in 2012 whereas construction businesses reported to be the lowest with 20% of implementation. It is common to note that construction industry is often criticised for its slow adoption of ICT and innovation. However, a different survey of top 15 construction companies in the UK reported that over 90% of the construction companies holding a Twitter account and LinkedIn account and 65% having a company Facebook account. Further, more than half of the companies have linked their website with social media accounts while all having the capability to analyse Internet traffic (Pauley, 2014). In addition to that another study reported that the most popular social platform used by construction professional is LinkedIn with approximately 91% users. Twitter captures the second place with 84% users followed by Facebook, YouTube, Blogs, Google+ and Pinterest with 83%, 68%, 47%, 40% and 26% respectively (Whiston Solutions, 2015).

Even though many studies suggest that large construction organisations are now becoming active in social media platforms, majority of the construction industry is composed of Small and Medium Enterprises (SME). Therefore, it is important to study the usage of social media within SMEs. Michaelidou *et al.* (2011) reported about the usage of social media in SMEs in the UK. Findings of Michaelidou *et al.* (2011) suggest that 27% of SMEs use social media for business purposes, while Facebook is identified as the most popular social media site with 77% SMEs using it followed by Twitter (55%), and LinkedIn (46%). Therefore, social media usage is evident in both large and SME construction organisations, with comparatively lower usage within SMEs.

Furthermore, social media adoption in construction industry is expected to add many benefits. This is reflected in CIC Group construction industry survey 2014. The survey findings suggest with majority of respondents agreeing that social media could improve the image of construction industry. However, it was also noticed that respondents had mixed views about the role of social media in construction. While some feel that social media should not be encouraged in a working environment as it is highly unregulated, unmanageable and perceived to be of high risk to reputation, others think that it is a good medium of marketing, attracting younger generation and disseminating knowledge through interactive forums, video blogs and the like. Nevertheless, CIC presumes that there is a huge scope for potential development of social media in construction that could facilitate collaboration, innovation and promotion (Construction Industry Council, 2014).

One of the important benefits of social media in construction industry could be knowledge management. Due to the fragmented nature of the industry knowledge management becomes even more important in construction industry (Dave and Koskela, 2009). Hence, social media could be a cost effective tool to manage knowledge within the industry. This is even more evident in case studies presented by Porkka *et al.* (2012) that demonstrates social media usage improves knowledge management and extends the use of traditional data in urban planning and construction projects.

Despite the benefits, reluctance of top management, lack of understanding and encouragement from top management (Brown, 2012), unfamiliarity of staff/lack of technical skills in handling social media platforms, perceived unimportance, lack of use by the competitors (Michaelidou *et al.*, 2011) are reported to be barriers of social media implementation in construction organisations.

In addition, some key points to consider when implement social media in a construction business are as follows (Pauley, 2014; Li and Solis, 2015):

- Commitment: poorly updated profile is considered to be more harmful than being present in social media platforms. Hence, decision to be made on what social media platforms that the company chooses to be present in and how it will be updated and how will it make a difference in the business.
- Having a solid social strategy: it is hard to believe sometimes that social media can reap greater benefits; however, many companies have started implementing it in a pro-active manner. Hence, it is important to pay attention to this new business strategy which will most likely influence the business to a greater extent in the near future.
- Know the strategy: it is important that social media usage does not conflict with business strategy and hence awareness of what is fed in to the social media is crucial. Therefore, it is important employees of the organisation are aware of this.
- Remain open-minded: many people assume that social media can only serve as a marketing tool while there are many other potential benefits identified.
- Measure and act: measuring the impact of social media implementation is important to take crucial decision about continuation and improvement. If social media implementation is not bringing any benefit to the company then no point in spending time and effort in maintaining.
- Allocate time and right people: it is important that a person is delegated for maintaining social media platforms for the company who understands the organisation's strategy and goals. Otherwise there could be misuse of the platform which can impact the organisation brand.
- A sound social media usage guideline and protocol to govern the usage.
- Executive support: this is considered to be extremely important and affects each point discussed above.

Nevertheless, there is a lack of reported case studies on social media implementation in construction organisations. Therefore, this chapter presents a case study of a construction organisation to understand social media implementation within the construction context.

16.7 Case study

16.7.1 Method

Case study method was adopted to explore the state of use of social media in a construction organisation as it allows an in-depth exploration (Fellows and Lieu, 2003) and holistic inquiry to be carried out to investigates a contemporary phenomenon within its natural setting (Harling, 2002).

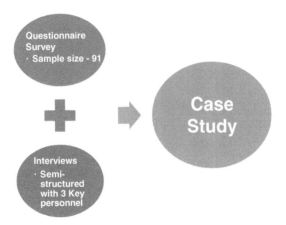

Figure 16.3 Research method.

The case study employed both qualitative and quantitative approaches as illustrated in Figure 16.3. While quantitative approach provides an overview of the state of organisation in terms of social media implementation, qualitative approach helps to gain deeper understanding of the situation. Initially, a questionnaire survey was conducted among the employees to understand employees' perspectives on social media and usage patterns. A survey sample size of 91 for a large organisation over 900 employees enables findings to achieve over 90% confidence level (Kotrlik and Higgins, 2001). Then, semi-structured interviews were conducted with three key personnel from the case study organisation to get insights into the survey findings and explore on other issues.

Key themes and patterns were identified through content analysis of interviews and questionnaire results were analysed using descriptive statistics. Finally, conclusions were arrived based on both the findings and comparisons made with the literature findings.

16.7.2 Case study

The case study organisation is a large Professional Quantity Surveying (PQS) organisation registered under the Royal Institution of Chartered Surveyors (RICS) with over 900 employees which can be considered to be a representative sample of large PQS organisations in the UK.

16.7.3 Survey findings

Survey findings are analysed in this section. Figure 16.4 illustrates the age profile of the respondents. Accordingly, more respondents lie between 20 to 29 age group (38%), representing younger generation in the sample. Also it is clear that sample represents fair proportion of employees in all age groups.

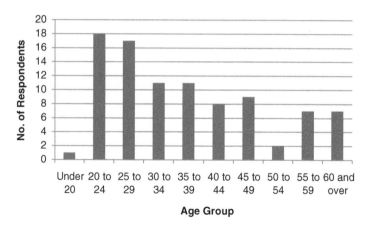

Figure 16.4 Respondent age profile.

16.7.3.1 Usage of social media within the organisation

16.7.3.1.1 PERSONAL PURPOSE

Figure 16.5 illustrates the usage of social media for personal purposes. Unsurprisingly, Facebook is the most used platform by respondents for personal purposes followed by LinkedIn and YouTube. Further, Flickr was

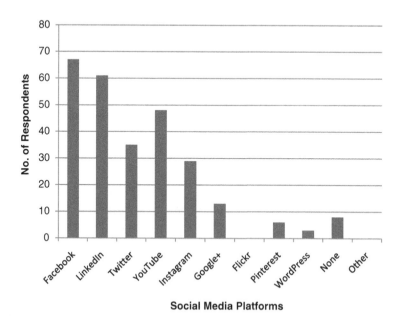

Figure 16.5 Usage of social media – personal purpose.

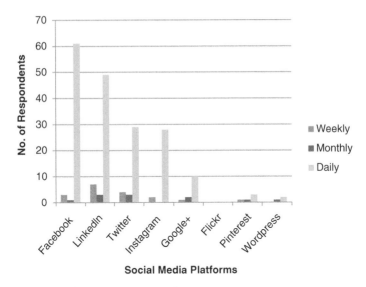

Figure 16.6 Usage frequency – personal purpose.

unpopular among the respondents resulting in non-usage within the case study organisation. Further, findings suggest that 91% of the respondents use social media platforms for personal purposes.

Figure 16.6 illustrates the frequency of usage of each platform. Findings suggest that the prominent social media platforms are frequently used. Especially, Facebook (67%) and LinkedIn (54%) are used on a daily basis by majority of the respondents while Twitter, Instagram, Google+, Pinterest and WordPress are not used by the majority. Furthermore, it is identified that 74% of respondents use social media platforms on a daily basis.

In addition, there is a claim that younger generation is considered to be active in Facebook than older generation though now older generations are also slowly picking up. Therefore, it is interesting to study the impact of age in social media usage. Figure 16.7 demonstrate the usage pattern of social media for personal purpose by age. It clearly illustrates that the younger generation uses social media mostly for personal purposes, especially, Facebook. Social media use for personal purposes declines as the age increases. Noticeable decrease is evident with Facebook usage. Again LinkedIn and Twitter seems to be popular among youngsters as well as seniors. Another important distinction is that as the age increases the diversity of social media platforms used decreases, where only two to four platforms are accessed out of the eight identified platforms by employees aged above 45.

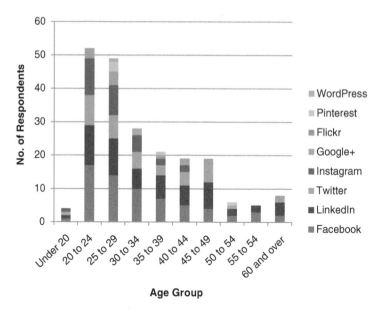

Figure 16.7 Social media for personal use by age.

16.7.3.1.2 BUSINESS PURPOSE

Figure 16.8 depicts usage of social media for business purposes. Accordingly, majority of the respondents use only LinkedIn for business purposes on a daily basis (see Figure 16.9). On the other hand, Twitter, Facebook, YouTube and google+ are also used by a few. Further, it can be observed that the use of social media for business purposes is relatively lower than personal usage resulting in only 56% of the respondents using it for business purposes indicating limited application of social media for business functions/purposes.

Figure 16.10 represents social media usage for business purposes by age category. It suggests that age groups of 20 to 24, 25 to 29 and 45 to 49 use social media the most for business purposes. However, an ambiguity arises as to what respondents perceive as business development as the age group between 20 and 29 is less likely to influence business development compared to senior people.

16.7.3.1.3 CAREER DEVELOPMENT

Figure 16.11 depicts usage of social media for career development. As expected LinkedIn is the dominant platform used by majority of the respondents (71%) while few others use Twitter, YouTube, Instagram and Google+ as means of career development. On the other hand more than a quarter of the respondents hardly use any platform for career development which is surprising.

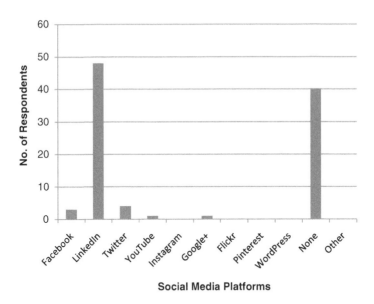

Figure 16.8 Usage of social media – business purposes.

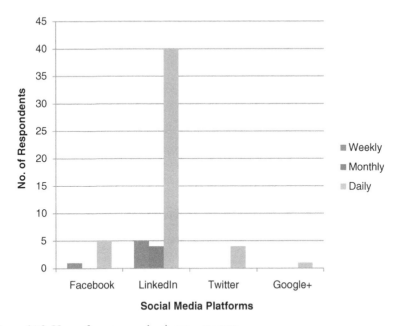

Figure 16.9 Usage frequency – business purposes.

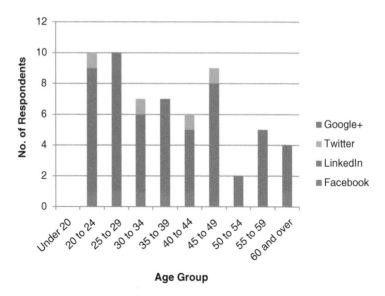

Figure 16.10 Social media for business development by age.

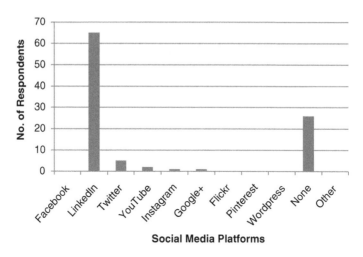

Figure 16.11 Usage of social media – career development.

Figure 16.12 illustrates the frequency of usage of LinkedIn and Twitter for career development. Accordingly, LinkedIn is accessed by the majority (54%) while Twitter is only accessed by only 8% on a daily basis. This reflects that personal use of social media overtaking usage for career development. This indicates a poor utilisation of social media platforms for professional and career enhancement which is not healthy in a business environment.

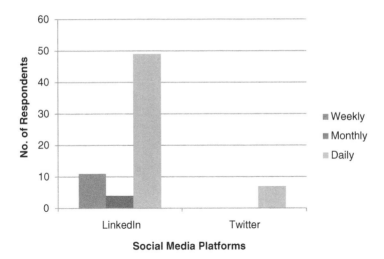

Figure 16.12 Usage frequency – career development.

Figure 16.13 conveys the message that employees at the age of 20 to 34 are very keen on using LinkedIn for personal development while people at the age over 55 also demonstrate fair usage of the platform. The reason for younger people being more active in LinkedIn platform is because they are in the stage of developing their career and looking for attractive opportunities to gain more valuable experience. Nevertheless, popularity of LinkedIn among senior professionals also showcases the increasing significance of social media in career development.

16.7.3.2 Social media usage policy within the organisation

Literature suggests that the use of social media brings huge benefits to the organisation if employees of the organisation are encouraged to use. Hence, the survey intended to capture whether it is encouraged within the case study organisation and majority (71%) confirmed that it is not encouraged while other respondents suggest that it is encouraged. Another important fact is that half of the respondents (50%) confirmed that they are unaware of the internet usage policy set by the employer which is a threat as literature suggests as there should be protocols in place to govern social media usage in a business environment. This is a key problem in many organisations which needs to be given serious attention where social media usage is allowed and encouraged. The knowledge of protocol and boundaries of usage is very important to prevent any damage to the brand image and to maintain the reputation of the organisation.

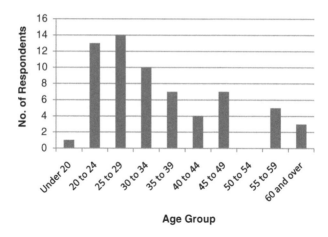

Figure 16.13 Usage pattern of LinkedIn by age.

16.7.3.3 Integration of social media with business goals and objectives

It is important that social media are integrated with the goals and objectives of the organisation for its successful implementation. However, responses suggest lack of understanding among employees in this regard resulting in majority agreeing on 'neither integration nor segregation' which is illustrated in Figure 16.14. It cannot be concluded form the findings whether social media are integrated or segregated form business goals and objectives, however, findings demonstrate lack of clear understanding among employees and lack of interest of top management to disseminate the social media implementation strategy.

16.7.3.4 Significance of social media in business activities

Social media are considered important in an organisation's activities for:

- Business development
- Knowledge management
- Marketing/advertising
- Competitive advantage
- Brand and industry awareness

Figure 16.15 illustrates importance of social media in various activities of the case study organisation from employees' perspective. Accordingly, employees consider that social media are most important to gain competitive advantage and knowledge management, followed by business development,

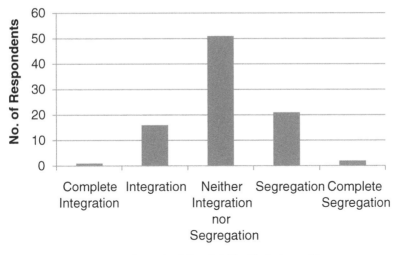

Level of Social Media Integration

Figure 16.14 Level of integration of social media with organisation goals and objectives.

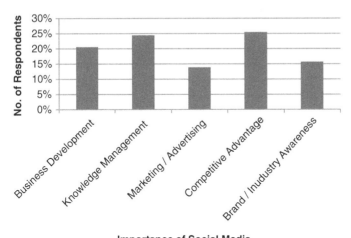

Importance of Social Media

Figure 16.15 Importance of social media for the case study organisation.

brand or industry awareness and marketing. Findings demonstrate evolving role of social media from primarily being a marketing tool to other areas. It is interesting to note that employees scored high for knowledge management. Therefore, the role of social media in organisation's knowledge management could be an interesting area to explore further.

Table 16.2 Interviewee profile

Interviewee	Industry experience
X1	20 years
X2	3 years
X3	1 year

16.7.4 Interview findings

Interviews were conducted with key personnel of the organisation with varying experiences which is listed in Table 16.2.

It is clear that the younger generation (20–29) demonstrates higher social media usage and awareness compared to the older generation. Hence, interviewees with varying experience tend to provoke different views about social media implementation in construction.

16.7.4.1 Use of social media within the organisation

The organisation maintains accounts in various social media platform such as: Blogs, LinkedIn, Twitter, YouTube and Facebook. All interviewees were aware of the active social media platforms used by the organisation partly as a result of email links being circulated among employees frequently about the news feeds. Further, the organisation is considered to be active in feeding information into the platforms mentioned above. This indicates the influence of social media in modern day businesses.

Interviewees feel that social media are a good tool that facilitates communications. It is also considered to be an influential marketing tool. Nevertheless, social media are perceived to be surrounded by negative stigmas making it difficult to satisfy everyone in the organisation, especially top management, regarding its usage and implementation within the organisation.

16.7.4.2 Usage of social media by interviewees

Usage of social media within the organisation is categorised in to three types namely:

- Personal purposes
- Business purposes
- Career development purposes

16.7.4.2.1 PERSONAL PURPOSES

It was interesting to note that Twitter is used by all three interviewees for personal purposes while one interviewee commented that Twitter is to be the most used platform among others. Second most used platform is Facebook, followed by Instagram and YouTube. Key fact to note here is that the interviewee with the most experience is using only Twitter for personal purposes while other two interviewees reported to have used other platforms.

While interview findings suggest that Twitter is the most used platform for personal use, survey demonstrates that Facebook is used mostly for personal purposes. This is because Facebook has become a viral in the present world attracting both younger and older generations. However, in a business environment this can have undesirable effects as well.

16.7.4.2.2 BUSINESS PURPOSES

Both YouTube and LinkedIn are used for business purposes and only used by the senior interviewee. To some extent this indicates that business feeds are preferred to be dealt by senior personnels of the organisation.

Similarly, survey also suggested that LinkedIn to be the dominant platform used for business purposes while Twitter, Facebook, YouTube and google+ are also used at a very lower rate.

16.7.4.2.3 CAREER DEVELOPMENT

All three interviewees identified Twitter and LinkedIn to serve as career development platforms. Survey also suggests that LinkedIn to be the dominant platform while Twitter, YouTube, Instagram and Google+ are also used for career development.

LinkedIn has become a very popular platform for job hunting where well-presented profiles attract competitive opportunities, leading to career development.

16.7.4.3 Social media usage protocol

Even though a clear record of social media platform usage in the case study organisation is evident it was noted that not all interviewees are aware of the protocol for social media usage within the organisation. The senior interviewee confirmed that there is a section specifically on social media usage protocol while another interviewee claimed that it should be made transparent and accessible to employees so that they are aware of it. Survey findings support this claim as half of the respondents stated that they are unaware of the usage policy. This conveys a message to the management that an effective communication is required in this regard. Further, this is a serious issue in the case study organisation which calls for a rapid action.

16.7.4.4 Implementing social media in construction organisations

Positive views on social media implementation within construction organisations are reported as it is believed that social media help to connect targeted audiences like clients, contractors as well as prospective employees and encourage co-operation within groups. In addition, social media,

- Is easily accessible and free
- Enable fast specialist feedback/better communication
- Has potential to reach to global audience
- Helps improve networks

However, one interviewee stated that social media are more suitable for other industries than construction as construction businesses are not really influenced by promotions. But potential benefits of social media are evident through survey findings in other branches of business like business development, knowledge management, and marketing.

Particularly, in the case study organisation social media help to improve the status of the organisation, the brand and help to manage knowledge.

Despite the advantages and financial benefits that social media bring in to the construction organisations, some barriers are also identified by the interviewees in social media implementation in construction businesses and identified below as follows:

- Age of employees – older generation lags behind the rapidly developing technology and thus, resists change in the organisation.
- More personal usage – use of social media for personal purposes rather than business purposes during business hours results in reduction of employee productivity.
- Lack of control – difficulty in exercising control over the usage.
- Conflict of interest – personal opinions may conflict with business interests.
- Lack of encouragement to use – lack of support from top management due to anticipated negative impacts. Sometimes by social media usage there is a risk of devaluing corporate image.

16.8 Discussion

16.8.1 Popular social media platforms among construction organisations

Table 16.3 compares the findings of the study with other studies which demonstrates a significant difference in some areas. Findings of this study exclude social media usage for personal purposes while Whiston Solutions (2015) does not clearly state the scope of the analysis. It is clear that LinkedIn is identified as the most popular social media platform by most of the studies. While Twitter is identified as another most popular platform, the study does not support the fact. Except for LinkedIn all other platforms lack popularity within the case study organisation. The reason

Table 16.3 Comparison of results of the study with other studies

	The study	Pauley (2014)	Whiston Solutions (2015)	Michaelidou et al. (2011)
	Construction professionals of the case study organisation (%)	Top 15 construction companies in the UK (%)	Construction professionals (%)	SME's in the UK (%)
LinkedIn	71	90	91	46
Twitter	5	95	84	55
Facebook	3	65	83	77
YouTube	2		68	
Blogs	0	Not analysed	47	Not analysed
Google+	1		40	
Pinterest	0		26	

for this could be that LinkedIn is more professionally aligned compared to other platforms.

Findings and literature review suggest that, generally, LinkedIn, Twitter and Facebook are being the most popular social media platforms within construction organisation. Nevertheless, YouTube, blogs, google+, Pinterest and Flickr are also used by a smaller percentage.

16.8.2 Drivers and barriers of social media implementation in construction organisation

Based on the review and findings, a list of barriers and drivers of social media implementation in construction industry are identified and grouped into different categories depending on the relevance of each factor.

Accordingly, barriers are categorised into three major types namely: employee related; management related and others. Similarly, drivers are categorised into six major types such as: societal and cultural change; business and clients; science, technology and innovation; sustainable development; work life; and markets which are depicted in Figure 16.16. Closer look at barriers suggests that most of these could be overcome by the top management as the power lies in the hands of the management. If the management decides to implement social business strategies then employee related barriers could be managed through proper training and monitoring of activities.

However, it is important to remember key points highlighted by Pauley (2014) and Li and Solis (2015) when implementing social business strategy including: commitment, having a social strategy, knowing the strategy, remain open-minded, measure and act, allocate time and right people, a sound social media usage guide and protocol and executive support. More importantly, a successful social business will have social media integrated

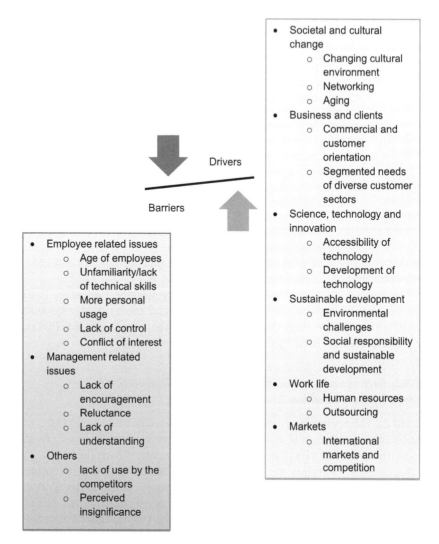

Figure 16.16 Drivers and barriers of social media implementation in construction organisation.

with business goals. When all these are combined together construction organisations can reap huge benefits through use of social media.

16.9 Conclusions

Social media are considered to be powerful tools due to its capability of enormous information transmission. It is also accessible by everyone mostly at no cost and sometimes at a low cost. Social media can be used for personal purposes, business purposes and career development purposes. It is clear that social media have influenced business activities to a great extent. As a result, many businesses are active in social media and maintain a company account in more than one platform. Few case studies witness the positive impact of social media implementation within a business environment despite the barriers. However, in a construction context social media implementation is still in its early stages of development. The case study of a leading construction firm in the UK revealed that the organisation is active in few popular social media platforms including: Blog, LinkedIn, Twitter, YouTube and Facebook. Facebook and Twitter are the most commonly used platforms by the employees for personal purposes while LinkedIn is mostly used for business purposes and career development. This conforms to the performance criteria identified by Kietzmann *et al.*, (2011) for social media viz identity. Further, it was identified that usage of social media within the organisation is not encouraged while the survey responses conveyed some form of unawareness of the usage policy of social media within the organisation. This could be a threat to the organisation's activities because unawareness of usage policy can lead to abuse of social media usage within a business environment. The information regarding protocol and boundaries of usage is crucial to prevent damages to the brand image of construction businesses and maintain the reputation of the organisation. As literature suggests integration of social media with business goals is important for a successful social business strategy. However, the case study organisation does not demonstrate a clear picture of its state in this aspect and lack of understanding about social media implementation strategy among employees was evident in the case study. This indicates poor communication of social media usage strategy within the organisation. However, significance of social media in business development, knowledge management, marketing/advertising, competitive advantage and brand and industry awareness are recognised by the employees. Despite the benefits, few other barriers are reported with regards to the social media implementation in the case study organisation which includes: age of employees, more of personal usage, lack of control, personal opinions conflicting with business messages and lack of encouragement to use. However, it is believed that social media can take the construction business to the next level if the barriers are overcome.

References

Ahlqvist, T., Bäck, A., Halonen, M. and Heinonen, S., 2008. *Social Media Roadmaps: Exploring the futures triggered by social media.* Helsinki: Edita Prima Oy.

Ahuja, V., Yang, J. and Shankar, R. 2009. Study of ICT adoption for building project management in the Indian construction industry. *Automation in Construction,* 18, 415–423.

Altimeter. 2015. *Social business strategy.* Available: http://www.altimetergroup.com/work-with-us/challenges-we-solve/social-business-strategy/. [Accessed 17 May 2015].

Andrews, C. 2012. Social media recruitment. *Applied Clinical Trials,* 21, 32–42.

Bennett, S. 2014. *The 13 most popular social networks (by age group).* Available: http://www.adweek.com/socialtimes/popular-social-networks-age/502497 [Accessed 20 May 2015].

Bradwell, P. and Reeves, R. 2008. *Network Citizens Power and Responsibility at Work,* London, Demos.

Broughton, A., Higgins, T., Hicks, B. and Cox, A. 2010. *Workplaces and Social Networking: The Implications for Employment Relations,* Brighton, The Institute for Employment Studies.

Brown, M. 2012. *Why the construction sector should engage with social media.* Available: http://www.theguardian.com/sustainable-business/construction-sector-social-media [Accessed 17 May 2015].

Coleman, V. 2013. Social media as a primary source: a coming of age. *EDUCAUSE Review,* 48, 60–61.

Construction Industry Council. 2014. *CIC 2050 Group Construction Industry Survey 2014,* London, CIC 2050 Group.

Dave, B. and Koskela, L. 2009. Collaborative knowledge management—A construction case study. *Automation in Construction,* 18, 894–902.

eBiz. 2015. *Top 15 most popular social networking sites|May 2015.* Available: http://www.ebizmba.com/articles/social-networking-websites. Accessed on 20 May 2015.

Fellows, R. and Lieu, A. 2003. *Research Methods for Construction,* UK, Blackwell Publishing.

Fuchs, C. 2014. Social Media: A Critical Introduction. London: Sage.

Golden, M. 2010. *Social Media Strategies for Professionals and Their Firms: The Guide to Establishing Credibility and Accelerating Relationships,* Hoboken, NJ, John Wiley & Sons.

Grahl, T. 2015. *The 6 types of social media.* Available: http://timgrahl.com/the-6-types-of-social-media/ [Accessed 17 May 2015].

Harling, K., 2002. Workshop. *Annual meeting of the American Agricultural Economics Association, Case studies: their future role in agricultural and resource economics.* 27July 2002. California.

Kietzmann, J. H., Hermkens, K., McCarthy, I. P., and Silvestre, B. S. 2011. Social media? Get serious! Understanding the functional building blocks of social media. *Business Horizons,* 54(3), 241–251.

Kotrlik, J. and Higgins, C. 2001. Organizational research: determining appropriate sample size in survey research appropriate sample size in survey research. *Information Technology, Learning, and Performance Journal,* 19, 43.

Li, C. and Solis, B. 2015. *The 7 success factors of social business strategy.* Available: http://www.briansolis.com/2015/04/7-success-factors-social-business-strategy-infographic/ [Accessed 17 May 2015].

LinkedIn. 2012. *Case study L'Oreal.* Available: http://www.slideshare.net/linkedineurope/loreal-case-studyv5 [Accessed 17 May 2015].

———. 2015. *About LinkedIn.* Available: https://press.linkedin.com/about-linkedin [Accessed 17 May 2015].

Michaelidou, N., Siamagka, N. T. and Christodoulides, G. 2011. Usage, barriers and measurement of social media marketing: an exploratory investigation of small and medium B2B brands. *Industrial Marketing Management,* 40, 1153–1159.

Moreau, E. 2014. *Top 15 social networking sites you should be using.* Available: http://webtrends.about.com/od/socialnetworkingreviews/tp/Social-Networking-Sites.htm. [Accessed 20 May 2015].

Myers, A. 2012. *13 types of social media platforms and counting.* Available: http://decidedlysocial.com/13-types-of-social-media-platforms-and-counting/ [Accessed 17 May 2015].

O'Reilly, T. 2005. *What is Web 2.0.* Available: http://www.oreilly.com/pub/a/web2/archive/what-is-web-20.html. [Accessed 20 May 2015].

Patten, J. 2007. *Blogging in the workplace: building a safe culture.* Available: http://www.personneltoday.com/hr/blogging-in-the-workplace-building-a-safe-culture/ [Accessed 17 May 2015].

Pauley, N. 2014. *How the top UK construction companies are using social media marketing in 2014.* Available: http://www.pauleycreative.co.uk/2014/01/how-the-top-construction-companies-are-using-social-media-in-2014/ [Accessed 17 May 2015].

Peansupap, V. and Walker, D. H. 2006. Information communication technology (ICT) implementation constraints: a construction industry perspective. *Engineering, Construction and Architectural Management,* 13, 364–379.

Porkka, J., Jung, N., Päivänen, J., Jäväjä, P. and Suwal, S. 2012. Role of social media in the development of land use and building projects, *Proceedings of ECPPM 2012,* Iceland, Reykjavik, July 25–27th.

Salcido, M. 2011. *Benefits and advantages of using social media\advantages of social media.* Available: http://www.organicseoconsultant.com/advantages-of-using-social-media/ [Accessed 17 May 2015].

Scott, M. 2014. *Understanding the basic categories of social media marketing.* Available: https://blog.ahrefs.com/understanding-basic-categories-social-media-marketing/ [Accessed 17 May 2015].

SEOPressor. 2015. *The 6 types of social media.* Available: http://seopressor.com/social-media-marketing/types-of-social-media/ [Accessed 17 May 2015].

Smith, N., Zhou, C. and Wollan, R. 2011. *The Social Media Management Handbook: Everything You Need to Know to Get Social Media Working in Your Business,* Hoboken, NJ, John Wiley & Sons.

Statista. 2015. *Number of monthly active Facebook users worldwide as of 1st quarter 2015 (in millions).* Available: http://www.statista.com/statistics/264810/number-of-monthly-active-facebook-users-worldwide/ [Accessed 17 May 2015].

Szkolar, D. 2012. Social networking for academics and scholars. *Information Space.* Available: http://infospace.ischool.syr.edu/2012/06/21/social-networking-for-academics-and-scholars/ [Accessed 10 February 2016].

Vasquez, E., Karely, F., Bastidas, C. and Enrique, C. 2015. Academic social networking sites: a comparative analysis of their services and tools. *iConference 2015,* California, USA, 24–27 March 2015.

Whiston Solutions. 2015. *Social media benefits for the property and construction industry.* Available: http://www.whiston-solutions.com/industry-developments/social-media-benefits-property-and-construction-industry [Accessed 17 May 2015].

Wikipedia. 2015a. *Facebook.* Available: https://en.wikipedia.org/wiki/Facebook [Accessed 17 May 2015].

———. 2015b. *HipHop for PHP.* Available: http://en.wikipedia.org/wiki/HipHop_for_PHP [Accessed 17 May 2015].

———. 2015c. *Twitter.* Available: https://en.wikipedia.org/wiki/Twitter#cite_note-10 [Accessed 17 May 2015].

———. 2015d. *YouTube.* Available: https://en.wikipedia.org/wiki/YouTube [Accessed 17 May 2015].

17 Mobile computing applications within construction

Zeeshan Aziz, Aizul Harun and Naif Alaboud

17.1 Introduction

In recent years, the information and communications technologies (ICT) have extensively been applied to address communication and collaboration challenges and to enhance productivity, performance and quality in design and delivery of construction projects. However, the construction industry is still lagging behind other industries such as aerospace and automobile, in terms of deployment of more advanced ICT systems for communication and integrated design and manufacturing. Construction clients are increasingly demanding shorter construction cycle times, access to up-to-date information, better value for money and improved quality. Recent advances in Information Communication Technologies in general and mobile technologies in particular offer new opportunities. Samuelson and Björk (2014) highlighted areas of information technology (IT) investment growth within building industry, including document handling, portable equipment/mobile systems, computer aided design systems, Information search via the Internet, accounting systems, project management, systems for technical calculations, project webs (EDM), product models/building information models (BIM), systems for costing/cost control, e-commerce, new business models and virtual reality.

Given the fact that construction sites comprise dynamic and unstructured work environments and hazardous work settings, it necessitates the use of intelligent ways to support mobile construction workers. In recent years, there have been significant developments in technologies such as information systems, cloud computing, wireless networking and mobile technologies, which can assist to better integrate mobile construction workforce. Recent advances in mobile information technology include improved wireless bandwidth, better quality of service, low cost, higher processing power and battery life, and hardware to support real-time connectivity. These advances enable better data sharing, which has become a trigger for growth in mobile computing.

This chapter reviews evolution and key drivers of mobile computing within construction. This is followed by a discussion on the need for integration between various project stakeholders is highlighted, to ensure smooth flow of information between project participants. Relevant research in the area and case studies on how various construction production management activities are supported using mobile computing is presented. Recent developments in the area of cloud computing and building information modelling and emerging application scenarios in field data collaboration, co-ordination, production support, building handover and facilities management are discussed. A prototype architecture and demonstrator illustrating use of mobile web services, Internet of Things for on-site environmental services monitoring is presented. Conclusions are drawn on future possible impact of emerging mobile technologies on the construction industry.

17.2 Drivers for mobile IT use in construction

Traditional methods of distributing construction information are labour-intensive and involve lot of manual intervention. These cause delays in key processes such as information gathering, processing and access. Also, there are problems in information redundancy, invalidity and conflicting or clashing data, resulting from manual data handling. This negatively reflects on the construction project, often leading to delays, cost overruns and quality issues, as a consequence of manual data handling. Mobile computing presents the missing link by allowing connectivity between the construction site operations and office-based operations. The potential advantages of mobile computing within construction context include provision of reliable and up to date information, decrease time and cost in construction operation, decline in faults, accidents, increase in productivity, better decision-making and better quality control. Over past few decades, a key focus of ICT developments within construction sector has been on office/fixed desktop-based clients. Mobile computing provides a vital link to better connect majority of the site-based staff to backend systems. This allows site based workers to access and share important construction project information from the point of work.

Construction project execution phase is often characterised by production of massive amount of information, in shape of project data, drawings, change orders, request for information, Health and Safety records etc. This information has to be updated and delivered to construction personnel in a way that supports decision making. Son *et al.* (2012) indicated that mobile computing devices enable the industry to seamlessly and effectively capture and transmit information and eliminate the kinds of errors and delays that are inherent in manual approaches. Kim *et al.* (2013) highlighted how the advent of smartphones, coupled with state-of-the-art mobile computing technology, provides construction engineers with unprecedented opportunities to improve the existing processes of on-site construction management and enable process innovation. Bowden *et al.* (2006) outlined potential advantages of mobile IT within construction, including reduction in construction time and capital cost of construction, reduction in operation and maintenance costs, reduction in defects, reduction in accidents and waste, increase in productivity and predictability.

Innovations in mobile hardware and software development mean that commercially available smartphones are fitted with high resolution colour touch screen, GPS navigation, gyroscope, wireless communication capability, digital cameras, sensors and high-speed data transfer with 4G. These powerful features 'enable a new generation of on-site management processes, such as high mobility, location-based customized work orders, real time information exchange, and augmented reality (AR)-based site visualization' (Kim *et al.*, 2013). Also, site supervisors and site managers have the needed information available for them to act and execute the tasks. This enables the participants to have a clear awareness of the current status of the work. Anumba and Wang (2012) summarised briefly the benefits of mobile computing for influencing the decision making:

- Mobile computing enable construction workers, many of whom are often nomadic, to remain connected to their offices while undertaking work at distributed locations.
- Using mobile and pervasive computing technologies, workers can access required from their point of work on as-needed basis.
- The tracking and management of workers at distributed locations is greatly facilitated, as project managers and others can monitor the progress of work in real time, enabling the re-direction of resources as appropriate to ensure timely completion.
- Distributed construction project team members stand to benefit from improved and more effective communication systems.
- The integration of context-awareness with mobile and pervasive computing technologies enable the timely delivery of the right information/ services to the right person as the right time.
- There is scope for greater efficiency in the delivery of information and services, as there can be tailored to the individual project team members.
- Improved collaboration (in terms of both quality of information exchange and the throughput of collaborative work) can be achieved using mobile and pervasive computing technologies.
- Real-time and context-specific retrieval of information directly addresses the problem of information overload and improves the throughput of project tasks.
- Mobile and pervasive computing technologies offer tremendous opportunities for enhanced training provision for construction workers.
- Tracking mobile construction personnel has major benefits with regard to safety and security of the worker, particularly in hazardous, challenging or remote work locations.

Table 17.1 provides summary of key benefits of mobile computing as discussed in literature.

Table 17.1 Summary of mobile computing benefits

The benefits	Description	Authors
Timely delivery of information	• Seamless and effective capture and information transmission • Awareness of the current status of the work	Chen and Kamara (2008, 2011), Anumba and Wang (2012), Kim et al. (2011, 2013), Nourbakhsh et al. (2012), Son et al. (2012)
Better decision making and reducing organisation fragmentation	Better decisions are taken because of the reliability and availability of the information from point of work and better integration of on-site staff	Anumba and Wang (2012)
Improved working efficiency	This includes better input and processing of information and provisioning of timely information	Bowden et al. (2006), Chen and Kamara (2008, 2011), Anumba and Wang (2012), Kim et al. (2011, 2013), Nourbakhsh et al. (2012)
Location-based construction information	The task information is associated with the corresponding location information which enabled the construction engineers to easily understand where the tasks	Kim et al. (2011, 2013), Wang et al. (2012)
User acceptance and perceived performance	The construction professionals perceive the advantages of mobile computing	Bowden et al. (2006), Son et al. (2012)
Increase productivity	The managers can reduce the time of organizing information, and the foremen can be sure that they are executing the work in accordance with the latest information	Bowden et al. (2006), Chen and Kamara (2008, 2011), Anumba and Wang (2012), Kim et al. (2011, 2013), Nourbakhsh et al. (2012), Son et al. (2012)
Construction process improvement	Improving processes can provides construction engineers with unprecedented opportunities to improve the existing processes	Bowden (2005), Chen and Kamara (2008, 2011), Anumba and Wang (2012), Kim et al. (2011, 2013), Nourbakhsh et al. (2012), Son et al. (2012)
High mobility of construction workers	It provides site engineers and foremen the ability to perform and move between site locations and site offices	Anumba and Wang (2012), Kim et al. (2013)
Tracking and management of workers	Monitoring the progress of work in real time, enabling the re-direction of resources as appropriate to ensure timely completion. Also, safety and security of the worker particularly in hazardous, challenging or remote work locations can be ensured.	Anumba and Wang (2012)
Improved collaboration	In terms of both quality of information exchange and the throughput of collaborative work.	Anumba and Wang (2012)
Enhance training provision for construction personnel	Exploration of new opportunities for training site personnel using emerging technologies such as Augmented Reality	Wang et al. (2012)

Table 17.1 summarised key benefits of mobile computing from construction industry perspective. However, to optimise mobile computing benefits for construction, it is important to explore convergence and synergy between different enabling technologies, to ensure technology is aligned to support construction worker needs. The following section provides a summary of key efforts in this area.

17.3 Evaluation of utilisation of mobile computing on construction sites

Using affordable mobile technology such as smartphones, tablets alongside with the latest generation of communication infrastructures such as touch screen, GPS, gyroscope, accelerometer and wireless communication capability, various software applications have emerged in recent years. Kim *et al.* (2013) categorised previous studies of mobile computing for construction into five different areas:

1   Development of a framework or platform to demonstrate how mobile computing should be used for construction.
2   Mobile computing as a tool for identification or general construction management.
3   Mobile computing for defect management.
4   Mobile computing for safety or disaster management.
5   Development of specific features of mobile computing.

The above provide a useful categorisation of mobile computing research. However, given fast pace developments in the area, including developments in context aware computing, augmented reality, use of drones for data collection and LIDAR, such categorisation needs to be constantly reviewed. Nourbakhsh *et al.* (2012) provided a summary of key published research on mobile application development within construction sector (Table 17.2). The review indicates wide range of applications in use, to serve wide range of objectives.

Chen and Kamara in (2011) presented a model for exploring mobile computing with information management on sites. There are two aspects of the model, including independent and dependent factors. The dependent aspects include mobile computers, wireless network and mobile application, perceived as essential components of mobile computing. Three independent factors include the user, construction information and construction site, that are fundamentals to govern mobile computing use in construction context. The independent elements help understand the construction environment in which mobile computing will be applied to manage information and determine the design of mobile computing systems (Figure 17.1) (Chen and Kamara, 2011).

Table 17.2 Developed mobile application systems for on-site information management

System function	Tools	References
Field inspection	PDA	
Inspection system, checklist and reference system, position check system Progress monitoring system	PDA	Kimoto *et al.* (2005)
PDA- and RFID-based dynamic supply chain management system	PDA, RFID	Wang *et al.* (2007)
Progress monitoring through monitoring status of precast concrete	PDA	Vilkko *et al.* (2008)
They proposed a model for process of selecting mobile computing technologies in construction	PDA	Chen and Kamara (2008)
Defect management: capturing digital images or movies of the defects, annotating a note regarding each defect, defining the locations of the defects, as well as sending the above information to an off-site database to be accessed from the design office	Nokia N80	Dong *et al.* (2009)
1 Construction equipment finder 2 Access to safety-related information	iPhone	Irizarry and Gill (2009)
A mobile collaboration tool which was functioned as a phone, fax, e-mail and canvas fordrawings.	N/A	Venkatraman and Yoong (2009)
A self-managed information system on mobile phone for informing workers to facilitate job openings	Nokia N73	Yammiyavar and Kate (2010)

Source: Nourbakhsh *et al.*, 2012.

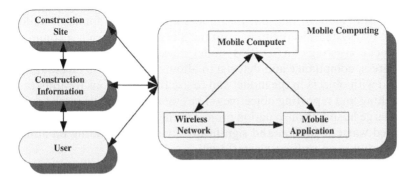

Figure 17.1 The application model for exploring mobile computing with information management on construction sites.

Source: Chen and Kamara, 2011.

Kim *et al.* (2011) proposed a location-based construction site management system, using a mobile computing platform. The developed system utilises iPhone SDK (software development kit) and is designed for the site management and construction drawing sharing. Kim *et al.* (2011) stated that the system showed a strong potential to implement a truly ubiquitous and intelligent construction site, by improving the current level of data sharing and communication practice in the construction industry. The system was further developed and tested on a real construction site. Bringardner and Dasher (2011) presented an adopted mobile computing system implemented in renovation of Dallas-Fort Worth Airport. The system was designed mainly to access the drawings, specification and other project documents and view the BIM model. Also, it constituted use of commercially available platform and low cost apps, cloud computing for storing and entering the project data, synchronisation of project information clouds and remote access of BIM models through tablet devices. Key benefits of the approach included low cost, much lighter, accessibility from tablets and desktops and more easily updated. As a result, the adopted system saved 1–2 h per day for each user because they can look at the drawings while standing at the issue. Nourbakhsh *et al.* (2012) developed a construction mobile application, comprising thirteen information groups (Figure 17.2). The functionality of the proposed application is first entering the user recognition page to authorize personnel's to get through the system. Different individuals access different sections of the application based on their roles and responsibilities. For example, supervisor managers can view, edit, and approve the information, but site managers can only view and edit. Davies and Harty (2013) presented 'Site BIM' which is a Table computer based application, to access design information and to capture work quality and progress data onsite. The system consisted of five components (Figure 17.3) including Document Management System, to allow for upload and receiving of information such as drawings. Coordinated 3D BIM models were located on site servers, maintained and coordinated by contractor document controllers. Due to capacity limitations, BIM models was split into floors and/or zones for each building. These two systems were linked with the Site dBase, which is a commercial product consisting of a 3D BIM model viewer and database functionality (progress, compliance and defects) to allow characteristic metadata to be linked with objects in the model and to use these relations for illustration, searching and reporting objectives. The system was implemented and test on a large hospital construction project in the United Kingdom. The result showed waste reduction and significant savings in spending on administrative and in co-ordinating staff time.

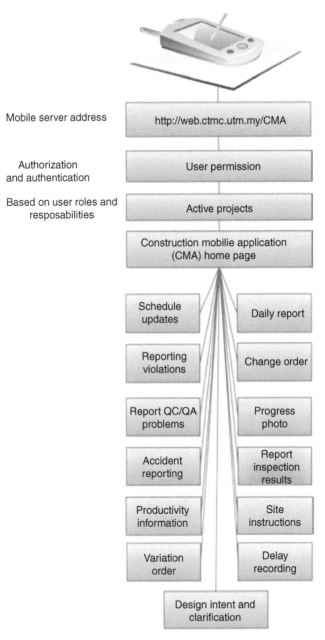

Figure 17.2 Construction mobile application.
Source: Nourbakhsh *et al.*, 2012.

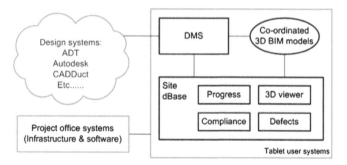

Figure 17.3 The components of SiteBIM systems.
Source: Davies and Harty, 2013.

17.4 The barriers and challenges of using mobile computing in construction site

There is a broader recognition that mobile computing can play a key role in management of information generated during a construction project and in construction process innovation. However, lack of supporting data and an understanding of hardware platforms give rise to some scepticism (Son *et al.*, 2012). Many construction professionals are unsure on how to implement mobile applications on their projects and concerned about its complexity (Bowden, 2005; Chen, 2008; Son *et al.*, 2012; Kim *et al.*, 2013). Furthermore, mobile computing has four key technical constraints including limitation of computational resources resulting from size and weight restriction, vulnerability to loss or damage concerning security considerations, variety of connectivity issues and reliability, and limited battery power (Chen, 2008). However, these limitations may be overcome due to rapid improvements in technology. Saidi *et al.* (2002) outlined two key barriers to the use of mobile computing in construction including (a) limitations of mobile devices including screen size, screen visibility, processing capability, and input method and b) construction industry characteristics including physical jobsite conditions (such as temperature, humidity, dust, etc.) and organisational issues such as the industry's fragmentation and low risk tolerance.

Moreover, several research studies have explored and discussed some of the emerging technology barriers and challenges which may affect the data exchange (Bowden, 2005; Chen and Kamara, 2011; Anumba and Wang, 2012; Wang, 2012; Son *et al.*, 2012; Nourbakhsh *et al.*, 2012). These barriers and challenges can be categorised into two broad categories of technological and organisational barriers.

Frequently, technological complexity is one of the main constraint to the use of mobile computing devices, related to screen size, input difficulties, battery life, etc. Kim *et al.* (2013) stated that the data input and output method of a mobile computer has an impact on the usability, which affect the information process on construction sites and should represent the information efficiently and effectively according to the user abilities. Size and weight of the mobile device may become too small to handle with larger hands and heavy to carry.

Also, there are various construction industry specific barriers for using mobile computing. In general, the construction professionals are aware and recognise the benefits and usefulness of the use mobile computing technologies on construction sites. However, they are clearly unsure how to implement them on their projects and concerned about its complexity (Bowden, 2005; Chen, 2008; Son *et al.*, 2012; Kim *et al.*, 2013). Such complexities arise because of the very nature of mobile computing which involves integration of various peripheral technologies, coupled with constraints of mobile devices such as limited size, interface etc. The success or failure of mobile computing or any technology provided by a firm is dependent on the readiness of the staff to utilise it in executing the tasks. Son *et al.* (2012) stated that the manager's return on investment in IT is often very low because of employees' refusal to use the IT that is available to them or their reluctance to use it to its full potential. Also, the training is another factor that affects the use of mobile computing. Moreover, strong support by the top management is vital for supporting mobile deployment. The extent to which top management is involved in IT implementation and understands its significance could make a difference. Even the technical support by the top management is needed to guaranteeing the utilization. Bowden *et al.* (2006) outlined four points about the cultural barriers which include:

- Attitudes towards the devices and IT in general could be a barrier.
- The devices were perceived by some participants in the usability trials as a gimmick or a toy, and it was thought that they should only be used when appropriate rather than as standard.
- Lack of awareness about information and communication technologies and the lack of exemplars demonstrating its successful use by others.
- Lack of IT literacy amongst site-based personnel.

17.5 Emerging technologies

Mobile computing has a large number of applications, across multiple disciplines. Emerging areas of growth include ubiquitous and location based services, augmented reality, sensor networking and profiling technologies. Key emerging technologies are discussed below.

- *Location-based services* refers to applications that utilise the user/object location knowledge to provide relevant information and services, with mobile phone working as a cursor to connect digital and physical world. An accurate and timely identification and tracking of construction components are critical to operating a well-managed and cost efficient construction project. Wide range of indoor and outdoor location tracking technologies can be used for facilitating construction operations.

- *Ubiquitous computing* is an emerging paradigm of personal computing, characterised by the shift from the dedicated computing machinery (requiring user's attention, e.g. PCs) to pervasive computing capabilities embedded in everyday environments. The vision of the ubiquitous computing requires a wide-range of devices, sensors, tags and software components to interoperate. The benefits of ubiquitous computing are perceived as ubiquitous access to information, seamless communications based on wireless technologies and computer mediated interaction with the environment through sensing, actuating and displaying. The key functionality to implement the ubiquitous computing functionality include context-awareness (the ability to capture user context such as location and other sensory data), service discovery (finding available service providers in a wireless network), awareness of user requirements/ preferences (making the user's desires known to other service providers), user-interface design (touch screen, voice input, speech output, etc.), the ability to match user requirements to services; and machine learning to improve performance over time, and adapt to better meet the user's needs. Relevance of the ubiquitous computing for the construction industry lies in the fact that these technologies have the potential to make construction collaborative processes and services sensitive to the data available in the physical world enabling a wide range of applications from field data collection, to materials management, to site logistics.
- *Sensor networks* – In the construction industry, sensor networks can be used to monitor a wide range of environments and in a variety of applications, including wireless data acquisition, machine/building monitoring and maintenance, smart buildings and highways, environment monitoring, site security, automated tracking of expensive materials on site, safety management and many others. In future, using different hardware technologies such as wireless communications, smart materials, sensors and actuator, it will be possible to add additional context dimensions, allowing for better mapping of the physical and virtual world.
- *RFID* – The term *RFID* describes technologies that use radio waves to identify individual items using tags. Being radio based, it has a non-line-of-sight readability. RFID technologies can be used for a wide range of applications to enhance construction processes including materials management, location tracking of tools/equipment, safety and security, supply-chain automation, maintenance and service provisioning, document control etc.
- Augmented reality technologies: This involves visualisation of built environment using photographs and videos captured through mobile devices, to develop a better understanding of spatial constraints. Mapping the camera pose (e.g. location, orientation, and field of view) allows photos captured from a mobile device into common 3D coordinates, to

allow virtual exploration of the site through integration with building data. This creates an augmented reality environment, where building data is superimposed on photo or video being captured through mobile device. The augmented photograph or video could be used to present key information to relevant site staff and to facilitate communication and reporting processes.

- Profiling technologies allow the delivery of personalised information to users, based on their profile and device capabilities. A W3C working group recommends the use of the CC/PP (Composite Capability/ Preference Profiles) (CC/PP, 2003) framework to specify how client devices express their capabilities and preferences to the server that originates content. Using the CC/PP framework, information collected from the terminal can be tagged with relevant context parameters (such as location and device-type). It is also possible to enable selection and content generation responses such as triggering alarms or retrieving information relevant to the task at hand.

17.6 Application of mobile web services for on-site environmental surveillance

The intensity of environmental impact of construction activities is significant and many of the activities throughout a construction project life cycle are not environment friendly (Tam and Tam, 2006). Although construction activities contribute to soil pollutions, key areas of concern with regards to environmental impact of construction activities include air, water and noise pollution (Gray, 2015). Similar observations were made by Zolfagharian *et al.* (2012), who, after interviewing industry experts, identified 'noise pollution' and 'dust generation by construction machinery' as the most risky environmental impact on construction sites.

Construction sites generate high levels of dust, resulting from use of materials such as concrete, cement, wood, stone, silica. Dust also results from activities such as land clearing, operation of diesel engines, demolition, burning, and working with toxic materials and this can carry for large distances over a long period of time (Gray, 2015). Construction dust is classified as particulate matter (PM_{10}). PM_{10} has an aerodynamic diameter of 10 microns or less and this allows such matter to be carried deep into the lungs where it can cause inflammation and a worsening in the condition of people with respiratory illness, asthma, bronchitis and even cancer. Noise pollution resulting from the construction sites result from use of vehicles, demolition activities and use of construction machinery such as pile drivers, cranes, rock drills, mixing machinery and other types of heavy duty equipment (Chen and Li, 2000). Noise also results from other site based activities. Excessive noise is not only annoying and distracting, but can lead to hearing loss, high blood pressure, sleep disturbance and extreme stress. Research has shown that high noise levels disturb the natural cycles of animals and reduces their usable habitat.

These adverse impacts have led to a growing realisation of the need to better implement environmental protection initiatives at construction sites. To ensure effective implementation of environmental management systems in the project development, periodic environmental surveillance through inspections and environmental quality monitoring should be conducted from time to time where non-conformity is identified (Environmental Protection Authority of Australia, 1996; CIRIA, 2010; Jabatan Kerja Raya Malaysia, 2013). Further corrective action and prevention action in the context of continuous improvement should be undertaken accordingly, to address the issues. However, existing approaches rely on manual and paper based inspection methods, which are time consuming, labour-intensive, produce limited information and can involve deficiencies and discrepancies (Vivoni and Camilli, 2003; Mohamad and Aripin, 2006; Damian *et al.*, 2007; Kim *et al.*, 2008; Harun *et al.*, 2015). Recent developments in mobile computing, sensor computing and the Internet of Things provide vital building blocks to enhance existing processes, leading to delivery of concise, accurate, timely and usable environmental information, crucial in the surveillance activities.

17.6.1 ENSOCS prototype

This Section presents a system architecture and prototype of a mobile web-based environmental information system. The system aims to improve environmental checking and the correction process, by providing a tool for environmental enforcement officer for managing their environmental surveillance activities on construction sites, to enhance their decision-making capabilities. This prototype application demonstrates the 'Internet of Things' concept, where the smartphone plays an intermediary role between environmental management teams, 'things' (wireless sensor networks and a weather station) and the Internet. The interactions between them have resulted in enrichment of and speedy information, enhancement in the delivery of reports, and enables the alerts of environmental non-compliances.

The proposed system was designed and developed for Internet browsing through a smartphone and works together with telemetry sensors (air and noise) and wireless weather station, to provide real-time environmental data monitoring, to support environmental surveillance undertaken by the inspection officer. The proposed approach demonstrates the interrelationship between activities and pollution in an innovative way, when compared to conventional paper-based methods. While maintaining the concept of a checklist, users may take a note of environmental observations using web forms. For air (PM_{10}) and noise (dBA) pollution, users can confirm their observational findings by referring to data transmitted by telemetry sensors in real time. Wireless weather station would also provide the indication of the potential polluter by providing the weather conditions (dry/wet) and wind speed and directions. In addition, in any events of non-compliance due

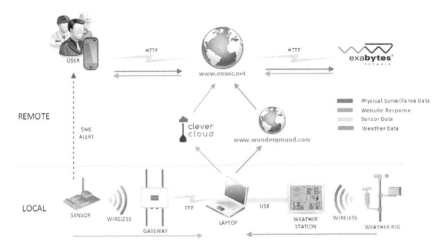

Figure 17.4 The ENSOCS architecture.

to the readings of air and noise parameters exceeding the threshold, early warning alert system is enabled through Short Messaging System (SMS).

For the purpose of prototype demonstration and evaluation, previous researches argue that participants involved in a controlled laboratory setting may not experience the potential adverse effects that are caused by changing and unpredictable network conditions and other environmental factors (Zhang and Adipat, 2005, Kjeldskov *et al.*, 2005). Thus, robust environment will enable realistic assessment of users' acceptability, as it is directly applicable to a mobile environment. Due to these reasons, ENSOCS has been deployed in a real-life construction project, alongside direct engagement with practitioners as evaluators. The set-up of the ENSOCS system comprised one Smart-Cities sensor node, one weather station, one sensor gateway and one Internet-ready laptop. System architecture is illustrated in Figure 17.4. ENSOCS architecture follows the common four-layer structure of the Internet of Things (IoT) (see Figure 17.4), with details of each layer presented as below.

Key layers of the architecture are explained as below:

a) Sensing and Control Layer – This includes the Wireless Sensor Network for capturing the reading of PM_{10} concentrations and noise levels and the weather station to obtain weather conditions data. The Wireless Sensor Network delivers the Short Messaging System (SMS) to the users as an early warning alert in events where the reading of PM_{10} concentrations and noise levels exceed or nearly exceed the threshold. Users may take a note of their environmental observations using web forms in the ENSOCS, which is accessible via the smartphone. Built-in smartphone

GPS and WLAN are also being used for the determination of the current location of the user;

b) Networking Layer – This includes wireless local area network (WLAN) and local area network (LAN) to provide connectivity between sensing and control layer and middleware layer;

c) Middleware Layer – This includes the server system which processes the location data gathered from the smartphones' built-in GPS, the sensor data, the data input from the user and the weather data. The server then intelligently maps the right information and services from the servers available in the system. The server systems of the ENSOCS prototype consist of the MySQL Database server and the web server.

d) Application Layer – This includes the mobile client, which allows for input from the user and, at the same time, allow for receiving the Short Messaging System (SMS), while ENSOCS mobile web-based environmental information system will display the environmental observation reports, data from the wireless sensor network (WSN), weather station data and location coordinates from the built-in smartphone GPS.

Smart Cities sensor node and the weather rig were separately placed inside an enclosure and powered by an external power supply connected to the solar panels. This allowed for installation of both units in an outdoor environment. For security reasons, they were installed at a secure location on a construction site, which was well guarded and monitored by construction personnel. The Smart Cities sensor node and weather rig were put in a location 1.4 m above the ground and at a position of 4 m from the construction hoarding and approximately 6 m from the nearest sensitive receptor. The installation took about 1½ h.

Smart Cities sensor node was equipped with XBee-ZigBee-Pro wireless radio communication, to enable communication with the sensor gateway within a distance up to 7 km. Whereas for wireless weather station, the weather data is transmitted every 48 s to the Weather Station from the weather rig via the radio frequency of 915 MHz within a distance of 300 ft (without any obstacles in the open field). Both Smart Cities sensor node and weather station make measurement and periodically sends the results to the server application for further analysis and database storage. ENSOCS mobile web-based environmental information system then collects information from the sensor node and weather station and displays the reading of data on the system dashboard (see Figure 17.5).

Besides the display of the real-time sensor and weather station data, ENSOCS mobile web-based environmental information system also provides the web form (Figure 17.6) to enable the user to record their environmental observations, while conducting the environmental surveillance. The integration between observations and real-time data would provide holistic view of the environmental problems at site. For instance, the environmental inspector may feel that the mufflers and noise-shielding were not working

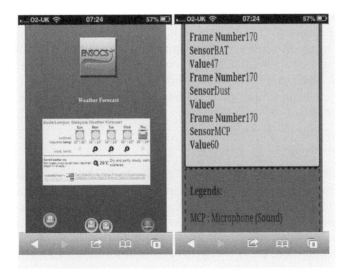

Figure 17.5 The display of weather data, air and noise data on ENSOCS' dashboard.

effectively during his environmental surveillance. However, only with the support of the real-time noise data would help to justify the findings and permits environmental inspectors to issue out the environmental surveillance report to address the non-compliance on the construction noise control.

The above case study demonstrates that the concept of a mobile environmental information system is feasible, and emerging technologies opens up

Figure 17.6 The web form (checklist) and the sample of a surveillance report (Left: Online report; Right: Emailed report) of ENSOCS.

huge potential to provide support for environmental surveillance on construction sites in real time. The validation process confirmed relevance of prototype demonstrator for supporting site based operations, for accelerating information delivery and for enhancing decision making capabilities through services provisioning and real-time environmental quality monitoring. However, need for further work to better align prototype system to take account of prevailing industry standards in field environmental management were identified.

17.7 Conclusions

Given the mobile and dynamic nature of construction production work, there is a need to integrate advances in mobile computing in the work environment to provide user-friendly and mobile access to construction product and production process. There are several issues that must be considered in utilising mobile computing regarding mobility, application abilities, services, integration of current systems and inputs and output methods. This is important to accommodate site personnel requirements and behaviours in the diversified construction site environment. Mobile computing system needs to be considered, applied and managed, while recognising the end user needs. Also, any deployment of mobile IT must follow a careful study of existing processes and organisational requirements, to ensure better alignment of technology to business processes. There is need to redesign existing information system to help achieve higher values of information integration and smooth information flow between site and office based operations. However, given significance of mobile computing for construction, necessitates appropriate investments in research and development, by both industry and academia.

References

Anumba, C. and Wang, X. (2012). *Mobile and Pervasive Computing in Construction.* West Sussex, UK: Wiley-Blackwell.

Brandon, P.S., Li, H., and Shen, Q. (2005). Construction IT and the 'tipping point'. *Automation in Construction, 14*, pp. 281–286.

Bringardener and Dasher (2011). 'Tablets! Gain an Edge by Connecting to Project Information and BIM Models in the Field or on the Go' Autodesk University 2011 [Online] http://aucache.autodesk.com/au2011/sessions/5226/class_handouts/v1_CR5226_Bringardner.pdf (last accessed on 16th Nov, 2016)

Bowden, S., Dorr, A., Thorpe, A., and Anumba, C. (2006). Mobile ICT support for construction process improvement. *Automation in Construction, 15* (5), pp. 664–676.

Chen, Y. (2008). 'Using Mobile Computing for Construction Site Information Management.' (PhD thesis), Newcastle, UK: Newcastle University.

Chen, Y. and Kamara, M. (2011). A framework for using mobile computing for information management on construction sites. *Automation in Construction, 20*, pp. 776–788. doi:10.1016/j.autcon.2011.01.002

Chen, Z. and Li, H. (2000). Environmental management of urban construction project in China. *Journal of Construction Engineering and Management*, *126*, p. 320.

CIRIA (2010). *Environmental Good Practice on Site*, (3rd Edition). London: CIRIA.

Damian, C., Fosalau, C., and Zet, C. (2007). Wireless Communication System for Environmental Monitoring. In *Proceedings of the 1stIMEKO-TC19 International Symposium on Measurement and Instrumentation for Environmental Monitoring: 19–21 September 2007*, (pp. 108–112). Iasi, Romania.

Davies, R. and Harty, C. (2013). Implementing 'Site BIM': A case study of ICT innovation on a large hospital project. *Automation in Construction*, *30*, pp. 15–24. dx.doi.org/10.1016/j.autcon.2012.11.024

Dong, A., Maher, M.L., Kim, M.J., Gu, N., and Wang, X. (2009). Construction defect management using a telematic digital workbench. *Automation in Construction*, *18* (6), pp. 814–824.

Environmental Protection Authority of Australia. (1996). *Environmental Guidelines for Major Construction Sites*. Melbourne, VIC: Environmental Protection Authority of Australia.

Gray, J. 2015. *Pollution From Construction*. Available: http://www.sustainablebuild. co.uk/pollutionfromconstruction.html [Accessed 9 July 2015].

Harun, A.N., Bichard, E., and Nawi, M. (2015). Traditional vs technological based surveillance on construction site: A review. *Applied Mechanics and Materials*, Transtech Publications Ltd, pg. 247–252.

Irizarry, J. and Gill, T. (2009). Mobile applications for information access on construction jobsites. In: O'Brien, W.J. and Caldas, C.H. (Eds.). *Proceedings of the International Workshop on Computing in Civil Engineering*: 24–27 June 2009, (pp. 176–185). Austin, US.

Jabatan Kerja Raya Malaysia. 2013. *Langkah-Langkah Tebatan Bagi Kawalan Pencemaran Alam Sekitar*. Available: https://www.jkr.gov.my/cast/index.php? action=cms&id=B2QCMAQ1 [Accessed 1 May 2015].

Kim, Y.S., Oh, S.W., Cho, Y.K., and Seo, J.W. (2008). A PDA and wireless web-integrated system for quality inspection and defect management of apartment housing projects. *Automation in Construction*, *17*, pp. 163–179.

Kim, C., Lim, H., and Kim, H. (2011). Mobile computing platform for construction site management. In *Proceedings of the 28th Annual Conference of the International Symposium on Automation and Robotics in Construction (ISARC): 29 June—2 July 2011, 28*, (pp. 280–287). Seoul, South Korea.

Kim, C., Park, T., Lim, H., and Kim, H. (2013). On-site construction management using mobile computing technology. *Automation in Construction*, *35*, pp. 415–423. doi:dx.doi.org/10.1016/j.autcon.2013.05.027

Kimoto, K., Endo, K., Iwashita, S., and Fujiwara, M. (2005). The application of PDA as a mobile computing system on construction management. *Automation in Construction*, *14*, pp. 500–511.

Kjeldskov, J., Graham, C., Pedell, S., Vetere, F., Howard, S., Balbo, S., and Davies, J. (2005). Evaluating the usability of a mobile guide: The influence of location, participants and resources. *Behaviour and Information Technology*, *24*, pp. 51–65.

Mohamad, R.H. and Aripin, A. (2006). Issues and Challenges in Environmental Monitoring and Enforcement in Sabah. *In:* Fourth Sabah-Sarawak Environmental Convention 2006, 2006 Kota Kinabalu, Sabah, Malaysia: Chief Minister's Department, Sabah and Environment Protection Department, Sabah, Malaysia.

Nourbakhsh, M., Zin, R.M., Irizarry, J., Zolfagharian, S., and Gheisari, M. (2012). Mobile application prototype for on-site information management in construction industry. *Journal of Construction Engineering and Management*, 19 (5), pp. 474–494.

Saidi, K.S., Ilaas, C.T., and Balli, N.A. (2002). The value of handheld computers in construction. In: *Proceedings of the 19th International Symposium on Automation and Robotics in Construction (ISARC)*, (pp. 557–562). Gaithersburg, MD.

Samuelson, O. and Björk, B. (2014). A longitudinal study of the adoption of IT technology in the Swedish building sector. *Automation in Construction, 37*, pp. 182–190. dx.doi.org/10.1016/j.autcon.2013.10.006

Son, H., Park, Y., Kim, C., and Chou, J. (2012). Toward an understanding of construction professionals' acceptance of mobile computing devices in South Korea: An extension of the technology acceptance model. *Automation in Construction, 28*, pp. 82–90. doi:10.1016/j.autcon.2012.07.002

Tam, V.W. and Tam, C. (2006). A review on the viable technology for construction waste recycling. *Resources, Conservation and Recycling, 47*, 209–221.

Venkatraman, S. and Yoong, P. (2009). Role of mobile technology in the construction industry—a case study. *International Journal of Business Information Systems, 4* (2), pp. 195–209.

Vilkko, T., Kallonen, T., and Ikonen, J. (2008). Mobile fieldwork solution for the construction industry. In: *Proceedings of the 16th International Conference of the Software, Telecommunications and Computer Networks (SoftCOM): 25–27 September 2008*, (pp. 269–273). Dubrovnik, Croatia.

Vivoni, E.R. and Camilli, R. (2003). Real-time streaming of environmental field data. *Computers and Geosciences, 29*, pp. 457–468.

Wang, L.C., Lin, Y.C., and Lin, P.H. (2007). Dynamic mobile RFID-based supply chain control and management system in construction. *Advanced Engineering Informatics, 21* (4), pp. 377–390.

Wang, X., Kim, M., Love, P., Park, C.S., Sing, C.P., and Hou, L. (2012). A conceptual framework for integrating building information modeling with augmented reality. *Automation in Construction, 34*, pp. 37–44. doi:10.1016/j.autcon.2012.10.012

Yammiyavar, P. and Kate, P. (2010). Developing a mobile phone based GUI for users in the construction industry: a case study. In: Katre, D., Orngreen, R., Yammiyavar, P., and Clemmensen, T. (Eds.). *International Proceedings of Human Work Interaction Design: Usability in Social, Cultural and Organizational Contexts. 316*, (pp. 211–223). Springer, Boston and Pune.

Zhang, D. and Adipat, B. (2005). Challenges, methodologies, and issues in the usability testing of mobile applications. *International Journal of Human-Computer Interaction, 18*, 293–308.

Zolfagharian, S., Nourbakhsh, M., Irizarry, J., Ressang, A., and Gheisari, M. (2012). Environmental Impact Assessment on Construction sites. In: *Construction Challenges in a Flat World, Proceedings of the 2012 Construction Research Congress.* (pp. 1750–1759).

18 Envisioning buildings: advances in construction visualisations

Emine Thompson

18.1 Introduction

The objective of this chapter is to highlight recent technological developments in visualisation domain that influence the operational practices of the construction sector. It seems every day a new digital technology makes its way into our daily life. This steady stream of technological input is changing the way we process and practice things. As a result, traditional ways of conducting business in any sector is altering also.

Utilising ICT in Construction industry is not a new concept. Many academics and practitioners and the R&D departments of large construction companies are interested in the capability and applicability of Information and Commination Technologies (ICT) for built environment practices. A significant amount of literature on this area has been generated over many decades. According to Turk, (2006) 'the first uses of computers in construction are reported in 1960s' and, over the years, integration of ICT in the architecture, engineering and construction (AEC) industry became an independent subject area within the AEC research domain. Since this first usage, many different ICT tools have been utilized in different areas of the construction industry, including data creation in the form of drawings, building information modelling (BIM), scheduling and budget information, data sharing, storing and updating data and utilizing complex data management and coordination systems.

Although, at some level, construction industry is benefiting from the ICT tools more than half a century now, Digital Built Britain Report (DBB, 2015) states that 'repeated reviews of construction industry performance going back to the Banwell Report of 1964 have shown that the UK's transactional, tactical approach to designing and building infrastructure is sub-optimal. Instances sustained cost inflation and market volatility experienced during periods of growth demonstrate that compared to many other capital delivery industries, construction remains inefficient. The potential for new technologies to deliver business transformation has been demonstrated in many industries; however the built environment has been slow to adopt these technologies and remains one of the last major industrial sectors to adopt new ways of working'. Higher levels of adaptation of ICT tools in construction industry are essential and the drive to embed ICT tools in the construction industry should aim to make the industry more efficient, sustainable and profitable.

Visualisation of a proposed building or a structure can be one of the most noticeable ways of ICT integration in the construction industry. Virtually creating the buildings assist the construction industry in many ways and would lead the projects to be more efficient, sustainable and therefore profitable. It is not only far easier to create an *accurate* model – a virtual prototype of the proposed structure with the ICT tools, but these tools also, with a click of a button, enable us to perform clash detections, visual impact assessments, line of sight analysis, sun and shadow analysis, better construction planning and scheduling, etc., all of which can prevent costly on-site problems.

Heydarian *et al.* (2015) also point out that technological advancements have the potential to improve and revolutionize the current approaches in design (e.g. by involving end-user feedback to ensure higher performing building operations and end-user satisfaction), in construction (e.g. by improving safety through virtual training), and in operations (e.g. by visualizing real-time sensor data to improve diagnostics).

Over all visualisation plays a vital role in public engagement situations, property sales and marketing activities, and provides accurate representations to communicate more effectively with all the stakeholders within a project. On a recent survey titled 'The Future of Visualisation in Architecture' carried out by Unity with 967 respondents over 95 countries across the world, it is being revealed that 65% of architectural practices and design firms use technology often or very often during the pitch/bid process to communicate their ideas (e.g. rendering engines, real-time applications, mobile apps, VR etc.). Also, 97% of the respondents felt visualisation was either important or very important to the overall design process (Unity, 2014). Representational paradigm presented in this chapter does emphasis on the applications of different techniques and tools for different temporal platforms where we can envision the future, the existing and the past with precision and interaction in different media.

18.2 Representational paradigm

Using visual tools is one of the major parts of the communication that take place in construction industry. Applying ICT technologies to model, simulate and visualise scenarios in architecture, engineering and the construction (AEC) industry is not an end in itself but rather a process that helps to reach a shared understanding of the proposed building.

The design of a building begins with a concept, and an idea in the mind of an architect. This image must be conveyed to clients, who understandably need to know what their building will look like. Members of the public also have an interest in finding out what is planned for their neighbourhood. (Giddings and Horne, 2003). Visually communicating what is planned for a specific location is not a new concept. Communicating visual information from our surroundings is as important in the digital age as it was to the pre-historic artists who created the first cave paintings'.

Over the centuries many different techniques and tools were used in order to represent the architect's vision. Visualisation in the built environment, whether it is a perspective drawing, a hand-made physical model, a photomontage, a virtual reality (VR) model, a BIM or a 3D print model of a proposed scheme, focuses not only on what the proposed scheme is going to be and how it will impact on the existing surroundings, but also on how it can be built cost-effectively and sustainably. Current visualisations, as well as being accurate representations, also contain many other types of information. All of these digital technologies provide wider opportunities to the AEC industry to enhance the quality of the final product and enable all parties to understand and communicate effectively and efficiently.

Visual communication can be done for different temporal platforms. We can separate these different platforms in temporal groups: future reality, existing reality and past reality (Figure 18.1).

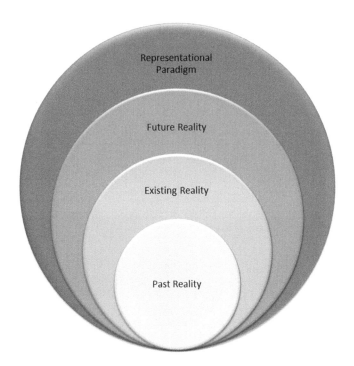

Figure 18.1 Temporal representation paradigm.

18.2.1 Future reality

This section covers some of the current *digital* tools and technologies that can be used to represent the proposed structure. For centuries before computer aided design/drafting (CAD) tools were available, paintings, sketches, plan drawings, artist's impressions, perspectives, and maps were utilized and these tools are still valid and useful. However, with the invention of CAD from the early 1960s, digital tools started to influence visual communications in the AEC industry. 2D CAD, as well as the traditional hand drawings and the specifications which provided the data, supported the communication between the members of building team. 2D CAD worked on the same principle of drafting and technical drawings which manipulated lines, circles, rectangles and text. One of the major benefits of using CAD systems in those days was being able to edit the drawings faster. This was ground breaking when compared with traditional hand drawings. Eventually 2D CAD became a norm in the industry and many other digital tools followed.

Nowadays visual communication tools have become more interactive and all stakeholders can easily understand the proposed scheme and utilize these new tools from their perspective, rather than just accepting what the designer or the architect would like to show.

We can talk about many different visualisation tools which can be used for representing the built environment, for example; 3D modelling, animation/walkthrough, virtual reality, augmented reality, building information modelling, and others. Many of these are directly related and interconnected with each other (Figure 18.2).

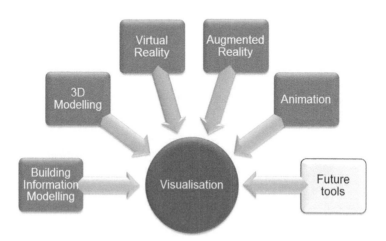

Figure 18.2 Visualisation tools for built environment.

18.2.1.1 Three-dimensional modelling in built environment

During the 1990s, instead of only being able to represent buildings in 2D CAD drawings, built environment professionals got to show their building in digital 3D models and since then 3D modelling became a significant tool within the industry. In the AEC industry, standard 3D modelling software such as AutoCAD, 3DsMax, Revit, Rhino, Maya, and SketchUP are some of the many 3D modelling software that can be used. These models can be exported to generic 3D file formats and can be utilized in various platforms.

3D models can be represented in different ways, such as wireframe, surface and solid models. As Sun and Howard (2004) explain, wireframe models represent the objects in a similar way to 2D drawings by using straight and curved lines, and there is no concept of surface in the wireframe model. On the other hand, a surface model provides both visual and mathematical information about the surface shapes of the object. A solid model represents a building using both mass and its boundary surfaces. Photorealistic visualisation where textures, lighting and environment accurately modelled can be a very effective way of communicating.

During recent decades, the AEC industry benefited from the developments in the gaming industry, which had the financial power to push computer technology by demanding better graphics cards, and greater computer power for ordinary personal computers so that their games could be played across the world by millions. Smith (2007) presents a chronological outline of the utilization of game engines in various industries and points out that this technology is being adopted for defence, medicine, architecture, education, city planning, and government applications. Trenholme and Smith (2008) notes that the current generation of computer games present realistic virtual worlds featuring user friendly interaction and the simulation of real-world phenomena.

Game engines such as Unreal Engine, CryEngine, Quake, Unity and hundreds of others offer vast possibilities to the AEC industry. Nowadays, game engines are compatible with many AEC standard software – for example, Autodesk Revit, ArchiCAD and SketchUP Pro – and models can be effortlessly shared with clients and other shareholders via various platforms, not only on desktop computers but on the Web and smartphones and tablets. Game engines' immense capabilities are being utilized by modellers in the AEC industry to take the 3D modelling to extremely realistic visualisations. However these visualisations, without the support of BIM, are still unintelligent models that do not contain any data that can be extracted and utilized.

18.2.1.1.1 ANIMATIONS AND WALKTHROUGHS

Enhanced 3D modelling and rendering techniques not only allowed the production of realistic still images but also enabled AEC industry specialists to create animations and walkthroughs of the proposed building. By

Figure 18.3 Animation of construction process.
Source: Student: Mark Cronin, 2009, Software 3D Studio Max.

producing series of images and putting them together in specialist software, designers were able to show the building before construction took place. This allowed the building to be explored in a more realistic view with accurate light, texture, and geometry. Initially the whole process was not easy and required sophisticated software and knowledge of cinematography. These were expensive techniques and initially only a few big companies were able to utilize these tools for the final design option. Nowadays however even the basic 3D modelling software offers an animation/walkthrough option as an output from the created 3D model, where initial ideas can be modelled and turned into animations and walkthroughs with ease (Figure 18.3).

18.2.1.2 Virtual reality in built environment

Concepts of stereoscopic vision have been around for more than two thousand years and in 1838, the stereoscope was invented by Sir Charles Wheatstone. Daguerreotype photography was invented the following year.

Together, they allowed people to view places, people and events as they exist in three dimensions by creating *a heightened sense of reality, of actually being there*. Stereoscopic images were the forerunner of virtual reality, providing the viewer with a single moment of artificially created three dimensional experience. (Ager and Sinclair, 1995)

Virtual reality is a term that is used frequently in the building industry, which conveys different meanings to different people (Greenwood *et al.*, 2008). Indeed, many different descriptions of VR can be found in the literature. According to Whyte (2002), 'the term *virtual reality* has become used to describe applications in which we can interact with spatial data in real-time'. However, other visualization methods, as Whyte (2002) points out, 'can describe the same or overlapping groups of technologies'. These include virtual environments, visualization, interactive 3D (i3D), digital prototypes, simulations, urban simulations, visual simulations and 4D-CAD. Greenwood *et al.* (2008) suggest that VR in the building construction context is a 'real-time visualization of a computer-generated model of a built environment'.

VR technology enables designers to create interactive and immersive environments. The users (whether this is the client or the designer or the planning officer) of a VR model can themselves interact with the model. This freedom, not staying within the boundaries of designer's view point, is a strong feature which empowers users to be in control and assist them to explore the proposed structure without the restrictions of a set walkthrough. This exploration can take place not only in a fully-immersive environments like a *CAVE* setting or in a semi-immersive environment with a big-wide screen but also on a desktop computer – desktop VR. Many other display technologies such as head-mounted display units and workbenches can also be utilized.

Since Whyte's appraisal of VR in built environment in 2002, many other applications and tools have appeared that help us to interact with our environment in real-time. Many are taking virtual reality one step further and making it more user-friendly and more affordable. The Oculus Rift – a VR headset which was originally designed for people to interact with their favourite computer games is now also used by AEC industry (Figure 18.4).

Recently Google brought out a very cheap cardboard tool which utilizes smartphone technology to give users similar interaction possibilities to the Oculus Rift in an immersive environment; they simply call this the Google Cardboard. Many different manufacturers now produce and supply cardboard viewers and one can actually download the template and create one's own viewer by using cardboard, a rubber band, magnets, lenses and some Velcro (Figure 18.5).

Bringing down the cost of VR applications is a very important factor for the AEC industry since using an extremely expensive kit on a construction site is not always an option or is simply not possible, but a standard smartphone with some free apps and a cheap kit will make this type of technology

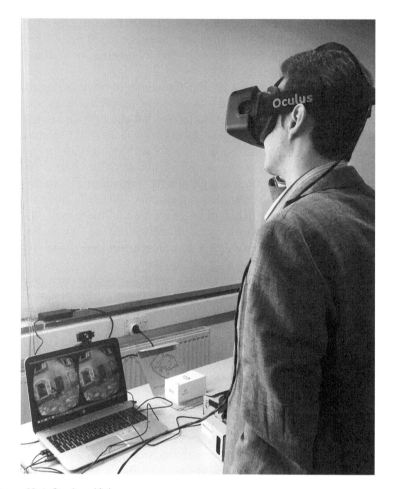

Figure 18.4 Oculus rift in use.

practical for all. Furthermore, combinations of technologies and tools are becoming possible and all these factors are making the visualisation of built environment richer and more informative and interactive.

18.2.1.3 Augmented reality (AR)

From the AEC industry point of view, augmented reality (AR) helps envisioning the proposed structures in context with in the real surroundings. AR, in short, is the augmentation of live video inputs with the digital visualisations. Rather than completely replacing the reality as in VR, AR compliments it by allowing virtual and real objects to exist together in the same environment, simultaneously in real-time. As Wang *et al.* (2013a) explain that AR creates

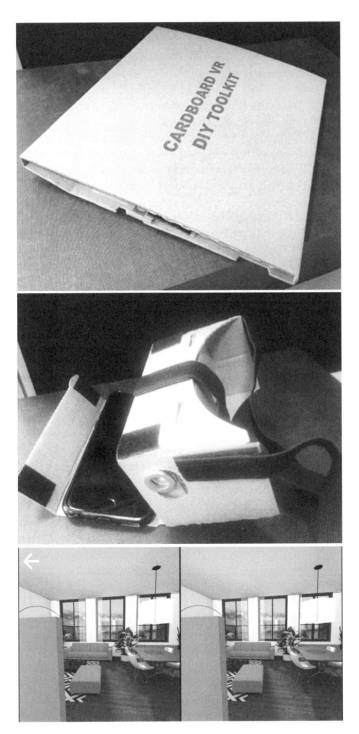

Figure 18.5 Cardboard VR tool, and two slightly different images which makes up the VR environment.

an environment where digital information is inserted in a predominantly real world view. It originated from marker-based tracking toolkits (e.g. ARToolkit, ARTag) which are used to determine tracking and registration (where to display the digital contents) and the media contents (what digital content to display). They also explain that AR can be applied to address a plethora of problems throughout a construction project's lifecycle, for example, planning, design, safety, and training.

Jiao *et al.* (2013) point out that huge progress has been made on AR techniques such as registration, tracking, and display hardware, but a construction AR system should be more convenient and combined with in-use applications to support multi-disciplinary users throughout the construction lifecycle.

Currently, both research and industry players are increasingly interested in utilising AR to facilitate better understanding of buildings. Substantial efforts have been made to mainstream building practice (Wang *et al.*, 2014a). Because AR technologies are becoming more mature and well established, AR applications are becoming more varied and popular. The trend of utilizing AR technologies in practical applications, such as education, design, manufacturing, construction, and entertainment reveals great potential for improving existing technologies and providing a better quality of life (Chi *et al.*, 2013).

In recent years, many different applications have been developed that can be used in conjunction with standard AEC software and tools such as 3D Interactive STHLM AB, AR-Mdedia, Augment, CityViewAR, ViewAR, Visuartech, SightSpace 3D, Smart Reality, and many others. Some tools also enable us to combine BIM and 4D models with AR.

18.2.1.4 *Building information modelling – BIM*

The concept of 3D object-based building systems, recognisable as a building information model, has existed since 1975. Eastman (1975) explains in detail how this system will work and will be beneficial for the construction industry. Since then the concept has evolved and become what we now know as Building information model (BIM). BIM is defined by the International Organization for Standardization (2010) as the 'shared digital representation of physical and functional characteristics of any built object'.

Visualisation is one of the most important characters about BIM (Wang *et al.*, 2014a). CAD uses lines, curves and rectangles in order to visualise whether it is 2D or 3D, whereas BIM related software utilizes the actual building components which have various sorts of information attached to them. 'BIM has been described as a game-changing information and ICT and cultural process for the construction sector. BIM processes are now mainstream for both new buildings and infrastructure and have great value in retrofit and refurbishment projects, where complimentary technologies such as laser survey techniques and rapid energy analysis are employed' (DBB, 2015).

BIM is utilized in many different areas of AEC, such as building performance, building analysis, building record, code compliance, cost estimation, design review, design, disaster planning, maintenance scheduling, sequential planning, site modelling, site utilization, space and assessment management, spatial planning, temporary structure design, etc. Various companies established their versions of software where the BIM process can take place, for example, Autodesk Revit, Graphisoft ArchiCAD, Bentley Architecture, Gehry Technologies Digital Project Designer, Tekla, VectorWorks and others.

18.2.1.4.1 IMMERSIVE AND INTERACTIVE BIM

Integration of BIM and AR is envisaged to become an innate feature of architectural visualisation within the building industry (Wang *et al.*, 2014a). As BIM becomes extensively used in practice, it is apparent that interaction with models becomes more important. As Dalton and Parfitt (2013) explain, interactive real-time 3D display are important as construction clients and project teams need new ways to visualize and interact with their data. Although some software, for example Autodesk Showcase and Autodesk Navisworks, allows interactive and immersive visualisation of BIM, more user-friendly and easier applications are required in order to keep the process simple, achievable and approachable.

A recent publication by Johansson *et al.* (2014), explains that the integration and use of immersive VR within the actual design process, and the current adaptation in the AEC field still suffers from a number of limitations; such as high cost, limited accessibility and limited BIM support. In order to overcome these limitations Johansson *et al.* (2014), developed a new system with three components: (1) the Oculus Rift Head Mounted Display (HMD), (2) an efficient real-time rendering engine supporting large 3D datasets that is (3) implemented as a plug-in in a BIM authoring software which becomes a portable system for immersive BIM visualisation (Figure 18.6).

Figure 18.6 System overview; the Oculus HMD, the revit viewer plug-in and the PowerPoint remote control.

Source: Johansson, Roupé and Viklund Tallgren, 2014.

To support an integrated design environment this rendering engine has been implemented as a viewer plug-in in Autodesk Revit. Because of this, immersive design review sessions can be performed directly in the BIM authoring software, without the need to export any data or create a separate visualization model. As the technology is portable, clients and design team members can take advantage of immersive visualization sessions without the need to travel to a specific location (Johansson *et al.*, 2014).

In similar research, Wang *et al.* (2014b) pointed out integrating BIM and Augmented Reality (AR) tools in order to display an immersive view right into the real environment where the participants can the see the as planned data onto the as built environment right in place. They explain four different integration options; BIM and AR Walkthrough, BIM and AR Context-aware Mobile System, BIM + AR for Onsite Assembly, BIM and AR for Way-finding.

AR can also visualize as-planned BIM facility information right in the context of the real workspace to enable project managers, subcontractors and other stakeholders to review the as-built progress against as-planned (Wang *et al.*, 2013b). In the same research, Wang *et al.* (2013b) looked into different areas in order to demonstrate BIM and AR integration and on-site use. The areas they focused upon are: interdependency, spatial site layout collision analysis and management, link digital to physical, project control, procurement (material flow tracking and management) and visualization of design during production. They explain that BIM and AR should be used together to overcome the BIM's inability to take into account the inherent uncertainty associated with design changes and rework, which prevail during construction, particularly in complex projects. They note that the use of an inbuilt context awareness and intelligence layer provides a platform that is able to couple BIM and AR so that information about 'as-built and as-planned progress' and 'current and future progress' can be obtained and presented visually.

Jiao *et al.* (2013) offer another solution that is based on a cloud BIM engine, a cloud BSNS (business social networking services) application, and an on-line AR system which utilizes web3D to render virtual objects, which originate from in-use BIM models. This proposed framework and system support special requirements such as usability, sourcing virtual objects from BIMs, and collaboration which are important factors if AR and BIM combination technology becomes a standard tool in AEC domain.

Using Game Engine technology also is another way of giving interactivity to BIM (for example: Pauwels *et al.*, 2011; Bille *et al.*, 2014; Edwards *et al.*, 2015). Edwards *et al.* (2015) created a prototype based around a designer creating a structure using BIM and this being transferred into the game engine automatically through a two-way data transferring channel. This model is then used in the game engine across a number of network connected client ends to allow end-users to change/add elements to the design, and those changes will be synchronized back to the original design conducted by the professional designer.

Bringing the interactive visualisation close to the construction site has immense benefits, and experiments on mobile visualization environment, which replicates a set virtual reality suite, are already on-going as explained on the paper titled *Developing a Mobile Visualization Environment for Construction Applications* by Mallia-Parfitt and Whyte (2014) which 'explores the design of a fully-immersive visualization prototype, which uses ultra-short throw projectors without the need for extensive set-up time. This solution offers great flexibility as two users can assemble and start using it at a new location within an hour'.

18.2.1.4.2 4D, 5D, ND BIM FOR VISUALISATION

*n*D visualisations and simulations are useful tools to support many aspects of construction project planning, for example collaborative planning, coordination and communication. Originally, as McKinney *et al.* (1998) explain, construction planners are forced to abstract CAD model building components into schedule models representing time. 4D-CAD (3D-CAD+time) removes this abstraction by linking a 3D building model and schedule model through associative relationships.

Application of BIM can be described as a process that expands 3D data into an nD information model, which allows dynamic and virtual analysis of scheduling, costing, stability, sustainability, maintainability, evacuation simulation and safety (Lieyun *et al.*, 2014). Providing time information by utilizing 4D BIM techniques for construction scheduling, workspace planning, constructability analysis, site monitoring has become an essential part of an improved, more reliable and efficient building process across the world. Koo and Fisher (2000) note that one of the major advantages of using 4D models as a tool for visualisation, simulation and analysis is their ability to show spatio-temporal and logical information through a single medium. In their recent publication Fadi *et al.* (2014) explains that visualization of construction schedules has much improved through the development of 4D modelling tools that offer an effective method for schedule planning and management. 4D models provide virtual visualization of the construction process, BIM applications for the construction management, specifically the schedule, cost, quality, and safety control, should be based on 4D models (Lieyun *et al.*, 2014).

Lieyun *et al.*, (2014) argue that 4Dmodels need to depict, visualize and analyse the constantly changing variables that occur as the construction phase proceeds. They also point out that some of the main challenges for implementing BIM applications are generating the 3D models, retrieving the job site environmental information and updating actual data from the job site, within the 3D models, as the construction process moves forward. Their answers to these challenges are using laser scanning and image processing to generate 3D models, utilizing augmented reality to retrieve actual environmental information from the job site, and Radio Frequency Identification

(RFID) to collect jobsite data. Clearly combinations of tools and techniques will be very useful however this type of technology is currently costly for smaller companies and smaller projects.

Since the mid-90s, research in the areas of 4D modelling (Collier and Fischer, 1995) has developed and nowadays there are many software applications which allow us to use *n*D tools and techniques. Current software, such as Autodesk Navisworks, Bently Navigator, Gehry Technologies GTeam, Solibri, Tekla BIMsight, combines visualisations and coordinates building projects though models. These tools are able to integrate 2D and 3D models with data, allowing visual analysis and helping to improve data visibility and communication; providing valuable clash detection options. There are plug-ins for SketchUP software that combine 3D models to time and/or cost data and can provide animated illustrations of construction sequence against time and/or cost. Many of these software provide interactive visualisation possibilities with real-time navigation and photorealistic renders. Many also offer on-site support with smartphone/tablet applications.

18.2.2 Existing reality: gathering digital data

The tools and techniques covered on the previous sections help us to envision the proposed construction. However as-built data is important and Zhu and Donia (2013) point out that as-built building information, including building geometry and features, is useful in multiple building assessment and management tasks. As-built data capturing is an important task, and traditional surveying techniques can be time consuming and more open to human error that what can be achieved with digital tools. Volk *et al.* (2014) in their recent research, summarised the data capturing tools. They have divided these techniques into two sub-categories of 'none-contact techniques' and 'contact techniques'. The latter covers the manual techniques, and the former covers many digital data capturing techniques – for example, photogrammetry, videogrammetry as imaged-based techniques and laser scanning (LADAR, LiDAR), laser measuring as ranged-based techniques. This section will specifically focus on 3D Laser scanning and its use in the AEC industry.

The AEC industry benefits from the use of 3D Scanning tools and techniques in many areas, from gathering as-built data to monitoring the construction process, to maintenance, to many others (Figures 18.7 and 18.8). As Gong and Caldas (2008) point out, 3D imaging has provided the construction industry with rapid, accurate and safe methods to collect detailed 3D spatial information. And the integration of direct geo-referencing (using GPS and inertial systems) into laser scanning technologies has given a further boost to 3D modelling (Zlatanova, 2008).

Laser scanning technology is typically classified as either phase-shift based or as pulsed time of flight (TOF). Phase-shift scanners send out waves of varying lengths to the target object and measure the phase shift in the

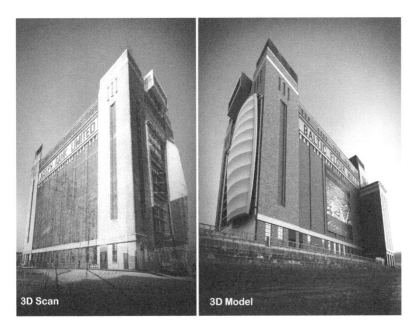

3D Scan   3D Model

Figure 18.7 3D point cloud data and render from the 3D model generated from scan
data, Baltic Art Centre, Gateshead.

returned waves of infrared light to calculate data points with X, Y, Z and I
(intensity) information. Phase-shift laser scanners are able to capture large
datasets quickly but have a limited range (10–70 metres typically). TOF
scanners have a much further range (up to 2000 m) but aren't able to capture
high density scans as quickly (Kimpton *et al.*, 2010).

A laser scanner provides distance measurements of surfaces visible from
the sensor's viewpoint, which can be converted to a set of 3D points known
as a point cloud. Individual point measurements can be accurate to a few
centimetres or less than a millimetre depending on the sensor, the range,
and the surface being scanned (Xiong *et al.*, 2013). This process may collect
millions of 3D points from multiple locations, this makes the point-cloud
data or 3D Scan data which than will be transferred into processing software were it will be registered and then the building elements will be recognised and finally can be displayed as a digital image (Figures 18.8 and 18.9).
El-Omari and Moselhi (2008) notes for 3D modelling purposes, many scans
are required from different positions and with reasonable accuracy to get
enough information pertinent to the geometry of the scanned object.

Zlatanova (2008) points out that there are many operational sensors for
3D data acquisition and they are readily available on the market (optical,
laser scanning, radar, thermal, acoustic, etc.). There are also many commercial pieces of software that are able to convert point cloud data for
usable format in 3D modelling software. For example 3D Reconstructor 2,

Figure 18.8 (a) 3D point-cloud data from Gateshead Old Town Hall scan, (b) 3D model created using point-cloud data from the Gateshead Old Town Hall scan, (c) Photorealistic render from the 3D model of the Gateshead Old Town Hall.

3D Reshaper, Cloud2max, Geomagic Studio, GOMinspect, Leios, Poly OWrks, Polycould, Rapidform XOS/Scan, Silverlining, Trimble, VRmesh Studio, and others (Kimpton *et al.*, 2010; Qian *et al.*, 2014). Apart from these specific softwares, there are now specific functions in many 3D modelling softwares in which point cloud data can be utilized, such as Autodesk 3D Studio Max, Autodesk Navisworks, Autodesk Revit, Bentley Pointools, CATIA, Rhinocero, SketchUp, Soildworks. The Autodesk's ReCap also offers a multitude of possibilities for working with point cloud data.

18.2.3 Past reality: making virtual history

Techniques and tools available today not only help us to envision the present or the future but also help us to visualise the past. 3D modelling, VR, AR and BIM tools provide opportunities to reconstruct the past with precision from maps, photographs, scans and other information sources. Pauwels *et al.* (2013) explain that our built heritage is increasingly described, managed and analysed using information technology, thereby creating a considerable amount of digital heritage.

Reconstruction can be for a heritage building that is still standing or for buildings, streets, settlements that no longer exist. Being able to have access to such information is very valuable for investigation, change monitoring, restoration, conservation, management and educational activities. Pauwels *et al.* (2013) explain that by using the right IT tools, one is able to document and combine information that would otherwise be very hard to combine, for instance, because some information is in a digital spreadsheet and other information is in a book. In contrast, when information is digitally combined, one is able to more easily show the history of a particular building and/or of its constituent building elements. Having this information at your disposal is useful for communicating and explaining the technical and historical value of a building to an audience. Second, it is also useful for experts to manage, maintain and potentially renovate a building.

For example, in a recent work Wyre (2015) studied utilizing VR technologies for the advancement of knowledge in interpretation and representation of the built heritage (The Legacy of Medieval Newcastle upon Tyne). With the cooperation of Newcastle City Council Historic Environment Section, he investigated many historical data sources in order the represent medieval architecture in Newcastle upon Tyne (Figures 18.9 and 18.10).

The generation of 3D cultural heritage models has become a topic of great interest in recent years. One reason for this is the more widespread use of laser scanning and photogrammetry for recording cultural heritage sites. These technologies have made it possible to efficiently and accurately record complex structures remotely that would not have been possible with previous survey methods (Dore and Murphy, 2012).

As explained in previous sections, laser scanning gathers very precise and useable data in many formats if the buildings, structures are in existence.

Figure 18.9 3D model construction of historical buildings.
Source: Ian Wyre, 2015.

Figure 18.10 Virtual medieval Newcastle.
Source: Ian Wyre, 2015.

The data can be used to create VR or AR environments where the virtual heritage can be experienced for educational or scientific purposes.

Another option is using point cloud data and transferring this information to a BIM platform to create a building information model. Brumana *et al.* (no date) point out that historic building information modelling (HBIM) aims to provide a three-dimensional parametric representation, enabling the user to draw object models and manage related data on historic architectural elements, within a common exchange format (IFC, industry foundation classes, and gbXML, green building XML) in order to support full interoperability between different software (Figure 18.11).

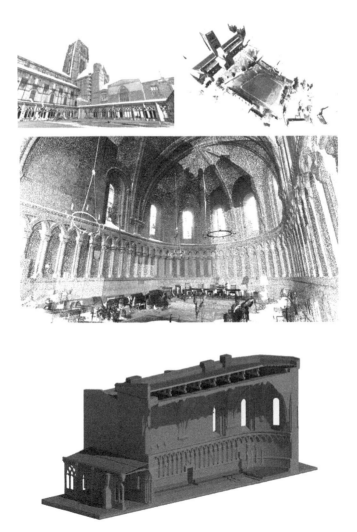

Figure 18.11 Point-cloud data examples from Durham Cathedral scans and Revit model.

18.2.4 *Virtual to real – the full circle: Three dimensional print*

The traditional way of model making is a well-established practice within the design and architecture sectors. These physical models are used as visual aids during the design process to envision the design concepts. Traditional methods of creating these physical models such as carving, castings, using moulds, and joining a variety materials together require craftsmanship and this process can be time consuming and expensive. New digital technologies and advancement in 3D modelling and digital fabrication have brought valuable solutions for model making in the form of the three dimensional (3D) print. Taylor and Unver (2014) date the origin of 3D printing to 1984 when Charles Hull (3D Systems) invented stereolithography, a printing process that enables a tangible 3D object to be created from digital data. They continue to explain that this technology is currently used for both prototyping and direct manufacturing in jewellery, footwear, industrial design, architecture, engineering, automotive, aerospace, dental and medical industries, education, geographic information systems, military armour and many other industries, and that this technology is presently being extended to build titanium aircraft parts, human bones, complex, nano-scale machines. Roebuck (2011) points out that since 2003 there has been large growth in the sale of 3D printers with the declining cost of these devices. Interestingly Pierrakakis *et al.* (2014) argue that 3D printing has the potential to do for manufacturing what the Internet did for information.

3D printing is an evolution of bi-dimensional printing, which allows one to obtain a solid object from a 3D model, realized within a 3D modelling software. The final product is obtained using an additive process, in which successive layers of material are laid down, one over the other. A 3D printer allows realizing, in a simple way, very complex shapes, which would be quite difficult to produce with conventional facilities (Dima *et al.*, 2014).

There is a variety of 3D printers in the market nowadays, including: Air-Wolf 3D HD2x, Beeverycreative – Beethefirst, CubePro Trio, SeeMeCNC – Orion, EOS, LulzBot Mini 3D Printer, MakerBot Replicator, Object, Project, Renishaw, Ultimaker 2, UP! Plus 2, Wanhao D5S, XYZprinting Da Vinci 1.0 AiO, ZPrinter, and many others. Taylor and Unver (2014) categorise 3D printers thus: personal (from £200 to £3000), professional (from £3000 to £80,000) and production (more than £80,000). Figures 18.12 and 18.13 show 3D print examples.

18.3 Future directions

Technological developments in visualisation are providing exceptional opportunities. Unity's (2014) survey showed that most impactful technologies in visualisation in the future will be the real-time rendering, GPU (graphical processing unit) rendering engines, 3D printing, VR (i.e. Occulus), BIM, cloud based rendering, CPU rendering engines, AR, 3D stereoscopic technologies and mobile/tablet based 3D graphics technologies which came in the top ten in this survey. And 77% of respondents to this survey believe that 3D real time visualisation will be a part of everyday workflows in the future.

For a long time the AEC industry has been labelled as one of the most inefficient and unsustainable industries. While there is an urgent need for new housing and other construction projects to take place, communities and cities around the world are confronted by acute challenges. Cities are already consuming 78% of the earth's resources and produce 60% of its carbon-dioxide (UN-Habitat) and it is predicted that by 2050, seven out of

Figure 18.12 Detail image from the 3D print virtual NewcastleGateshead model.

ten people will be living in cities (WHO, 2010). The AEC industry in many areas is making progress not to contribute to this unsustainable trend by utilizing new tools and technologies.

Supporting this, Wong *et al.* (2014) point out that promising innovations such as cloud and mobile technologies are expected to initiate the next wave of technological development, which will take the construction industry to the next level of technological advancement. Likewise Chen *et al.* (2015) point out the requirements of using a combination of technologies to en-hance the AEC industry by arguing that BIM is at risk of being 'blind and

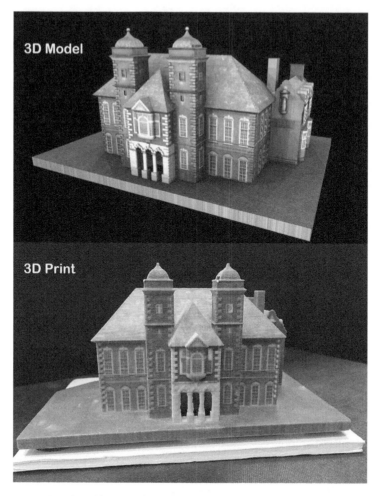

Figure 18.13 Rendered image of the college house (Newcastle upon Tyne) and 3D print model.

deaf' if its contained information cannot be synchronized with ongoing building processes in a real-time manner. However, as Azhar and Cox (2015) summarise, there are many mobile solutions (tablets, smartphones, cloud technologies, RFID, wearable devices) which can be used to monitor project progression, to gather construction information based on GPS, to create real-time visual documentation and sharing, to track compliance, workers, equipment and materials on site and to overlay 3D models with project information. They continue to explain that coordination, safety, and quality are the areas which are benefiting the most from the use of these mobile solutions.

Another example of this fusion is the Social-BIM (SBIM), a sociotechnical mode of BIM that enriches the co-creation process for Levels 2 and 3 BIM. It enables 'shared situational awareness' by empowering remote participants with visual and remote control of BIM models (Adamu *et al.*, 2015).

3D printing is another tool that will influence the construction industry in many areas. 3D printing can be greener than traditional manufacturing with reduced material waste, limited energy use, efficient use of raw materials, with a reduced carbon footprint for transportation of products (Campbell *et al.*, 2011). Since a more sustainable and greener construction industry is the target, then utilization of 3D printing should be considered. Research into taking 3D printing further, to see if building components can be constructed by these technologies, might reveal important solutions. There are already experimental 3D printed houses in China (3ders) the Netherlands (3DprintCH), and the USA (3DPrint) and Prof Behrokh Khoshnevis's Contour Crafting (CC) technology (Contour Crafting) is promising full automation in the construction industry which can bring a breakthrough in custom made and portable housing. Some of these 3D printing experiments are targeting zero energy houses and some using bio materials.

18.4 Conclusion

Digital technologies, data and knowledge integration in the AEC industry is leading to a synthesis of knowledge that transcends mere combination. Although the cultural change resulting from this fusion is inevitable and already taking place, this integration is helping the AEC industry to become efficient, productive and sustainable.

It is clear that digital technologies are everywhere and that it is only logical to use these ICT tools in the AEC industry. However it is also clear that human beings remain the most creative power behind the design and planning of efficient and effective places. When imagination, knowledge and deep understanding of the available tools come together, it will generate long lasting buildings that are sustainable through their entire life cycle. This becomes more apparent when we consider that 'none of the great architects of history, neither Sinan, chief architect of Sulemaniye Mosque in Istanbul, nor Michelangelo, architect of St Peter's Basilica in Rome, nor even Le Corbusier had such technologies available. However we can only marvel at what they might have achieved today, when digital technologies are able to help built environment professionals to design complex structures more efficiently and effectively' (Thompson, 2008).

18.5 Acknowledgement

Thanks to the Northumbria University Department of Architecture Built Environment's Specialist Support Team who attended various site visits and carried out the 3D scanning and photography of the scanning projects presented here. They are also responsible for the accurate 3D print outputs seen in this chapter. Acknowledgement also goes to students from BSc Architectural Technology programme at Northumbria University; Mark Cronin and Daniel Lipton and PhD Student Ian Wyre for their hard-work and dedication to produce great outputs. Also past and present colleagues of Northumbria University Virtual Reality and Visualisation Studio, namely Dr. Danilio Di Mascio, Dr. James Charlton, Graham Kimpton, and Iwan Peverett, who have worked on many of the projects mentioned in this chapter and their continuous efforts on trying new technologies. Also big thanks to BIM Academy for their collaboration on the Scan to BIM project presented here.

References

Adamu, A.Z., Emmitt, S., and Soetanto, R. (2015) 'Social BIM: co-creation with shared situational awareness', *Journal of Information Technology in Construction*, 20, pp. 230–252. Available at: http://www.itcon.org/papers/2015_16.content.07536. pdf. (Accessed: 1 June 2015).

Ager, M.T. and Sinclair, B.R. (1995) StereoCAD: three Dimensional Representation, in (ed.) *Sixth International Conference on Computer-Aided Architectural Design Futures*. Singapore: National University of Singapore, pp. 343–355.

Azhar, S., and Cox, A.J. (2015) Impact of Mobile Tools and Technologies on Jobsite Operations, in (ed.) *51st ASC Annual International Conference Proceedings Associated Schools of Construction*. College Station, TX: Texas A&M University.

Bille, R., Smith, S.P., Maund, K., and Brewer G. (2014) Extending Building Information Models into Game Engine, in (ed.) *Proceedings of the 2014 Conference on Interactive Entertainment*. New York, NY: ACM, pp. 1–8.

Campbell, T., Williams, C., Ivanova, O., and Garrett, B. (2011) Could 3D printing change the world. *Technologies, Potential, and Implications of Additive Manufacturing*, Washington, DC: Atlantic Council.

Chen, K., Lu, W., Peng, Y., Rowlinson, S., and Huang, G.Q. (2015) 'Bridging BIM and building: from a literature review to an integrated conceptual framework', *International Journal of Project Management*, 33(6), pp. 1405–1416.

Chi, H., Kang, S., and Wang, X. (2013) 'Research trends and opportunities of augmented reality applications in architecture, engineering, and construction', *Automation in Construction*, 33, pp. 116–122.

Collier, E. and Fischer, M. (1995) Four-dimensional modeling in design and construction, Stanford, CA: CIFE Technical Report #101.

Dalton, B. and Parfitt, M. (2013) Immersive visualization of building information models, *Design Innovation Research Centre working paper*, 6, [1.0], Reading: University of Reading.

DBB. (2015) *The Digital Built Britain*, London: HM Government. https://www.gov. uk/government/publications/uk-construction-industry-digital-technology.

Dima, M., Farisato, G., Bergomi, M., Viotto, V., Magrin, D., Greggio, D., Farinato, J., Marafatto, L., Ragazzoni, R., and Piazza, D. (2014) From 3D view to 3D print, *Proceedings of SPIE*, 9143: p. 91435E.

Dore, C. and Murphy, M. (2012) Integration of Historic Building Information Modeling (HBIM) and 3D GIS for recording and managing cultural heritage sites, in (ed.) *18th International Conference on Virtual Systems and Multimedia (VSMM)*. Politecnico Di Milano, Italy. IEEE, pp. 369–376.

Eastman, C.M. (1975) 'The use of computers instead of drawings in building design', *AIA Journal*, 63(3), pp. 46–50.

Edwards, G., Li, H., and Wang, B. (2015) 'BIM based collaborative and interactive design process using computer game engine for general end-users', *Visualization in Engineering*, 3(1), pp. 1–17.

El-Omari, S. and Moselhi, O. (2008) 'Integrating 3D laser scanning and photogrammetry for progress measurement of construction work', *Automation in Construction*, 18(1), pp. 1–9.

Fadi, C., Lee, S., Nikolic, D., and Messner, J.I. (2014) Visualization in 4D Construction Management Software: a Review of Standards and Guidelines, in (ed.) *Proceedings of the International Conference on Computing in Civil and Building Engineering*. Orlando, FL: American Society of Civil Engineers, pp. 315–322.

Giddings, B. and Horne, M. (2003) *Artists' Impressions in Architectural Design*, London: Taylor & Francis.

Gong, J. and Caldas, H.C. (2008) 'Data processing for real-time construction site spatial modeling', *Automation in Construction*, 17(5), pp. 526–535.

Greenwood, D., Horne, M., Thompson, E.M., Allwood, C.M., Wernemyr, C., and Westerdahl, B. (2008) 'Strategic perspectives on the use of virtual reality within the building industries of four countries', *Architectural Engineering and Design Management*, 4(2), pp. 85–98.

ISO 29481-1:2010(E). (2010) *ISO 29481–1:2010(E): Building Information Modeling— Information Delivery Manual—Part 1: Methodology and Format*, Geneva: International Organization for Standardization.

Jiao, Y., Zhang, S., Li, Y., Wang, Y., and Yang, B. (2013) 'Towards cloud augmented reality for construction application by BIM and SNS integration', *Automation in Construction*, 33, pp. 37–47.

Johansson, M., Roupé, M., and Viklund Tallgren, M. (2014) From BIM to VR-Integrating immersive visualizations in the current design process. In: Thompson, E.M. (ed.) *Fusion-Proceedings of the 32nd eCAADe Conference-Volume 2*. Newcastle upon Tyne: Northumbria University, pp. 261–269.

Kimpton, G., Horne, M., and Heslop, D. (2010) Terrestrial laser scanning and 3D imaging: heritage case study—The Black Gate, Newcastle upon Tyne, in (ed.) *ISPRS Commission V Mid-Term Symposium 'Close Range Image Measurement Techniques'*. [online] ISPRS, pp. 325–330. Available at: http://www.isprs.org/proceedings/XXXVIII/part5/papers/223.pdf [Accessed 25 Jan. 2017].

Koo, B. and Fischer, M. (2000) 'Feasibility study of 4D CAD in commercial construction', *Journal of Construction Engineering and Management*, 126(4), pp. 251–260.

Lieyun, D., Zhou, Y., and Akinci, Y. (2014) 'Building Information Modeling (BIM) application framework: the process of expanding from 3D to computable nD', *Automation in Construction*, 46, pp. 82–93.

Mallia-Parfitt, M. and Whyte, J. (2014) Developing a mobile visualization environment for construction applications. Available at: http://centaur.reading.ac.uk/37543/ (Accessed: 18 December 2015).

McKinney, K., Kim, J., Fischer, M., and Howard, C. (1996) Interactive 4D-CAD, in (ed.) *Proceedings of the Third Congress on Computing in Civil Engineering*. ASCE: Anaheim, CA, pp. 383–389.

Pauwels, P., De Meyer, R., and Van Campenhout, J. (2011) 'Linking a game engine environment to architectural information on the semantic web', *Journal of Civil Engineering and Architecture*, 5(9), pp. 787–798.

Pierrakakis, K., Kandias, M., Gritzali, C., and Gritzalis, D. (2014) 3D Printing and its regulation dynamics: the world in front of a paradigm shift, in (ed.) *Proceedings of the 6th International Conference on Information Law and Ethics.* Available at: http://www.cis.aueb.gr/Publications/ICIL-2014%203D%20Printing.pdf. Accessed on 1 June 2015.

Qian, Z., Agnew, B., and Thompson, E.M., (2014) Simulation of Air flow, Smoke Dispersion and Evacuation of the Monument Metro Station based on Subway Climatology, In: Thompson E.M. (ed.) *Fusion—Proceedings of the 32nd eCAADe Conference—Volume 1.* Newcastle upon Tyne: Northumbria University, pp. 119–128.

Roebuck, K. (2011). *3D Printing: High-impact Emerging Technology-What You Need to Know Definitions, Adoptions, Impact, Benefits, Maturity, Vendors.* Dayboro, QLD: Tebbo.

Smith, R. (2007) Game impact theory: five forces that are driving the adoption of game technologies within multiple established industries, *Games and Society Yearbook*, Available at: http://www.modelbenders.com/papers/Smith_Game_Impact_Theory.pdf (Accessed: 1 June 2015).

Sun, M. and Howard, R. (2004) *Understanding IT in Construction*, Abington OXON: Routledge.

Taylor, A. and Unver, E. (2014) 3D Printing-Media Hype or Manufacturing Reality: Textiles Surface Fashion Product Architecture, Textiles Society Lecture, 17th February 2014, Huddersfield, UK: Textile Centre of Excellence, Available at: http://eprints.hud.ac.uk/19714/ (Accessed: 1 June 2015).

Thompson, E.M. (2008) 'Is today architecture about real space, virtual space or what?', *The Northumbria Working Paper Series: Interdisciplinary Studies in the Built and Virtual Environment*, 1(2), pp. 171–178. Available at: https://www.north-umbria.ac.uk/static/5007/bepdf/vol1no2june08.pdf (Accessed: 1 June 2015).

Trenholme, D. and Smith, S. (2008) 'Computer game engines for developing first-person virtual environments', *Virtual Reality*, 12(3), pp. 181–187. Available at: http://dro.dur.ac.uk/5274/1/5274.pdf?DDD4+dcs0sps. (Accessed: 1 June 2015).

Turk, Z. (2006) 'Construction informatics: definition and ontology', *Advanced Engineering Informatic*, 20(2), pp. 187–199. Available at: http://www.sciencedirect.com/science/article/pii/S1474034605000911 (Accessed: 1 June 2015).

Unity. (2014) *The Future of Visualisation in Architecture*, Available at: http://try.unity3d.com/report/?utm_source=CE&utm_medium=link%20click&utm_campaign=try%20unity%20DR: Unity and CGarchitect. Accessed on 1 June 2015.

Volk, R., Stengel, J., and Schultmann, F. (2014) 'Building Information Modeling (BIM) for existing buildings—Literature review and future needs', *Automation in Construction*, 38(2), pp. 109–127. Available at: http://www.sciencedirect.com/science/article/pii/S092658051300191X (Accessed: 1 June 2015).

Wang, J., Wang, X., Shou, W., and Xu, B. (2014a) 'Integrating BIM and augmented reality for interactive architectural visualisation', *Construction Innovation*, 14(4), pp. 453–476.

Wang, X., Kim, M.J., Love, P.E., and Kang, S. (2013a) 'Augmented Reality in built environment: classification and implications for future research', *Automation in Construction*, 34(2), pp. 37–44. Available at: http://www.sciencedirect.com/science/article/pii/S0926580512002166 (Accessed: 1 June 2015).

Wang, X., Love, P.E., Kim, M.J., Park, C.S., Sing, C.P., and Hou, L. (2013b) 'A conceptual framework for integrating building information modeling with augmented reality', *Automation in Construction*, 34, 37–44.

Wang, X., Truijens, M., Hou, L., Wang, Y., and Zhou, Y. (2014b) 'Integrating augmented reality with building information modeling: onsite construction process controlling for liquefied natural gas industry', *Automation in Construction*, 40, pp. 96–105. Available at: http://www.sciencedirect.com/science/article/pii/S092658051300215X (Accessed: 1 June 2015).

WHO (World Health Organization). (2010) *Hidden Cities: Unmasking and Overcoming Health Inequities in Urban Settings. Centre for Health Development, & United Nations Human Settlements Programme,* ISBN 978 92 4 154803 8. Available at: http://www.who.int/kobe_centre/publications/hiddencities_media/who_un_habitat_hidden_cities_web.pdf (Accessed: 1 June 2015).

Whyte, J. (2002) *Virtual Reality and the Built Environment*, Abington OXON: Routledge.

Wong, J., Wang, X., Li, H., Chan, G., and Li, H. (2014) 'A review of cloud-based BIM technology in the construction sector', *Journal of Information Technology in Construction (ITcon)*, 19, pp. 281–291. Available at: http://www.itcon.org/2014/16 (Accessed: 1 June 2015).

Wyre, I. (2015) A methodology for spatial archaeology: visualising the legacy of Medieval Newcastle Upon Tyne using digital and virtual toolsets. PhD thesis, Northumbria University.

Xiong, X., Adan, A., Akinci, B., and Huber, D. (2013) 'Automatic creation of semantically rich 3D building models from laser scanner data', *Automation in Construction*, 31, pp. 325–337. Available at: http://www.sciencedirect.com/science/article/pii/S0926580512001732 (Accessed: 1 June 2015).

Zhu, Z. and Donia, S. (2013) 'Spatial and visual data fusion for capturing, retrieval, and modeling of as-built building geometry and features', *Visualization in Engineering*, 1(1), pp. 1–10. Available at: http://rd.springer.com/article/10.1186/2213-7459-1-10. (Accessed: 1 June 2015).

Zlatanova, S. (2008) Working Group II—Acquisition—Position Paper: data collection and 3D reconstruction, in (ed.) *Advances in 3D Geoinformation Systems*. Berlin and Heidelberg: Springer, pp. 425–428.

Web pages

3ders (2015). *WinSun China Builds World's First 3D Printed Villa and Tallest 3D Printed Apartment Building*, Available at: http://www.3ders.org/articles/20150118-winsun-builds-world-first-3d-printed-villa-and-tallest-3d-printed-building-in-china.html (Accessed: 1 June 2015).

3DPrint (2015). *Andrey Rudenko Plans to 3D Print a 2-Story 'Zero Energy' House in 5 Days with Advanced 3D Printer*, Available at: http://3dprint.com/40154/3d-printed-house-rudenko/ (Accessed: 1 June 2015).

3DprintCH (2015). *3D Print Canal House*, Available at: http://3dprintcanalhouse.com/smart-building (Accessed: 1 June 2015).

Contour Crafting (2014). *Contour Crafting*, Available at: http://www.contourcrafting.org/ (Accessed: 1 June 2015).

UN-Habitat. *Climate Change*, Available at: http://unhabitat.org/urban-themes/climate-change/ (Accessed: 1 June 2015).

19 The multi-agent paradigm in construction e-business and its use in the next generation of data-driven decision-making tools

Esther Obonyo and Chimay Anumba

19.1 Introduction

Collaborative technologies, proliferating sensors, Big Data analytics, cloud computing, mobile devices, and the Internet of Things (IoT) have resulted in an unprecedented increase in data availability volume, and diversity. These technologies through providing faster, cheaper and more accurate information can potentially streamline the workings of the senior leaderships. This is notwithstanding there is still a disproportionately large 'potential to practice' gap in efforts directed at improving performance through the use data-driven decision making tools. There are several explanations for the status quo. The data required to perform meaningful analysis may be missing or insufficient. Where the data exists a significant proportion of is presented in a non-standardized or and non-digital format. In some cases, the intended user may have to work with poorly defined procedures and mechanisms for extracting, processing, and analyzing data. The users may also lack awareness and/or interest in the existing of the data. This paper reviews the ways through which the agent paradigm has been used to address similar challenges in construction e-business applications. It starts with background information in the domain and proceeds to describe classical agent-centered models that were used to develop APRON, a construction e-business prototype. The paper then identifies the main drawbacks of the agent-centered approach and demonstrates how the organisational metaphor can be used to address these limitations providing an exemplary procurement scenario. This culminates into a discussion on the need for an agent-centric approach in the next generation of data-driven decision making tools.

The construction industry embodies professionals in information-intensive activities. Decisions have to be made based on the information available, but there are heterogeneous, distributed, dynamic, semi-structured and unstructured data and knowledge sources. The existing information sources have an open architecture with structures exhibiting dynamism. The components change and cannot be predetermined because they are implemented by different people, at different times, using different tools and different techniques. This problem can now be rigorously addressed using emerging paradigms in information technology that facilitate the modelling of the fragmentation in the construction industry. Distributed artificial intelligence (typically implemented in the form of intelligent agents) offers an innovative approach to overcoming this problem. The phrase 'intelligent agents' is used here to refer to systems capable of autonomous, purposeful action in the real world whereas 'multi-agent system' (MAS) refers to a computational system in which two or more (homogenous or heterogeneous) agents interact or work together to perform a set of tasks or to satisfy a set of goals (Lesser, 1999). Such a system comprises (1) an environment, (2) a set of passive objects that can be associated with a position in the environment, (3) an assembly of agents, which are specific objects representing active entities of the system, (4) an assembly of relations linking objects (and thus agents), (5) an assembly of operations with which agents perceive, produce, consume, transform and manipulate objects and (6) operators representing the assembly as well as reaction modifications (Ferber, 1999). A multi-agent system is therefore a consolidation of autonomous 'problem solvers'.

The notion of a community of agents cooperating to fulfill a complex task is the fundamental benefit of deploying the agent-based technologies in a system. Interestingly, most of the existing applications are generally based on models for closed systems thereby limiting the potential for exploiting agents in systems deployed by different developers. This can be largely attributed to the fact that agent technology is a relatively new field and the enabling infrastructure is still maturing. In the subsequent sections proceeds to provide an overview of the classical agent-centered models that were used to develop APRON, a construction e-business prototype. The paper then discusses the main drawbacks of the agent-centered approach and demonstrates how the organisational metaphor can be used to address these limitations providing an exemplary procurement scenario.

19.2   Deploying agent-centered multi-agent systems

APRON, a construction e-business prototype was deployed to demonstrate the potential of using the agent paradigm to enhance information management. The simulated use case in the implemented prototype was based on processing product information for the specification and procurement of light bulbs from the Philips Lighting website (URL1). The site hosts close to two hundred catalogues in Adobe Acrobat PDF format. The information that would be of interest to an end-user such as wattage, cap size and voltage is presented in a semi-structured format. The website does not have a search facility that would support guided navigation based on attributes such as wattage and voltage. It is also not possible to query the information directly from any another application. Furthermore, relevant data has to be re-keyed for reuse in a different application. The MaSE approach was used to design this prototype. MaSE is a further abstraction of the object-oriented paradigm where agents are a specialization of objects (DeLoach *et al.*, 2004a). Unlike simple objects whose methods can be evoked by other objects, agents in this framework coordinate with each other via conversations and act proactively to accomplish individual and system-wide goals. Further details on agent classes, agent conversations and the resulting MAS architecture can be found in Obonyo (2004), Obonyo *et al.* (2004, 2005a,b). An important parameter in the definition of agent-based systems is their reliance on a society of agents interacting and cooperating to achieve some collective goals. Much of the pioneer research work in this paradigm has dwelt on deploying 'agent-centered multi-agent systems' (ACMAS) such as APRON.

ACMAS-driven research focuses on the internal mental state of an agent the relationship between these states and its overall behaviour. Ferber *et al.* (2004) pointed out that in this view communications become speech acts whose meaning may be described in terms of the mental states of an agent as is evident in agent communication languages such as the KQML and FIPA ACL. This has resulted in the design and development of agent-based systems in which agents can only communicate with one another in a closed system. Because agents generally exist in the context of multi-agent software systems with some defined global behaviour being derived from the interaction of constituent agents, having a closed system greatly undermines the potential benefits of cooperation (Zambonelli *et al.*, 2000).

The main benefits of defining a societal structure for MAS include reducing the system's complexity, increasing the system's efficiency, and enabling more accurately model the problem being tackled (Jennings and Wooldridge, 2000). Without this societal structure the patterns and the outcomes of the interactions are inherently unpredictable and predicting the behaviour of the overall system based on its constituent components is extremely difficult (sometimes impossible) because of the high likelihood of emergent (and unwanted) behaviour (Jennings, 2000). As Ferber *et al.* (2004) points out, in such a scenario it is not possible for agents designed by different designers to interact unless one makes some assumptions about the primitives of communications and the architecture of agents. As agents based on the ACMAS lack have access to these constraints, they are constrained to using the same language and have to be built using very similar architectures. Moreover, such an approach lacking generality and being tuned to specific systems and agent architectures exploits abstractions that are unsuitable for modeling agent-based systems (Bussmann, 1998; Ferber and Gutknecht, 1998; Ferber *et al.*, 2004).

There are three other major weaknesses of the ACMAS model (Ferber *et al.*, 2004). Firstly, since all agents' communication is without any external control, an applications designer often has to balance between: (1) allowing free interaction of agents thus making it easy for an agent to act as a pirate and use the system fraudulently, and; (2) implementing too strong security measures that could prevent the system from working efficiently in domains where speed and response is of critical concern. Secondly, with AOMAS all agents are accessible from everywhere, grouping entities that work closely together into 'packages' that may or may not be hidden as is done in software engineering. The actual challenge within a MAS model would be coming up with a dynamic framework for grouping agents that work together. Finally, since in the ACMAS model the platform is the only supported framework, it is not possible to fully exploit the component concept that is used in classical engineering as an abstract architecture.

19.3 Organisational-centered MAS and construction e-business

There has been an emerging interest in the use of macro-level concepts such as 'organisations', 'groups', 'communities', 'roles' in designing multi-agent systems as a possible solutions to the limitation of ACMAS models. Ferber *et al.* (2004) has shown that the OCMAS approach can altogether dispense with the use of mental states used in classical ACMAS models. An agent-based system such as the one designed for construction e-business can actually be viewed as several interacting organisations and it is in fact possible for an agent to be part of multiple organisations (Zambonelli *et al.*, 2000). It is therefore not surprising that organisational constructs have been generally perceived as being the first-class entities in MAS. Such agent-based systems can thus be deemed to have computational mechanisms for flexibly forming, maintaining and disbanding organisations. This implies that the notion of a primitive component can be varied according to the needs of the observer. Additionally, such structures provide a variety of stable intermediate forms in which individual agents or organisational groupings can be developed in relative isolation and then added into the system in an incremental manner (Jennings, 2000). Since the behaviour of agents within a system is based on the behaviour and structure of human organisations, each agent has a clearly defined role within the system. Consequently, interactions are no longer mere expressions of classical object-oriented, interdependencies (Booch, 1994), but are also a characterization of the position occupied by an agent within the organisation. The organisational metaphor also simplifies the design of the system by separating the component level (intra-agent) design dimensions from the system level (inter-agent). The organisational metaphor also makes it easier for the designed MAS to closely mirror the real-world organisations they are intended to support (Zambonelli *et al.*, 2000).

To explore the OCMAS notion further, an exemplary architecture extending the implemented APRON prototype was designed leveraging on the Agent/Group/Role (AGR) model (Ferber and Gutknecht, 1998; Ferber *et al.*, 2004). In the AGR model an agent is only specified as an active communicating entity playing (multiple) roles within (several) groups while organisational concepts such as groups, roles, structures and dependencies, are first-class citizens of a design methodology. A key attribute of the AGR model is its minimalist structure-based view of organisations as a role-group structure imposed on the agents. AGR also says that agents can have their joint behaviour orchestrated by interaction protocols, but the nature and the primitives to describe such protocols are left open. Based a simple use case such as the one shown in Figure 19.1 the AGR model can be used to model a MAS organisational structure for construction e-business with interactions between three group structures: a procurer group structure (ProcurerGS); the manufacturers group structure (ProviderGS), and a contracts group structure (ContractGS) used when a client decides to buy a product from the provider.

The 'broker' is the same agent with dual roles in the client group and the manufacturer group. When an agent enters a client group, the client asks the broker for a product. Then the broker sends a call for proposal to

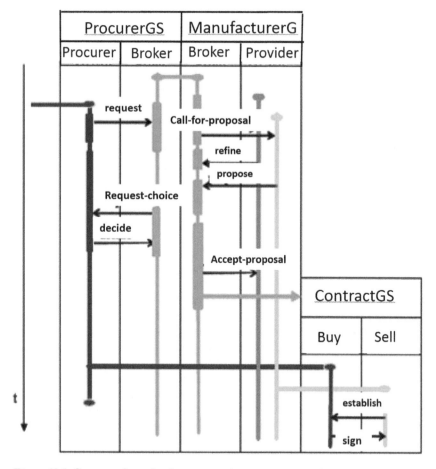

Figure 19.1 Construction e-business example.

manufacturers. The resulting proposals are presented to the client for product selection. In case of a suitable match, a contract group is created with both the client and the chosen manufacturer, taking the respective role of 'buyer' and 'seller'. This process can be repeated for all the other organisational concepts including 'organisation rules' and 'organisation patterns'.

19.4 Discussion and further work

Given that the construction industry still lags behind other industry sectors in the use of information and knowledge systems (Woldesenbet, 2014), the authors reviewed trends and emerging best practice among the world's leading 'information processors' to envision the next generations of agent-based systems for construction e-business. The 2008 financial and economic crisis promoted researchers in decision making to unpack the 'black box' that has shrouded executive choices for many years (Davenport, 2009; McElheran and Brynjolfsson, 2016). The research efforts here have focused on examining the accessible components of the 'black box' such as the nature of the decisions being made, the type of information supplied and key stakeholders and their roles in the process. Through performing a systematic analysis of the 'black box', decisions that affect the strategic direction that the studied organisation takes can be reengineered (Davenport, 2009). The leading 'information processors' are increasingly achieving goal through using data analytics and automated systems. The use of the terms 'analytics' in this discussion refers to 'the use of data and quantitative analysis to support decision making' while automation means 'using decision rules and algorithms to automate decision processes' (McElheran and Brynjolfsson, 2016).

In a study of 500 managers and executives, Larson (2016) established that only 2% of them used improved practices and technologies in decision making. The limited impact of conventional DDDs can be attributed to complex and ambiguous nature of the resulting information. When presented with this type of information, many people default to using mental shortcuts and cognitive biases that both distort perceptions and limit the available options when making choices (McElheran and Brynjolfsson, 2016). There are some knowledge and information management challenges within the highway infrastructure sector that are essentially a 'potential to practice' gap. Highway agencies across the globe have made significant investments in the use of computer-enabled technologies for the collection, storage and management of different types of data ranging from roadway inventory to pavement condition data during the life cycle of a highway infrastructure project. The benefits accruing from this investment remains limited largely because the available data is not being fully utilized to inform decision making (Woldesenbet, 2014).

Any organisation, irrespective of the nature of business it performs, is fundamentally an 'information processor' (McElheran and Brynjolfsson, 2016). The existing information and knowledge management technologies through providing faster, cheaper and more accurate information can potentially streamline the workings of the senior leaderships. When coupled with principles of behavioral economics, such technologies can illuminate biases and areas of irrationality, thereby increasing the likelihood that actions that end up being pursued result in the desired outcome. A 3-month study of 100 leaders established that managers who made better decisions by using practices and technologies based on behavioral economics achieved they results 90% of the time with as many as 40% of them exceeding their expectations (Larson, 2016). Another survey of an estimated 50,000 US manufacturing enterprises revealed that the use of data-driven decision making tools (DDDs) grow from 11% to 30% of manufacturing plants between 2005 and 2010 (McElheran and Brynjolfsson, 2016). This trend suggests that 'information processors' are increasingly recognizing the important role DDDs can play in efforts directed at improving decision making.

This is notwithstanding there is still a disproportionately large 'potential to practice' gap. In a study of 500 managers and executives, Larson (2016) established that only 2% of them used improved practices and technologies in decision making. The limited impact of conventional DDDs can be attributed to complex and ambiguous nature of the resulting information. When presented with this type of information, many people default to using mental shortcuts and cognitive biases that both distort perceptions and limit the available options when making choices (McElheran and Brynjolfsson, 2016). There are some knowledge and information management challenges within the highway infrastructure sector that are essentially a 'potential to practice' gap. Highway agencies across the globe have made significant investments in the use of computer-enabled technologies for the collection, storage and management of different types of data ranging from roadway inventory to pavement condition data during the life cycle of a highway infrastructure project. The benefits accruing from this investment remains limited largely because the available data is not being fully utilized to inform decision making (Woldesenbet, 2014). There are several explanations for the status quo. The data required to perform meaningful analysis may be missing or insufficient. Where the data exists a significant proportion of is presented in a non-standardized or and non-digital format. In some cases, the intended user may have to work with poorly defined procedures and mechanisms for extracting, processing, and analyzing data. The users may also lack awareness and/ or interest in the existing of the data.

The existing DDDs also have in-built complexity seeing that they are: (1) Populated using several data collection methods including automated systems and developments such as smartphones, camera, sensor, bar code, radio frequency identification, voice recognition and satellite navigation; (2) Underpinned with diverse data storage mechanisms ranging from electronic and digital systems such as database to data warehouses, ontology frameworks and non-relational databases; (3) Processed using a broad range of data analytical tools such as data mining, knowledge discovery in database, machine learning and business intelligence tools, knowledge bases, and expert systems, and; (4) Accessed using different management approaches such as enterprise resource planning, total quality management, cloud computing, lean manufacturing, and business process management throughput their business cycles.

It is important to bear in mind that systems cannot totally replace human decision makers. 'It takes an expert human being to revise decision criteria over time or know when an automated algorithm no longer works well (Davenport, 2009)'. The human decision maker must therefore fully understand the DDDs as well as the inputs/outputs and the models. To bridge the gap between potential and practice, DDDs must make explicit the assumptions underpinning decision modelling – such assumptions may not remain valid as time progresses. There is therefore a need for a feedback loop that informs both the system and the human user when the initial assumptions are no longer applicable. It is also prudent to work with a set of models. There are usually multiple pathways to the desired outcome and the most optimal pathway can change depending on the prevailing factors in the external environment.

Clearly, there is a need for systematic integrated, trans-disciplinary and in depth knowledge of the feedbacks between the different parts of knowledge and information systems that underpin decision making as well as the emergent properties of the entire decision making landscape. These complex requirements can be best characterized using an agent-based approach. The authors will in future efforts assess the feasibility of adapting Verburg *et al.*'s (2016) agent-based framework for operationalizing modelling for the Anthropocene to characterize the business models for the next generation of DDDs for highway infrastructure management. The framework addresses the challenge of moving from science-discovery modelling where the focus is the functioning of specific systems to generating models based on solution-driven questions related to management. Examples of parameters modelled in the former include process-response relationships, thresholds, tipping points, early warning signals and connectivity and those in the latter include adapting to future climate change, identifying the unintended consequences of specific actions, or maximizing social-ecological resilience.

In previous efforts, the authors investigated the use of agent-based modelling applications at the micro-level and deployed prototypes for end user applications. At the Macro level, highway infrastructure management is a policy problem. Financing the design, construction and maintenance of infrastructure is investing in a public good to enhance sustainable economic growth. Stakeholders such as the researchers and private practice players are exposed to uncertainties, inefficiencies and market failures, both in everyday operations management generating immediate value from the current products as well as in innovation management to generate new products for future value. In the follow-up research activities, will also investigate the feasibility of using Vermeulen and Pyka's (2015) agent-based modeling for decision making in economics under uncertainty approach to bridge the 'potential to practice' gap in highway infrastructure management through: economic theorizing to understand the factors that contributed to the observed outcome in the existing real-life systems, and; 2) policy experimentation and evaluation to provide decision makers with a range of interventions that can be used to yield a desired outcome (Figure 19.2). There are some outstanding issues in the design and development of agent-based tools. Investigating the use of agent-based modelling in the specified use case will contribute to the evolving body knowledge on challenges of challenges of validating and operationalizing agent-based models in data-driven decisions models (Vermeulen and Pyka's, 2015). It will also generate additional empirical data that can be used to further eliminate the existing ambiguities in interpreting simulation outcomes because of factors such as onset behavior, learning, and predefined temporal patterns.

Before agent-based systems can be deployed in large-scale real life applications within the context of construction e-business, it is necessary for researchers to reduce the gap between the real world and the design models used in agent-oriented software engineering. In particular, MAS need to be designed for the deployment of open systems that can truly exploit the flexibility and autonomy inherent in intelligent agents. The main limitation of most (if not all) other agent-based construction applications is the fact that they were not implemented as open systems. Deployment of open agent-based applications for construction e-business has not been possible largely because the existing design and development platforms focus on the

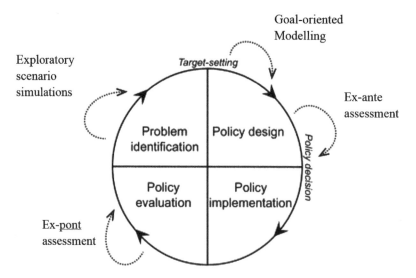

Figure 19.2 Roles of agent-based models in the highway infrastructure management systems.

Source: Verburg *et al.* (2016).

internal mental states of an agent, the relation between these states and the agents overall behavior using agent communication languages such as the KQML and FIPA ACL. Consequently, the resulting applications are based on agent-centered multi-agent systems in which agents can only communicate with one another in a closed system.

One aspect of the research discussed in this paper explored the use of the organisational metaphor to encapsulate macro concepts in the design of MAS for deploying applications that are more open and dynamic. There has been a growing interest among researchers in agent-oriented software engineering in developing methodologies for modelling the organisational abstractions. Recent examples of such efforts include Girodon *et al.*'s (2015) proposed DOCK methodology which enhances knowledge management systems through modelling human organisations using their roles, collaborations, skills, goals and knowledge. In Rodrígueza *et al.* (2015), a MAS architecture which incorporates social behaviors is used to manage data from sensors. This approach was tested in a residential home for the elderly. El Habib Souidim *et al.* (2015), Rahmanzadeh (2015), Mahani and Aga (2014), Billhardt *et al.* (2014), DeLoach and Garcia-Ojeda (2014) and Case (2014) also use a similar approach.

Defining open and dynamic models implies that the agents roaming on the Web will inevitably join groups designed by other developers using different terms. As the organisational metaphor becomes truly global and models interaction between agents in different groups implemented by different developers, there will be a need for reconciling differences in the use of concepts and terms. Before the organisational metaphor can take root as a superior approach to analyzing and designing agent-based system, the agent community will have to address the semantic complexities inherent in the use of the OCMAS model. This problem can be addressed through defining an ontology for the MAS organisational concepts. Preliminary reviews of emerging trends in this area has revealed that an example of such ontology has been implement in a different domain by Coutinho *et al.* (2005). Further research at the micro-level will focus on extending this ontology for use in defining an OCMAS model for construction e-business.

References

Billhardt, H., Julián, V., Corchado, J.M. and Fernández, A. (2014). An architecture proposal for human-agent societies. *Highlights of Practical Applications of Heterogeneous Multi-Agent Systems*. The PAAMS Collection: PAAMS 2014 International Workshops, Salamanca, Spain, June 4–6, 2014. Springer International Publishing, pp. 344–357.

Booch, G. (1994). *Object-Oriented Analysis and Design*. 2nd edition. Addison-Wesley, Reading, MA.

Bussmann, S. (1998). Agent-oriented programming of manufacturing control tasks. In *Proceedings of Third International Conference on Multi-Agent Systems (ICMAS'98)*, IEEE Computer Society, Washington, DC, pp. 57–63.

Case, D.M. (2014). Engineering multi-group agents for complex cooperative systems. In *Proceedings of the 2014 International Conference on Autonomous Agents and Multi-Agent Systems*, International Foundation for Autonomous Agents and Multiagent Systems, Paris, pp. 1707–1708.

Coutinho, L.R. Jaime, S., Sichmam J.S. and Boissier, O. (2005), Modeling organisation in MAS: a comparison of models, *First Workshop on Software Engineering for Agent-oriented Systems, SEAS*. Brazilian Computer Society, Uberlância, Brazil.

DeLoach, S.A. and Garcia-Ojeda, J.C. (2014). The O-MaSE methodology. In *Handbook on Agent-Oriented Design Processes*. Springer, Berlin and Heidelberg, pp. 253–285.

El Habib Souidim, M. Songhao, P., Guo, L. and Chang, L. (2015). Multi-agent co-operation pursuit based on an extension of AALAADIN organisational model. *Journal of Experimental & Theoretical Artificial Intelligence*, 07 July 2015. http://dx.doi.org/10.1080/0952813X.2015.1056241. Accessed 30 January 2016.

Ferber, J. (1999). *Multi-Agent Systems, An introduction to Distributed Artificial Intelligence*, Addison-Wesley, An imprint of Pearson Education, Harlow, UK.

Ferber, J. and Gutknecht, O. (1998). Aalaadin: A meta-model for the analysis and design of organisations in multiagent systems. In *Third International Conference on Multi-Agent Systems*, IEEE, Paris, pp. 128–135.

Ferber, J., Gutknecht, O. and Michele, F. (2004). From agents to organisations: An organisational view of multi-agent systems. In Giorgini, P., Muller, J. and Odell, J., (eds), *Agent-Oriented Software Engineering (AOSE) IV, LNCS 2935*, July 2003, Melbourne, VIC, pp. 214–230.

Girodon, J., Monticolo, D., Bonjour. E. and Perrier, M. (2015). An organisational approach to designing an intelligent knowledge-based system: Application to the decision-making process in design, projects. *Advanced Engineering Informatics*, 29(3), pp. 696–713.

Jenning, N.R. (2000). On agent-based software engineering. *Artificial Intelligence*, 117(2), pp. 277–296.

Jennings, N.R. and Wooldridge, M. (2000). Agent-oriented software engineering. In Bradshaw, J. (ed.), *Handbook of Agent Technology*, AAAI/MIT Press, Cambridge, MA.

Larson, E. (2016). A checklist for making faster, better decisions. *Harvard Business Review*, 7 March, 2016.

Lesser, V. (1999). Cooperative multi-agent systems: A personal view of the state of the art. *IEEE Transactions on Knowledge and Data Engineering*, 11(1), 133–142.

Mahani, M.N. and Agah, A. (2014). Strategic reorganisation in multi-agent systems: Inspired by intelligent human organisations. *International Journal of Cooperative Information Systems*, 23(4), 1450009.

McElheran, K. and Brynjolfsson, E. (2016). The rise of data-driven decision making is real but uneven. *Harvard Business Review*, 3 February, 2016.

Obonyo, E.A. (2004). APRON: Agent-based specification and procurement of construction products. Doctoral thesis, Department of Civil and Building Engineering, Loughborough University.

Obonyo, E.A., Anumba, C.J. and Thorpe, A. (2004). Specification and procurement of construction products, a case for an agent-based system, *International Journal of IT in Architecture, Engineering and Construction*, 2(3), 204–215.

————. (2005a). APRON: An agent-based specification and procurement system for construction products, *Engineering, Construction and Architectural Management*, 12(4), 329–350.

————. (2005b). Specification and procurement of construction products using agents, In Anumba, C.J., Ugwu, O.O and Ren. Z. (eds), *Agents and Multi-Agent Systems in Construction*. Taylor & Francis Group, London and New York.

Rahmanzadeh, A. and Nazemi, E. (2015). Fhorganization: New organization model for multi-agent systems, *International Journal of Computer Networks and Communications Security*, 3(8), 337–342.

Rodrígueza, S., De Paza, J.F., Villarrubiaa, G., Zatoa, C., Bajob, J. and Corchadoa, J.M. (2015). Multi-agent information fusion system to manage data from a WSN in a residential home, *Information Fusion*, 23, 43–57.

Verburg, P.H., Dearing, J.A., Dyke, J.G., Leeuw, S., Seitzinger, S., Steffen, W. Syvitski, J. (2016). Methods and approaches to modelling the Anthropocene. *Global Environmental Change*, Volume 39, July 2016, pp. 328–340, ISSN 0959-3780, http://dx.doi.org/10.1016/j.gloenvcha.2015.08.007. Accessed 6 June 2016. (http://www.sciencedirect.com/science/article/pii/S0959378015300285).

Vermeulen, B. and Pyka, A. (2015). Agent-based modeling for decision making in economics under uncertainty. Economics Discussion Papers, No 2015-45, Kiel Institute for the World Economy. http://www.economics-ejournal.org/economics/discussionpapers/2015-45.

Woldesenbet, A.K. (2014). Highway infrastructure data and information integration and assessment framework: A data-driven decision-making approach. Graduate theses and dissertations. Paper 14017.

Zambonelli, F., Jennings, N.R. and Wooldridge, M.J. (2000). Organisational abstractions for the analysis and design of multi-agent systems. *Workshop on Agent-oriented Software Engineering ICSE 2000*. Limerick, Ireland, June, 2000.

20 Conclusions – Summary, the status quo and future trends

Bingunath Ingirige, Srinath Perera,
Kirti Ruikar and Esther Obonyo

20.1 Introduction

This chapter provides the concluding highlights of this book. Many past studies conducted in the construction industry confirm the positive impact of utilising ICTs, with an improving effect on costs, scheduling and quality. The various sections and chapters of this book presented some of the realisation and leveraging of the benefits of new advances in construction ICT and e-Business through discussion of theory, contextualising practical case studies, policies and regulation framing advancements and identifying some of the growing trends in the future.

20.2 General summary

The book covers issues of new advances in construction ICT and e-business. The issues discussed are not purely technological, but also include, social, process, and people issues that enable a coordinated and a cohesive approach to advance ICT and e-Business use in construction organisations. Along with contributions from leading academics in the field of ICT in construction, the book also includes contributions from industry practitioners. This is in the form of case studies of successful ICT and e-business implementations and the lessons learnt thereof. The book is therefore of relevance to both, academics and industry alike.

Chapter 1 defined the subject area, by providing a general introduction to enabling technologies as well as defining e-business. Chapters 2–4 cover procurement issues ranging from e-commerce to e-procurement. Chapters 5–8 focus on building information modelling (BIM). Under this section, Chapter 5 provides a theoretical perspective on BIM with some of the essential contextual requirements to enable its effective and efficient use. Chapters 6 and 7 provide valuable insights from case studies to support BIM use. Chapter 8 gives an outlook of the use of extranets in construction. Chapters 9–11 address numerous areas of cloud computing and discuss issues of data exchange standards, collaboration tools and city level information modelling. Chapters 12–14 address the important process issues of how effectively the ICT technologies can be assimilated within the construction industry with formalising and rethinking of the process. Under this category Chapter 12 considers maturity of processes for e-business use and Chapter 13 is about strategic considerations of IT infrastructure. Chapter 14 concentrates on the holistic industry structure by focusing on issues faced by micro and SME organisations in construction, often an area not receiving much attention. Chapters 15–17 address socio-cultural issues in advancing construction ICT and deal with technological issues such as the use of social media and mobile computing applications in construction and how these are best delivered within a good socio-cultural atmosphere. The chapters also discussed case studies that address implementation issues. Chapter 18 provides a detailed review of virtual reality and its use in envisioning buildings. It is followed by Chapter 19 that explores the use of AI agents in building design and management. The book concludes with this final chapter that summarises the status quo with respect to ICT use in the construction industry providing a review of the trends and future directions for e-business in construction.

20.3 Key advancements and enablers

As with any other industry, the issues in advancing ICTs in construction are not merely covered under the technology umbrella. They not only encompass all aspects of the technology, but also include process and cultural aspects. All throughout the book the same sentiment resonates; i.e. an appropriate process and culture will nurture the technologies within a context, and only then can the technology be truly assimilated into the context and society and industry can benefit from it. This section highlights such key advancements and enablers of ICTs within the construction context.

- One of the key advancements in construction operations is associated with the Internet of Things (IoT). The IoT is gradually emerging as a pervasive concept with its presence being developed to interconnect a variety of things or objects such as radio-frequency identification (RFID) tags, sensors, actuators and mobile phones, through unique addressing schemes that are able to interact with each other and cooperate with their neighbours to reach common goals. This interconnectivity adds new dimensions to data, not previously possible. As such, a richness of 'connected' context, new opportunities and new challenges; all of which will shape the future ICT landscape. It will bring about contextual awareness to the built environment with numerous possibilities of connected and responsive buildings that adapt to changing micro environment.
- BIM has been a primary focus of the UK industry for good part of the last decade, even though its origins date back much further. It is seen as a major advancement in construction ICT and e-business that is

prompting a rethink of the way in which the industry operates from the traditional inward looking, silo-based approach to a model that is more collaborative and 'singular'. What will eventually be achieved is largely dependent on effectiveness of processes and the capability of stakeholders to communicate a shared understanding through the lifecycle of a building. It concerns the re-branding of the industry in a move to put right the adversarial image it inherits.

• Given the UK Government's mandate for a fully collaborative 3D BIM (Level 2) as a minimum by April 2016 on publically procured projects, much of the recent aspirations (and subsequent efforts and developments) have been focused on attainment of this Level 2. But, as the 2016 deadline looms the shift in focus to Level 3 and beyond becomes inevitable and necessary. Simply adopting a UK centric view in an industry that is increasingly globally dispersed and culturally diverse is perhaps too narrow a view to possess. Thus, a shift in perspective towards an all-encompassing 'global' context becomes necessary.

• The European Union has mandated the adoption of e-procurement by 2016, which will work as an enabler to achieve advancements. Similarly, the UK Government has mandated the adoption of BIM by April 2016. These apply to Construction Works, in addition to Goods and Services procurement. Both the European Union and the UK Government, hope to harness the capabilities of these enabling technologies in a bid to modernise and improve performance of the construction sector of the economy (see Chapter 3).

• As the adoption of BIM catches momentum and this is then matched with capability and capacity to handle the large volumes of interconnected data generated over a project's lifecycle; we will begin to uncover opportunities that were previously not possible. For example, the predicted technological adoptions that flow from the BIM revolution will include *automated digital decisions* (i.e. those that do not require human input) and *predictive digital solutions* (i.e. solutions that are automatically predicted based upon digital information) (see Chapter 5). This change will impact on how a human would interact with the construction decision-making processes. In such instances the humans would be gatekeepers of information inputs, examining trends and identifying 'outliers' that fall outside the expected norm.

• Another revolutionary trend in the construction industry has been that of 3D printing. It crosses boundaries that have historically and traditionally existed. With advancements in 3D printing, the traditional boundaries are beginning to haze. For example, it bridges gaps between industries (e.g. construction and manufacturing), between lifecycle stages of processes (design to assembly) and between processes and products. Its capability to print objects (i.e. components) in controlled environments has a range of benefits. For example, 3D printing can be greener than traditional manufacturing with reduced material waste, limited energy use, efficient use of raw materials, with a reduced carbon footprint for transportation of products; indicative of a revolutionary future (see Chapter 19).

• Historically, advancements in ICTs have been in response to the 'emergent' need of the industry to support intra-(CAD, BoQ) and inter-disciplinary processes and functions (groupware, extranets, SaaS). The construction industry's move towards a collaborative mode of working (driven by Latham, 1994; Egan, 1998) was supported by developments in tools that enabled partnering and collaboration. Extranets evolved, firstly as part of a local area network, then expanded into a secured area network; and are now only naturally extending their capabilities into cloud computing to accommodate the demands of the 'connected' users. These emerging channels of data storage and communications create new opportunities (platform independent global connectivity) and present new threats and risks (security) that need constant monitoring and consideration.

• The 2008–2009 financial and economic crisis sparked a growing interest among researchers in decision-making to perform a systematic analysis with the goal of reengineering decisions that affect the strategic direction that the studied organization takes (Davenport, 2009). These efforts examine accessible components of decision-making such as the nature of the decisions being made, the type of information supplied, the key stakeholders and their roles in the process. The finding from such studies coupled with principles of behavioural economics are being used to enhance the functionality of the next generation of decision support systems (see Chapter 19). Automated systems and data analytics being used to catalyse the use of the large volume of resulting information. There has been a steady progress towards the realisation of a systematic integrated, trans-disciplinary and in-depth knowledge of the feedback between the different parts of knowledge and information systems that underpin decision-making as well as the emergent properties of the entire decision-making landscape.

20.4 Key barriers

There are many barriers to the advancement of ICTs and e-business in construction. Some of these barriers are due to the legacy systems that still operate in construction, some because of the inherent culture and others are due to the emergent threats. These barriers are as follows:

• Despite some of the advancements and uptake of new technologies, the construction industry still has deep-rooted traditional practices and paper-based processes that remain in use. Until their use is phased out and effort is made to educate the laggards, issues of integration, seamless document exchange and interoperability will remain; and constrain some of the advancements.

- The advent of e-procurement has allowed the time taken for tendering to be greatly reduced in the Public Contracts Regulation 2015. It is evident that e-procurement provides a viable electronic alternative to the more traditional paper-based process, such as tendering, associated with the procurement process. Although new framework agreements and state-of-the-art procurement systems have emerged, there are adversarial relationships between construction industry participants. Much needs to be done to address the adversarial culture, if the benefits are to be truly realised.
- Cloud computing is emerging as a viable option for hosting and delivering services over the Internet. It offers a multitude of benefits including ease of access, scalability and lower operating costs, as examples among several other benefits. However, in spite of the benefits the move towards advancements such as cloud computing will bring a fresh set of challenges in the areas of data security, new cyber legislation and procurement; and contract changes in the future.
- BIM adoption requires a shift in focus from an internal, organisational perspective towards an external, project's lifecycle perspective. To facilitate this change, in the first instance efforts have been made to educate the professions and improve their awareness. However, for benefits of BIM-enabled collaboration to be truly realised, the whole 'project' team would require to pull together to achieve success thus the issue of economies of scale. Also, attempts in isolation will not take a project or by extension the whole industry on a course of improvement. More the number of users, more the opportunities of collaboration and cost effective performance.
- The current strategies of implementing BIM and e-procurement seem to be disjointed. Streamlining the processes could provide a win-win solution to improving these facets of a construction project.
- The advancement and proliferation of technologies bring about new legal challenges. These relates to issues of validity within construction contracts, copyright and data ownership issues, communication protocols, issues related to data storage, liabilities related to loss of data among others.
- Success in ICT in construction is largely dependent on the effectiveness of processes and the capability of stakeholders to communicate. This primary consideration seems ill-conceived among industry participants.
- Data availability, reliability and privacy are matters that require specific solutions. Smart cities will require an open protocol to communicate between different sensors, actuators and meters. Such standard open protocols are still under development and in their infancy.
- Small businesses in supply chains find integrating the supply chain difficult due to poor interoperability issues resulting from advanced ICT sophistication.

20.5 Key benefits of advancing ICT and e-business

Construction industry continues to be faced with problems of communication and coordination that result in experiencing cost and time overruns at different scales. Advancements in ICT and e-business could potentially revolutionize many areas in construction and reduce or mitigate cost and time overruns. Some of the key benefits are as follows:

- Simplified procurement processes via e-procurement.
- More transparency of transactions via e-business and e-procurement.
- Caters to auditing requirements in relation to tender evaluation procedures in e-tendering and e-purchasing.
- Realising the potential of web-enabled project management in construction.
- Further advancements such as web-based BIM.
- Transaction cost savings achieved through digitalizing and efficient communication.
- Improved process efficiency.
- Shortened internal and external communication cycle times.
- Convenience of archiving completed work.
- Better customer relationships through improved processing.
- Managing the 'information overload' problem and increased use of collected data in executive decision-making.
- Incorporating a feedback mechanism in the existing information and knowledge processing systems will increase relevance and statistical reliability.

20.6 Future directions

The book identifies that the construction industry has advanced in its introduction and implementation of ICT and e-business and that the industry has several pockets of excellent innovation. However, it has been established that certain processes, practices, legislations and regulations are constraining the movement in a big way. Therefore despite the industry's key developments, more positive strategies should continue to energise ongoing advancements and key debates. Some of the key areas forming the future agenda are as follows:

- The full potential of Internet of Things (IoT) and I4.0 is yet to be fulfilled, but that being said the entire legacy of extranets is said to be the underlying technology that will develop the IoT for the future.
- The revolution and evolution of extranets has now interwoven into mainstream construction and this trend will continue in the future. Extranet systems could now be customised to provide a more flexible and secure systems for users.

- It is also pertinent to recognise that advancements in ICTs such as BIM should be grounded within the contexts of societal, economic, environmental and political changes that will shape the overall future together with the emerging BIM technologies. Such a grounded approach will set the foundation for a move towards advanced, collaborative and intelligent BIM (Level 3 and beyond in the UK).
- Cloud solutions, developed in the last decades are ready to face issues of open protocols, which are essential in the case of city planning. The market probably will force the proprietary protocol company to shift some applications to open protocol. The approaches taken with SaaS will further advance and become mainstream method of ICT service deployment.
- The move towards cloud computing will bring another set of challenges in the areas of data security. An interesting conflict is highlighted, namely that between the movements towards free and open information, connectivity and collaboration and the need to secure that information. Such debates are likely to continue in the future.
- There are many initiatives currently in place that integrates GIS applications with construction processes. One such area is known as district information modelling (DIM). GIS has a decennial history of success that makes it essential in civil engineering and territorial planning hence using the DIM concept in the case of smart cities where, as an example, the information concerning energy flows, traffic flows and building stock, will become widely used in many construction industry applications.
- The Strategic e-Business Framework developed (Chapter 13) hinges on the recognition that e-business is a collaborative effort, requiring consideration of both the internal and the external environments in six key phases of situation analysis, vision establishment, critical success factors definition, action-plan development and implementation, and strategy review. More recently, CeB-CMM (Chapter 12) was developed as a construction-specific e-business capability maturity modelling (CMM) tool. Using such construction-specific frameworks and with time as a critical mass of lessons learnt from practical case studies enhance the use of e-business in the construction industry in the future. It will facilitate construction organisations to effectively plan and use ICT for augmentation of their construction processes.
- There will be several initiatives in advancement of e-business practices in developing countries. For instance, Ghana uses their new e-government initiatives to provide incentives for SMEs to go online by simplifying administrative procedures, reducing costs and allowing them to enter new markets. Such initiatives are followed by many developing countries around the globe.
- Technological developments in visualisation are providing exceptional opportunities. A significant number of key industrialists believe that

3D real time visualisation will be a part of everyday workflows in the future.

- As postulated in Chapter 15 and exemplified in Chapter 16, social media tools and philosophies are now increasingly moving from the periphery of construction industry communications to the mainstream of recruitment, marketing and customer relationship management. Many organisations that used to ban social media use from employees are now reverting their policies and procedures to formally working with social media. With Internet-born generations gradually becoming the mainstream workforce, the construction industry will no doubt move towards embracing social media. *It will become* an intrinsic part of everyday working and professional interaction. In a similar way to embracing CAD or GIS in to construction practices it will gradually be part of the normal working ethos of the project teams. Automation, robotics and intelligent systems in construction will become common place on construction sites. The pace of advancement of these technologies is significant and these will be part of the twenty-first century digital working environment. Many of the existing systems have been deployed as stand-alone entities. In future systems, there will be a need for macro-level modelling which better reflects the critical interdependences with emerging cyber-physical systems. Chapter 19 identified specific examples of these requirements based on checkpoints in information processing within highway infrastructure management.
- The construction industry is massively data rich. It generates large amounts of data every minute of its operations. The whole of the built environment generates data in all its phases of planning, design, construction and a lifetime of operation. The hidden relationships of the 'blackhole of built environment data' are yet to be explored. Big data, data analytics and data mining all have a huge potential for the future in exploring and codifying these potential relationships of the society and its political and economic environment and how it interacts with the built environment.

References

Davenport, T.H. (2009). *Make Better Decisions*, Harvard Business Review, Harvard Business School Press.

Egan, J. (1998). Rethinking construction: The report of the construction task force. London: Department of the Environment, Transport and the Regions.

Latham, M. (1994). Constructing the team: Joint review of procurement and contractual arrangements in the United Kingdom construction industry. London: HMSO.

Index

Page numbers followed by *f* indicate figures; those followed by *t* indicate tables.